Fixed-Bed Reactor Design and Diagnostics

Fixed-Bed Reactor Design and Diagnostics

Gas-Phase Reactions

Howard F. Rase

The W.A. Cunningham Professor
Department of Chemical Engineering
The University of Texas at Austin

Butterworths

Boston London Singapore Sidney Toronto Wellington

Library of Congress Cataloging-in-Publication Data

Rase, Howard F.
Fixed-bed reactor design and diagnostics: gas-phase reactions / Howard F. Rase.
p. cm.
Bibliography: p.
Includes index.
ISBN 0-409-90003-6
1. Chemical reactors. I. Title.
TP157.R343 1990
660′.283—dc20 89-33538

British Library Cataloguing in Publication Data

Rase, Howard F.
Fixed-bed reactor design and diagnostics: gas-phase reactions.
1. Chemical engineering equipment. Reactors. Chemical reactions
I. Title
660.2′844
ISBN 0-409-90003-6

Butterworth Publishers
80 Montvale Avenue
Stoneham, MA 02180

10 9 8 7 6 5 4 3 2 1

Printed in the United States of America

To my wife
Beverly

Contents

Preface

This book is written for professionals who seek recommendations for modeling, scale-up, and diagnostics of gaseous reactions catalyzed by solid catalysts in fixed beds. It is not a review of the literature. It provides brief but clear and essential theoretical and conceptual background (Part I), presents recommended mathematical models, scale-up procedures, and diagnostic methods (Part II), and a generous number of practical examples (Part III). The material is written and cross-referenced so that the busy user can enter the book at any desired subject area of interest and gain the needed information. To this end, although nomenclature is defined at the time of first use, a complete listing is provided at the end of the book. Further, a very thorough author-prepared index is included, and references to background material and examples appear throughout the book. The writing style is designed to appeal to the user who seeks brevity, clarity, and rapidly scannable material. To this end, specially designed tables are used whenever possible to present concepts and procedures in a concise and readily understandable form.

The book has been restricted to a single reactor type and fluid phase because gaseous reactions over solid catalysts constitute a unique group of commercially important reactions. This uniqueness can be explained by two significant facts.

1. Gaseous reactions over solid catalysts are the most common reactions in the production of commodity chemicals and petroleum products and are, therefore, subject to rather continuous efforts to improve design methods, process efficiency, and process and quality control.
2. Thermodynamic and physical property data for the ideal gaseous state are precise, copious, and readily available in computer data bases and well-known volumes of compiled data. Since many gaseous reactions are conducted at elevated temperatures, the ideal gas assumption often holds even at elevated pressure. In those cases where it does not, excellent equations of state for gases make possible reasonably accurate corrections to reflect nonideal behavior.

Both reasons create a favorable climate for the development and application of mathematical models for use in improved design of new units, revamping of existing units, and diagnosis and control of operating units. For these same reasons, predictive model control has the greatest chance to be applied successfully to gaseous reactions, particularly since excellent proprietary reaction-rate equations are available to companies interested in applying this emerging tool for fine-tuning a process to maximize profit and minimize chances for undesirable reactor excursions. Clearly, the discussion of methods in these contexts can be realistically applied in tone and scope.

The ability to predict equilibrium compositions and adiabatic reaction temperatures without requiring laboratory studies is a great boon to the development of a model and the selection of reasonable operating conditions for producing a desired result. Even when time constraints are such that model development is precluded, thermodynamic calculations are most helpful in selecting the best conditions for scale-up.

It is my hope that this book will encourage the use of reasonable models, reliable scale-up methods, and systematic diagnostics initially based on development-laboratory data and constantly updated by plant experiences.

I am especially indebted to Doyle Broom, Jerome DeHey, and Patrick Havens, who have served as teaching assistants in my courses. Doyle Broom wrote the original interactive programs for fixed-bed reactors described herein, and Jerome DeHey and Patrick Havens made valuable suggestions and were most helpful in implementing much of the professional software described in this book.

Howard F. Rase
Austin, Texas

CHAPTER 1

Introduction

Fixed-bed catalytic reactors have been aptly characterized as the workhorses of the process industries. For economical production of large amounts of product, they are usually the first choice, particularly for gas-phase reactions. Many catalyzed gaseous reactions are amenable to long catalyst life (1–10 years); and as the time between catalyst changeouts increases, annualized replacement costs decline dramatically, largely due to savings in shutdown costs. It is not surprising, therefore, that fixed-bed reactors now dominate the scene in large-scale chemical-product manufacture.

The first commerical application of a fixed-bed reactor occurred in 1831 when Peregine Phillips, a British vinegar maker, patented a process for making sulfur trioxide by passing air and sulphur dioxide, made by burning sulfur, over a hot bed of platinum sponge. The sulfur trioxide, formed by the catalytic oxidation of the sulfur dioxide in the air, was subsequently dissolved in water to form sulfuric acid, which, to this day, is the most widely used and important commercial acid. Phillips saw the advantage of a solid catalyst that could be held in a container while reactants were passed through it to produce the desired product. Since the catalyst is not consumed in the reaction and is held in the bed, it can be used repeatedly as a continuous flow of reactants is passed over the bed of catalyst, without the need for separating and recycling the catalyst. During the early 1800s, solid catalysts were just beginning to be studied and described by distinguished chemists. In the ensuing years numerous catalysts were discovered, and most were applied in commercial fixed-bed processes, making the fixed-bed reactor the dominant type for solid catalysts.

Since those early days, many special fixed-bed designs have been developed, but all types of fixed-bed reactors fall into one of two major categories: adiabatic or nonadiabatic. The adiabatic design is always preferred when acceptable conversion and selectivity to the desired product is possible. It is simpler and less costly than the nonadiabatic design, which requires heat transfer along the bed between the reaction side and a cooling or heating medium. For such heat transfer to be effective, a high ratio of heat-transfer surface-to-reactor volume is required, usually in the form of multiple small-diameter tubes containing the catalyst and with the heat-transfer medium flowing exterior to the tubes.

As would be expected, a number of reactor configurations have evolved to fit the unique requirements of specific types of reactions and conditions. Some of the more common ones used for gas-phase reactions are summarized in Table 1.1 and the accompanying illustrations. The table can be used for initial selection of a given reaction system, particularly by comparing it with the known systems indicated.

Table 1.1 Fixed-Bed Reactor Configurations for Gas-Phase Reactions

Classification	*Use*	*Typical Applications*	*Description*[a]
Single adiabatic bed	Moderately exothermic or endothermic nonequilibrium limited	Mild hydrogenation	Fig. 1.1
Radial flow	Where low ΔP is essential and useful where change in moles is large	Styrene from ethylbenzene	Fig. 1.2
Adiabatic beds in series with intermediate cooling or heating	High conversion, equilibrium limited reactions	SO_2 oxidation Catalytic reforming Ammonia synthesis Hydrocracking Styrene from ethylbenzene	Fig. 1.3
Multitabular nonadiabatic	Highly endothermic or exothermic reactions requiring close temperature control to ensure high selectivity	Many hydrogenations Ethylene oxidation to ethylene oxide, formaldehyde by methanol oxidation, phthalic anhydride production	Fig. 1.4
Direct-fired nonadiabatic	Highly endothermic, high temperature reactions	Steam reforming	Fig. 1.5

[a]See figures for detailed descriptions.

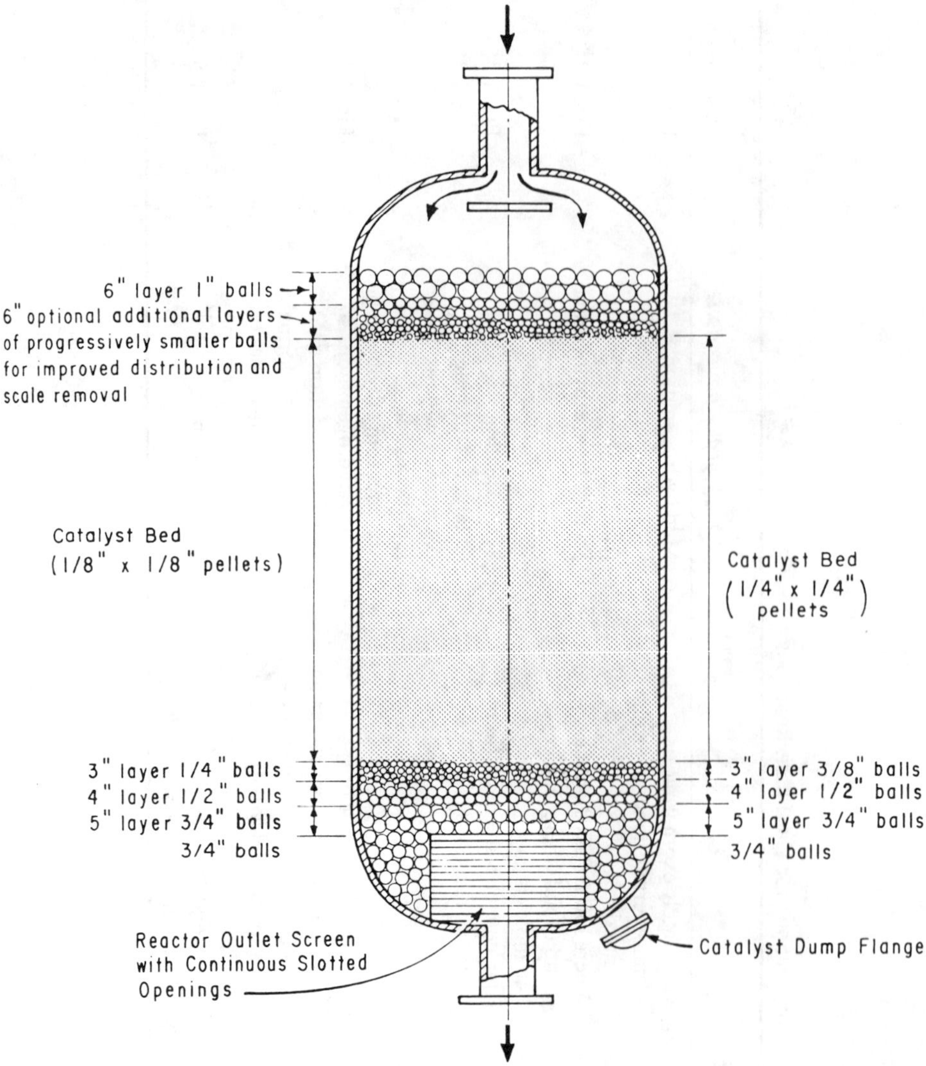

Figure 1.1 Single adiabatic bed
Catalyst bed is supported by inert ceramic balls or pellets. Slotted reactor outlet prevents flow from bypassing lower portion of catalyst bed. Inert packing at inlet of bed prevents movement of catalyst particles by high-velocity gas and also prevents scale from collecting in the bed. Single-inlet baffle is usually adequate for beds with an L/D of 10 or more. When flow rates are large and ΔP must be minimized, large-diameter short beds are indicated. In such cases spherical shells are used with the bed occupying the essentially straight portion of the large sphere. In such cases concentric cone distributors are used along with exit collector screens, as shown, to ensure good distribution. (Reproduced by permission: Howard F. Rase, "Chemical Reactor Design for Process Plants," copyright 1977, John Wiley & Sons, Inc., New York.)

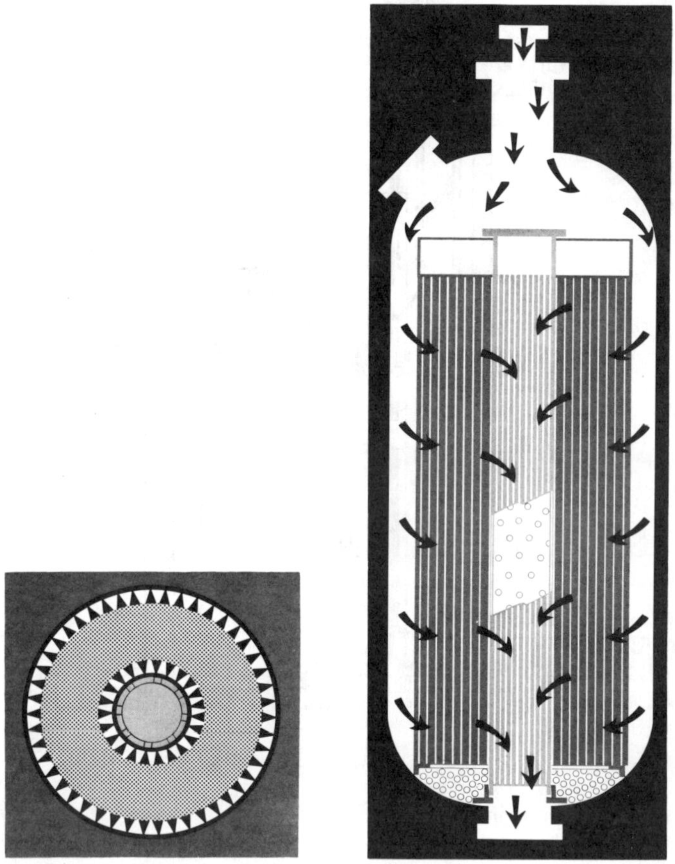

Figure 1.2 Radial-flow reactor
Diagrammatic representation of center pipe and distributor. In this arrangement flow is into the basket and out of the center pipe. Designs for flow from center pipe outward are also used. The smooth side of the slotted screen (smaller openings) is always arranged to contact catalyst bed. The perforated distributor in the center pipe is usually designed for 2 psi ΔP to ensure good distribution along the entire length. (Reproduced by permission: Johnson Filtration Systems, Inc., St. Paul, Minnesota.)

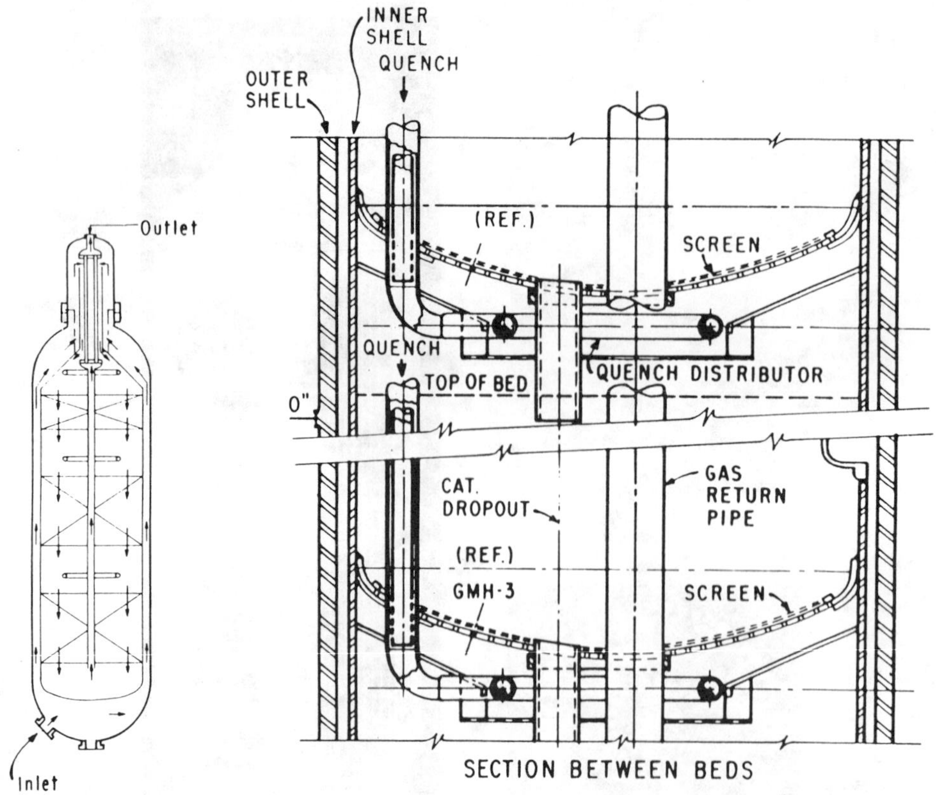

Figure 1.3 Adiabatic beds in series

Equilibrium-limited exothermic reactions and rapid endothermic reactions require multiple beds with, respectively, intermediate cooling or heating. This function can be accomplished with multiple beds in a single shell using external or direct-contact heat exchange. Alternatively, single separate reactors can be used in series with intermediate exchange. Illustrated is a Kellogg-type ammonia reactor. Overall view shows movement of gases. Inlet gas travels in the annulus between the stainless steel basket and the outer low-alloy reactor body, which is protected by refractory on its inside wall as well as the cooler inlet gas so that it can withstand the high pressure at a lower temperature. The inlet gas passes through an exchanger at the top to gain heat from the product stream. Flow then proceeds through the successive beds with quench gas (recycle reactants) injected between beds through a concentric ring. Beds are emptied from the bottom upward. As the top of a bed recedes below the connecting pipe, the bed above begins to discharge. Individual beds in series can range from cylindrical vessels as in Fig. 1.1 to shallow-bed spherical reactors. Intermediate heat exchange can be by shell-and-tube heat exchangers of low ΔP on the gas side, or, for endothermic systems at high temperature, low-ΔP furnaces can be used. (Note: A new design, with an internal exchanger instead of quench, eliminates the diluting effect of quench and increases efficiency.) (Reproduced by permission: M.W. Kellogg Co., Houston, and Eshenbrenner, G.P., Wagner, G.A., *Chemical Engineering Progress, 68*, 62 (1972) by permission of the American Institute of Chemical Engineers.)

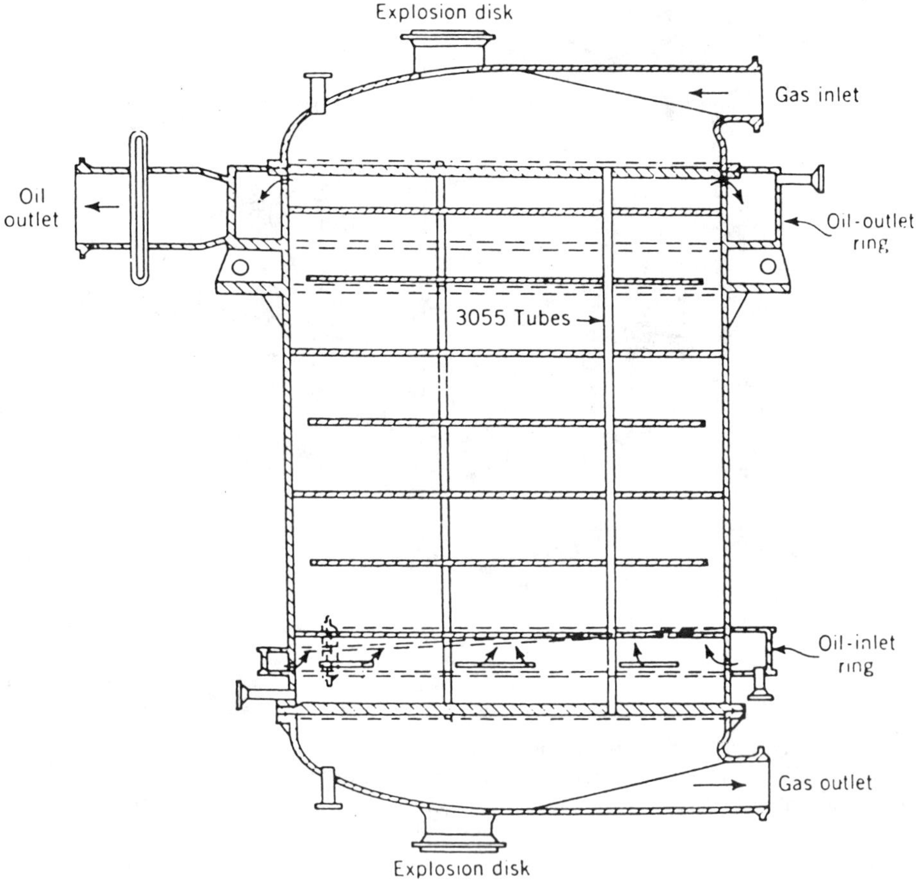

Figure 1.4 Multitubular reactor
Illustrated is a small 3000-tube reactor that is cooled by a heat-transfer fluid such as an oil or Dowtherm. The fluid is circulated through an exterior exchanger where it is cooled by boiling boiler feedwater. The stream formed is separated in a disengaging drum and then superheated in a stack-gas exchanger at some nearby furnace for subsequent release in the steam distribution system. The pressure of the steam generated corresponds to the saturated steam pressure at the temperature of the boiling water, which temperature is close to the entering hot-fluid temperature. Alternatively, when outlet reactor temperature, which is essentially the same as the cooling medium temperature, is lower ($<525°K$), boiling boiler feedwater is used in the shell side as the cooling medium. It passes to a disengaging drum, and new boiler feedwater is added at the drum. Circulation between the drum and the reactor shell occurs by a thermosiphon mechanism. In both types of cooling systems the back pressure on the drum serves to control the boiling temperature and, thus, the reactor shell-side temperature, which is the major reactor control variable. (Reproduced by permission: "Kirk-Othmer Encyclopedia of Chemical Technology," 3rd Edition, Vol. 9, copyright 1980, John Wiley & Sons, Inc., New York.)

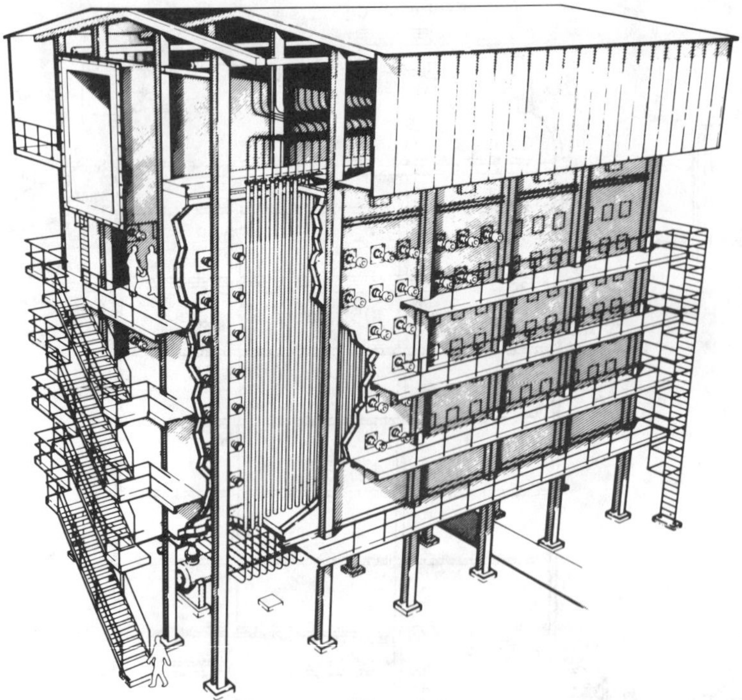

Figure 1.5 Direct-fired nonadiabatic reactor

This type reactor is used for high-temperature, rapid endothermic reactions requiring high fluxes obtainable by a radiant heat source. Various styles of furnaces are in general use with feed being preheated in the convection section of the stack and then flowing down through the vertical tubes that are packed with catalyst. Each type is designed for controlled heat flux along the tube. The radiant-wall type by Selas and Topsoe has multiple adjustable side-wall burners. The terrace-wall type by Foster Wheeler has large upward-burning burners placed in terraces at the site wall. Downfiring is practiced in the Kellogg and ICI furnaces, and upfiring in the Exxon design. The pictorial is of the Topsoe-type furnace for steam reforming service. Note the hairpin expansion sections at the top or inlet (low alloy) and at the outlet (high alloy). The unit has two furnace chambers with horizontal rows of burners on both sides of each chamber. The outlet gases (973–1173°K) pass to a brick-lined collection system. Preheat is provided by exchange with the hot stack gases. (Reproduced by permission: J.R. Rostrup-Nielsen, "Catalytic Steam Reforming" in *Catalysis, Science and Technology*, Vol. 5, edited by J.R. Anderson and M. Boudart, Springer-Verlag, Berlin, 1984.)

PART

I

THEORETICAL AND CONCEPTUAL BACKGROUND

Design, operation, and troubleshooting of fixed-bed reactors for gas-phase reactions require understanding of heterogeneous catalysis (reactions on solid catalysts), the laws of thermodynamics, reaction kinetics, and transport processes together with the associated empirical coefficients essential for quantitative work. In Chapters 2 to 4, each of these areas is treated in a manner that will enable the reader to recall and understand the knowledge essential for competent decision making related to fixed-bed reactors. The goal is to provide the busy professional all essentials in a condensed format. For more detailed treatment the reader should refer to standard works on each subject.

CHAPTER 2

Heterogeneous Catalysis and Kinetics

Knowledge of the broad concepts relevant to the behavior of solid catalysts and the usual types of rate expressions used to describe heterogeneous catalysis is essential to competent reactor design and analysis. These understandings can be expanded and quantified for a particular reaction system by original experimental observations and/or studies on the same system reported in the literature. Essentially all generalizations about solid catalysts are qualitative, but detailed behavior and useful quantitative formulations for specific catalysts have been and will continue to be presented in both the scientific and applied literature.

HETEROGENEOUS CATALYSIS

The term *heterogeneous catalysis* refers to catalytic phenomena occurring on a solid catalyst in a fluid (vapor, liquid, or vapor-liquid) reaction system. The literature in the area is vast, and the subject is complex and, at times, bewildering, but progress in understanding has definitely been made. What began as an intuitively satisfying explanation of heterogeneous catalysis now seems to be well established by increasingly sophisticated experimental observations, particularly those reported in the 1970s. Just as for homogeneous reactions, a totally generalized concept can only be stated in rather elementary terms.

General Concept of Heterogeneous Catalysis

Reactions catalyzed by solid catalysts occur through the bonding (chemisorption) of one or more reactants with surface chemical entities of the catalyst (active sites) to produce a catalyst-reactant complex. Such complexes decompose either directly or through interaction with other complexes (surface reaction) to yield the reaction products that will then appear in the fluid phase after chemidesorption from the surface. (This concept does not speak in terms of detailed events on the atomic scale, but understanding at this level is advancing for particular catalyst systems. See Fig. 2.1).

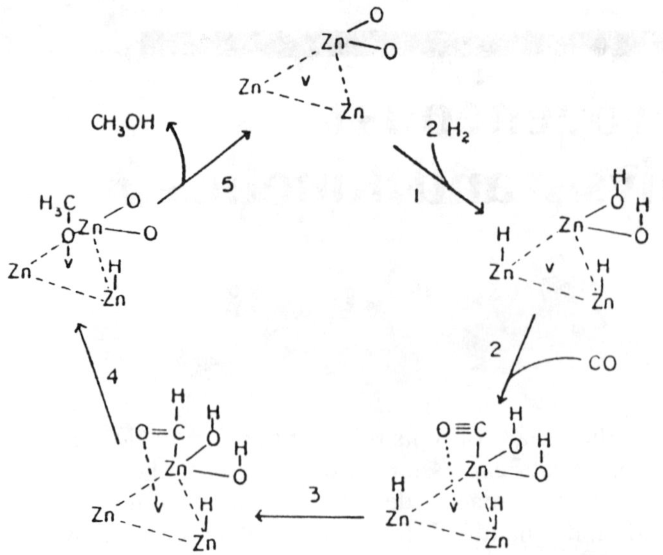

Figure 2.1 Proposed mechanism for methanol synthesis on ZnO catalyst
The mechanism illustrated is for methanol synthesis ($CO + 2\,H_2 \rightleftarrows CH_3OH$) using a ZnO catalyst. A typical modern low-pressure (50 atm) commercial catalyst consists of mixed ZnO and CuO supported on aluminum oxide or chromium oxide. Methanol synthesis from CO and H_2 has been commercially practiced for over 50 years, but the reaction is not yet fully understood. This situation is typical since new catalyst development always adds new complexities that present formidable challenges to researchers seeking understanding. Newer methanol catalysts are more active and useful at lower-temperatures and permit lower-pressure operation for this exothermic equilibrium-limited reaction. Steps illustrated:

1. Adsorption of H_2.
2. CO adsorption involving interaction of the oxygen end of the CO and the electron-deficient vacancy (v) to produce a strongly bonded activated complex. One unique quality of a good methanol catalyst is that it does not dissociate the CO.
3. Reaction proceeds by hydrogenation of the adsorbed CO by migration of an adsorbed hydrogen to form a surface formyl species.
4. Hydrogen of the formyl species to a surface methoxide.
5. Hydrogenation of the methoxide to methanol and desorption of the methanol to yield the original catalyst surface.

These steps have been confirmed by experimental evidence documented by the referenced material. Much of it involves the known behavior of analogous homogeneous reactions involving organometallic catalysts. As a new component, such as Cu in this case, is introduced into the catalyst, agreement on its function will remain controversial for years. The important point from a practical view is that any rational mechanism that makes chemical sense is highly useful for decision making relative to catalyst selection, reactor design, and troubleshooting of an operating unit. (Reproduced by permission: *Catl. Rev.-Sci. Eng. 22*, 235 (1980), by courtesy Marcel Dekker, Inc.)

Corollary 1: For maximum activity per unit mass of a catalyst pellet, the active components must be dispersed over as much accessible surface as possible. (It follows logically that most commerical catalysts will have the active component(s) dispersed on a highly porous carrier with a high fraction of usable area. See Fig. 2.2.)

Corollary 2: Since the heterogeneous catalytic process occurs at the surface of the catalyst, reactants will be transported to the surfaces of the solid catalyst from the bulk fluid phase to replace those consumed. Transport in the opposite direction by products will occur as the products of reaction move from the catalyst surfaces out to the bulk phase. Two distinct transport processes take

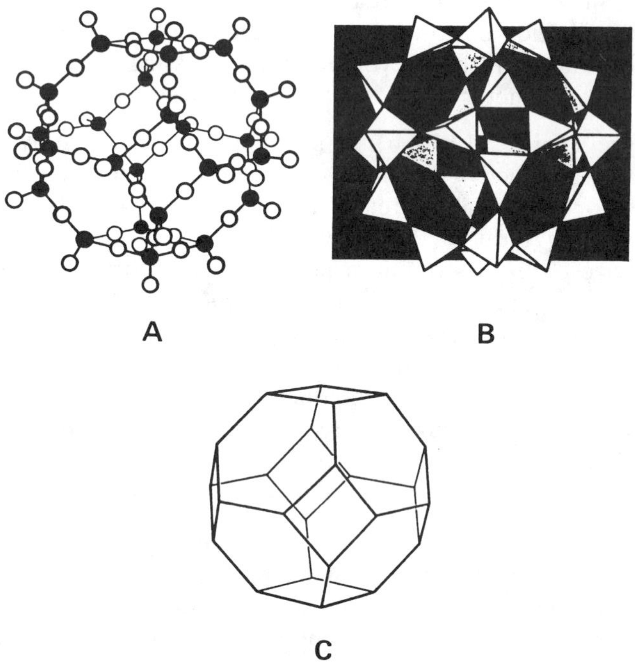

Figure 2.2 Porous catalyst
Three ways of depicting the sodalite unit in aluminosilicate frameworks. Most catalysts are amorphous, and detailed pore structures are not known. Zeolites are crystalline, and pore sizes and arrangements have been documented by x-ray measurements. As shown here, zeolites are made up of uniform openings or pores that vary, depending on the particular type of zeolite. These micropores make up individual crystallites that, when held together in a final pellet or multicrystal aggregate, contain larger pores as well (macropores) created by gaps between crystallites or aggregates. The network of pores and various pore sizes will also occur in amorphous catalysts, but the variety of micropores will be greater. (Reproduced by permission: "Zeolite Chemistry and Catalysis," J.A. Rabo, editor, ACS Monograph 171, copyright 1976, American Chemical Society, Washington, D.C.)

place. Between the bulk fluid and exterior surface of the catalyst, turbulent diffusion is the most likely mode, due to the flow of the fluid around the particles. Transport within the interior porous structure of the catalyst pellet occurs by a diffusional process that is affected by the nature and arrangement of the pores and the resulting interconnecting network (see Fig. 2.2) as well as by the molecular structure and concentration of each diffusing species.

Corollary 3: As reaction occurs on the surface, heat is generated or consumed, causing a transfer of heat between the catalyst and the bulk phase toward the lower-temperature sink. The heat is transferred by a turbulent process between the bulk and exterior surface and by a complex free convective and conductive process within the catalyst pellet.

Corollary 4: Since the surface area of the carrier depends on its porosity and the mode of distribution of pores, it is reasonable to conclude that certain large reactants or products will be prevented from accessing pores that are smaller than the effective diameter of the molecule in question.

Corollary 5 (see Tables 2.4 to 2.6): Since heterogeneous catalysis is a surface phenomenon, loss of activity (*deactivation*) can occur through loss of surface area of carrier or of available area of the active components. Loss of carrier surface by *sintering* reduces the accessibility of reactants to active sites by reducing the size or by destroying a portion of the pore openings. Available area of the active component can be lost when conditions are such as to encourage rapid *agglomeration* of the active components, strong irreversible bonding of products or impurities to active components (*poisoning**), and physical covering of the active components by deposits (*coking* being most common). Deposits that occur at pore entrances (*pore plugging*) and reduce the cross section at that point can also reduce the effective active area of the catalyst by preventing access to such pores.

Corollary 6: Just as for homogeneous reactions, the chemical reaction rate for an elementary catalytic step must be proportional to the product of the concentrations or partial pressures of the interacting components, each raised to a power corresponding to its stoichiometric coefficient in the elementary step.

The general concept of heterogeneous catalysis and its corollaries are based on evidence gathered over the years. It is so general, however, that it serves primarily as a general conceptual framework for reasoning about catalyst performance. There is no general theory of catalysis that is operational in the sense that would permit a priori detailed design of a catalyst for a particular need using fundamental principles of physics and chemistry. The literature, however, is rich in excellent studies of various catalysts of commercial interest. The data presented and the various hypotheses proposed in such

---

* If the poison can be made to desorb upon removing it from the feed or by purging the bed with inert gas, it is called a *temporary poison*. If it cannot be removed by purging, it is called a *permanent poison*.

studies are invaluable to the designer and to the use of catalysts. Although admittedly the hypotheses proposed are often tentative and subject to considerable debate, they have great value in developing a framework for reasoning that is so essential in the decision processes associated with design and operation.

Sources of much of the current knowledge to be found in the open literature can be found in journal articles abstracted by *Chemical Abstracts*, special publications of symposia, and book series dedicated exclusively to catalysis, as listed in Table 2.1.

Table 2.1 Journals and Book Series Exclusively Devoted to Catalysts

JOURNALS

Applied Catalysis

Industrial & Engineering Chemistry Research (section on Kinetics and Catalysis in each issue)

Journal of Catalysis

Journal of Molecular Catalysis

Kinetics and Catalysis (English translation of Kinetika i Katalız)

Preprints of American Chemical Society Petroleum Chemistry Division

BOOK SERIES

"Advances in Catalysis and Related Subjects" (annual), Academic Press, New York.

"Applied Industrial Catalysis," Academic Press, New York.

"Catalysis," The Chemical Society, London.

"Catalysis," P. H. Emmett, Reinhold, N.Y. (7 volumes, 1954-60).

"Catalysis Reviews - Science and Engineering" (annual), Marcel Dekker, New York.

"Catalysis, Science & Technology," Springer-Verlag, New York.

"Chemical Physics of Solid Surfaces and Heterogeneous Catalysis," Elsevier, New York.

"Conferences on Catalysis in Organic Synthesis," Academic Press, New York.

"Proceedings of the International Congress on Catalysis" (every 4 years), (various locations and publishers)

"Proceedings of the International Symposium on Catalyst Deactivation," Elsevier, New York.

"Studies in Surface Science and Catalysis," Elsevier, New York.

Summary of Data on Commerical Catalysts

The empirical nature of the practice of catalysis necessarily requires heavy dependence on the lore associated with individual practical catalysts. Of particular interest are the activity and selectivity characteristics of the catalyst and the type of reactor used. Tables 2.2 through 2.6 present a summary of some of this lore. These tables should serve as a useful reference in early planning and problem solving. Manufacturers of catalysts through their sales and technical service representatives supply invaluable details and assistance to both customers and potential customers. Licensors of proprietary processes offer similar services along with detailed assistance in planning, design, and troubleshooting.

Table 2.2 Characteristics of Typical Gas-Phase Industrial Solid-Catalytic Processes in Fixed Beds

Reaction	*Operating Conditions*[a]	*Catalyst*	*Typical Life (Years)*	*Process Affecting Life of Catalyst Charge*	*Catalyst Property Affected*
Ammonia synthesis $N_2 + 3H_2 \rightleftharpoons 2NH_3$	450–550°C 200–500 atm SA	Fe with promoters (K_2O) & stabilizer (Al_2O_3)	5–10	Slow sintering	Activity
Methanation (ammonia & hydrogen plants) $CO/CO_2 + H_2 \rightleftharpoons CH_4 + H_2O$	250–350°C 30 atm A	Supported nickel	5–10	Slow poisoning (e.g., S & As from plant upsets)	Activity
Acetylene hydrogenation ("front end") $C_2H_2 + H_2 \rightarrow C_2H_4$	30–100°C 50 atm A	Supported palladium	5–10	Slow sintering, increased bed temperature	Activity/selectivity
Sulfuric acid $2SO_2 + O_2 \rightarrow 2SO_3$	420–600°C 1 atm SA	Vanadium & potassium sulfates on silica	5–10	Slow physical deterioration (formation of dust)	Pressure drop & mass transfer
Methanol synthesis (low temperature) $CO + 2H_2 \rightleftharpoons CH_3OH$	200–300°C 50–100 atm SA or NA	Copper on zinc & aluminum oxides	2–8	Slow sintering	Activity
Low-temperature carbon monoxide shift $CO + H_2O \rightleftharpoons CO_2 + H_2$	200–250°C 30 atm A	Copper on zinc & aluminum oxides	2–6	Slow poisoning, & accelerated sintering by poisons	Activity
Hydrocarbon hydrodesulfurization (including natural gas) $R_2S + 2H_2 \rightarrow 2RH + H_2S$	300–400°C 30 atm A	Cobalt & molybdenum sulfides on aluminum oxide	2–8	Slow coking, metal deposits in some cases (feedstock important)	Mass transfer, pressure drop
High-temperature carbon monoxide shift $CO + H_2O \rightleftharpoons H_2 + CO_2$	350–500°C 30 atm A	Fe_3O_4 on chromia	2–4	Slow sintering, pellet breakage due to water condensation	Activity & pressure drop

Steam reforming natural gas $CH_4 + H_2O \rightleftharpoons CO + 3H_2$	500–850°C 30 atm DFNA	Nickel on calcium aluminate or alumina	2–4	Sintering, occasionally carbon formation & pellet breakage due to plant upsets	Activity pressure drop
Ethylene partial oxidation $2\,C_2H_4 + O_2 \rightarrow 2\,C_2H_4O$	200–270°C 10–20 atm NA	Silver on α-alumina with promoters	1–4	Slow sintering increasing bed temperature	Activity & selectivity
Benzene oxidation to maleic anhydride $C_6H_6 + O_2 \rightarrow C_4H_2O_3$	350°C 1 atm NA	Vanadium & molybdenum oxides with promoters on α-alumina	1–2	Irreversible formation of inactive vanadium phase	Activity & selectivity
Reduction of aldehydes to alcohols $RCHO + H_2 \rightarrow RCH_2OH$	220–270°C 100–300 atm A	Copper on zinc oxide	0.5–1	Slow sintering, pellet breakage (depends on feedstock)	Activity or pressure drop
Partial oxidation of methanol to formaldehyde $CH_3OH \rightarrow CH_2O + H_2$ $CH_3OH + \frac{1}{2}\,O_2 \rightarrow CH_2O + H_2O$	600–700°C 1 atm NA	Silver granules	0.3–1	Poisoning, e.g., iron, some coking (due to poisons)	Selectivity
Acetylene hydrogenation ("tail end") $C_2H_2 + H_2 \rightarrow C_2H_4$	30–100°C 50 atm A	Supported palladium	0.1–0.5	Coke & oil formation (regeneration possible)	Mass transfer
Catalytic hydrocarbon reforming	460–525°C 8–502 atm SA	Platinum alloys on treated alumina	0.01–0.5	Coking (depends on hydrogen pressure), frequent regeneration	Mass transfer
Catalytic cracking of oils	500–560°C 2–3 atm (fluidized bed)	Synthetic zeolites	0.000002	Very rapid coking (continuous regeneration)	Mass transfer

[a]A = adiabatic fixed bed; NA = nonadiabatic fixed bed (multitubular); SA = staged adiabatic with intermediate heating or cooling; DFNA = direct-fired nonadiabatic tubular reactor.

Reproduced by permission: Denny, P.J. and Twigg, M.V. in "Catalyst Deactivation," Intl. Symp. Antwerp, Oct., 1980, ed. Delmon, B. and Froment, G.F., Elsevier, Science Publishers, Physical Sciences & Engineering Div., New York and Amsterdam, 1980. Reactor designations added.

Table 2.3 Generalized Methods for Preventing or Moderating Catalyst Deactivation

Problem	*Cause*	*Method of Minimizing Problem*
Poisoning		
Loss of catalytic surface sites	Adsorbed impurities	Purify feed Use additives that adsorb poison Select conditions that lower adsorption strength Select mass-transfer regimes that reduce activity loss

Table 2.3 *(continued)*

Problem	*Cause*	*Method of Minimizing Problem*
Fouling by coke or carbon		
Loss of catalytic surface sites	Free-radical reaction in gas phase	Avoid free-radical formation Minimize free space for gas Use free-radical traps Use gasifying agents, diluents
	Free-radical reactions at reactor walls	Coat reactor with inert material
	Formation and growth on metals	Avoid coke precursors in feed Use gasifying agents, additives, diluents Passivate the metal
	Formation and growth on metal oxides and sulfides	Reduce acidity of oxide or sulfide Avoid coke precursors in feed Use shape-selective molecular sieves
Loss of catalyst; pore plugging; destruction	Formation of coke or v-carbon in large amounts	Avoid free radicals and coke precursors Use gasifying agents and additives that lower carbon solubility Use large pellets or large pores
	Hot spots in pellet or bed	Use small pellets Use slurry or fluid-bed reactor
Sintering		
Loss of metal area	Particle or crystal migration	Reduce temperature Use thermal stabilizers Avoid water
Loss of support area	Crystallization or collapse	Reduce temperature Use support stabilizers Avoid water

Reproduced by permission: Bartholomew, C.H., *Chem. Eng.* (*N.Y.*), p. 96 (Nov. 12, 1984).

Table 2.4 Catalyst Deactivation by Poisoning

DEFINITION OF CATALYST POISON:

A catalyst poison is an impurity or product in a reaction system that is significantly more strongly adsorbed than the reactants and which, when present in even small amounts, can rapidly reduce catalyst activity due to the inability of poisoned sites to participate in the desired reactions. Since the number of active sites per unit surface area of the support is not large, the cumulative effect of poisons can be devastating to viable operation over a relatively short period of time. If a poison is removed from the feed and the catalyst activity increases and is restored after time, the poison is said to be <u>temporary</u>. If such action does not improve the activity, the poison is called permanent since it is so strongly adsorbed that removal from the feed will not cause desorption at operating conditions.

PROTECTION OF REACTOR WITH GUARD BEDS:

Other than avoiding feeds with known poisons and preventing upsets in upstream units that mighty dump poisons into the reactor bed, the use of guard beds is recommended to protect reactors unduly subject to uncontrolled release of poisons into the feed system. A guard bed consists of a smaller vessel containing often partially deactivated catalyst that is capable of adsorbing poisons for some period of time. When breakthrough does occur, the bed can be bypassed and the catalyst replaced in a timely fashion. If the brief operation with poisoned feed in the main reactor in not acceptable, dual guard beds may be used so that one can always be in service.

GENERALIZATION ABOUT POISONS:

Much of the knowledge about poisons is experience oriented and unique for particular catalysts or types of catalysts. Vendors and licensors provide detailed information on poisons affecting their particular catalysts. Some generalizations are possible, however, for the broad classes of metallic, insulator, and semiconductor types as shown below.

Catalyst Type	*Poisons*	*Comments*
Metallic Catalysts* (Fe, Co, Ni, Ru, Rh, Pd, Os, Ir, Pt, and to a lesser extent Cu, Ag.)	Compounds of N, P, As, Sb, O, S, Sc, and Te with unshared election pairs. Examples: <u>Toxic</u> / <u>Nontoxic</u> H: S : H (toxic) (R) C: S : H (toxic) / (R) C: S(O)(O) : OH (nontoxic)	The elements listed are toxic when in molecules containing unshared electron pairs that can donate an electron to form a strong covalent bond with the metallic catalyst component.

*Portion on metallic catalysts based on Maxted, E. B., <u>Advan. Catal. Relat. Subj.</u> <u>3</u>, 129 (1951)

Table 2.4 *(continued)*

Catalyst Type	*Poisons*	*Comments*
	H O H: P̤ : H O: Ṗ̤:O O	
	Metallic ions and metal derivatives having all 5d orbitals occupied: Cu^{+}, Zn^{2+}, Cu^{2+}, Ag^{+}, Cd^{2+}, In^{3+}, Sn^{2+}, Au^{+}, Hg^{2+}, TI^{+}, Pb^{2+}, Bi^{3+}, Mn^{2+}, Fe^{2+}, Co^{2+}, Ni^{2+}	No toxicity is observed for ions with no d-orbitals or empty d-orbitals. The d-orbitals are strong electron donors for the intermetallic bond between the metallic ion and the metallic catalyst.
	Compounds containing multiple bonds: CO and cyanogen derivatives. Strongly bound and thus permanent poisons.	Other multivalent hydrocarbons are more strongly adsorbed than monovalent. Thus, benzene>cyclohexene, acetylene>ethylene. Enables selective hydrogenation, for example, of acetylene in mixture with ethylene.
Insulators (silica-alumina, zeolites, and activated alumina)	NH_3 and organic bases (quinadine>quinoline >pyrrole>peridine> decylamine>aniline)	High proton affinity causes poison to combine with active acid site thereby destroying the acid carbonium-ion activity.
Semiconductor Types (metal oxides and sulfides)	•Strongly electropositive or electropositive substances may cause poisoning.	•Less sensitive to poisoning especially because of high operating temperature.

	•Metal oxides that are partially reduced are affected by same poisons as metallic catalysts and also by halogens and in some cases small amount of oxygen, H_20 or oxygenated compounds.	•Partial reduction to a particular favorable oxidation state may be crucial for good activity, and small amounts of these poisons can destroy the delicate balance.

Table 2.5 Catalyst Deactivation by Sintering

DEFINITION OF SINTERING:

Sintering is a thermal process that produces a decline in surface area of the active ingredient and/or the support which results in a decline in the observed rate of reaction. The process occurs in various ways, but all mechanisms involve a decline in free energy produced by a decline in surface area and a concomitant decline in solid-fluid interfaces as low-energy solid-solid interfaces are formed. Sintering often becomes rapid above 500 - 650°C but will, of course, occur over a long period of time at lower temperatures.

Type of Process	*Mechanism*
<u>SUPPORTED METALLIC CATALYSTS</u>	
Atomic Migration	Occurs by diffusion and the process is often called agglomeration (coalescence). Two mechanisms have been proposed. Because of smaller curvature, small crystallites will have a higher concentration of atoms at the interface between crystallite and support than will large crystallites. Bulk individual atoms will have a yet higher interfacial concentration. Thus the movement that becomes significant at elevated temperatures is from bulk atoms to small crystallites, and from small crystallites to larger crystallites.

Table 2.5 (*continued*)

Type of Process	*Mechanism*
	Over an extended period, the length of which is inversely proportional to temperature, the average size of crystallites increases.
Crystallite Migration	Occurs with crystallites smaller than 5 nm in diameter. As they move over the surface, collisions with other crystallites occur and agglomeration results. Apparently this type of migration occurs during the early stages of sintering of metallic catalysts. As the catalyst ages and temperatures are increased, atomic migration probably dominates.
Redispersion of Agglomerated Catalytic Components	Sintered catalysts usually must be discarded when the activity declines to an unacceptable level. One commercial catalyst, is regularly reactivated by redispersion of the agglomerated platinum. This is accomplished by treating with an organic chloride that releases chlorine $Pt\ (s) + Cl_2 \rightleftarrows PtCl_2\ (g)$ endothermic The hotter agglomerates shift the equilibrium to $PtCl_2$, which vaporizes then condenses on cooler portions of the support where Pt is favored, thus redispersing the platinum. This chemical-vapor transport is a plausible but not totally documented mechanism.
SUPPORTS (aluminum and silicon oxides, silica-aluminas and zeolite)	
Diffusion (oxides)	In the case of oxides, vacancies in the defect solid migrate toward grain boundaries, and

	material is transferred from the grain boundaries to the neck area between particles constituting pores. This growth reduces pore size and available surface area.
Viscous Flow (silicates and silica-aluminas)	Glass-forming oxides at 600°C experience a major drop in viscosity, which facilitates material transfer by viscous flow, and this process is greatly enhanced by the presence of steam. This phenomenon is the basis of vitrification in the ceramic industry, but is not welcome in catalysis.
STABILIZING ADDITIVES (PROMOTERS)	
Components, particularly metals, and the support itself.	Various additives have been used to retard sintering. Supports in common use are effective in preventing agglomeration of active metals when used at recommended operating conditions. Irreducible oxides or certain elements such as other transition metals or Ba further enhance the stability. Small amounts of lower-valent oxides or chlorides are known to inhibit grain growth in alumina supports, perhaps by easily filling vacancies and thus inhibiting diffusion.

Table 2.6 Catalyst Deactivation by Fouling (Coking or Inorganic Deposits)

DEFINITION OF FOULING:

Fouling is the physical covering of active sites and/or the entire surface of the support, by carbonaceous (coke) or inorganic deposits. Severe fouling can lead to pore plugging which cuts off access to active regions and causes precipitous loss in activity. Moderate fouling causes a steady but slower decline in activity that can reach a rather steady and acceptable level for long-term operation. Carbonaceous deposits that are highly porous permit diffusion to the active sites beneath the deposits; or, under some conditions, the active component itself will migrate toward the exterior surface of the deposits.

Table 2.6 (*continued*)

Type of Fouling	*Description*
Coking	• Coke is a high molecular weight polymer with low hydrogen content. Its exact nature and structure vary greatly with the reaction system, conditions of operation, and age of the deposit. Formation of coke on the catalyst surface is thought to occur in many instances by polymerization of aromatic compounds present or formed, often via olefin polymerization. The polymer dehydrogenates and condenses to form large aggregates of polynuclear aromatics having a turbostatic graphite structure. Coke subjected to prolonged exposure to high temperatures will become more graphitic and more difficult to burn off. Coke can be removed by burning, usually with diluted air (regeneration). The burning rate must be adequate to proceed through the entire bed but not so fast that severe hot spots develop and destroy catalyst activity by sintering.
	• Hydrogen inhibits polymerization and dehydrogenation and thus the coking process. It is often used in excess to accomplish such a goal.
	• Purging with steam will remove heavy oils adsorbed on coke, and a hydrogen purge at elevated temperature can remove fresh coke in some cases by depolymering it.
Inorganic Deposits	• These include corrosion products from upstream equipment such as furnaces, salts

present in the feed that precipitate at the reaction conditions, and metal oxides resulting from decomposition at reaction conditions of metal-organics such as asphaltenes (porphyrin-type molecules in heavy cuts of petroleum). The organic structure surrounding the metal is removed at reaction conditions, leaving the metal on the catalyst, which is oxidized during regeneration.

- In most cases inorganic deposits are difficult or impossible to remove economically, and, if present in large amounts, plug catalyst pores, thus making it necessary to replace the catalyst.
- Guard beds, often employing partially deactivated catalyst, can be used to cause deposition of inorganics in the guard bed, thus protecting the main reactor.

Summary of Causes of Catalyst Deactivation

The important phenomenon of catalyst deactivation has an extensive literature that is most valuable when the behavior of a specific catalyst of interest has been investigated thoroughly. General concepts are also useful, however, and they are summarized for convenience in Tables 2.2 to 2.6, which provide a quick view of the issues.

The prevention of catalyst deactivation (Table 2.3) to some degree is essential for a successful process, but deactivation occurs, if only gradually, on all catalysts, as summarized in Tables 2.4 to 2.6. The ideal solution is to develop catalysts with greater resistance to sintering, poisoning, and fouling. It is this area of catalyst development along with the development of more selective catalysts that remains an art form. Though guided by rapidly evolving insights on the effects of pore structure, the known role of additives on specific catalysts, and some rational chemical explanation of a particular reaction system, improved catalysts continue to result from pure discoveries or from innovations based on experience with other catalysts. Clearly, when one is selecting catalysts or working with catalyst problems, knowledge of both commercial catalyst development and basic studies on various catalysts as reported in the literature is invaluable.

Physical Characteristics of Solid Catalysts Important in Design and Operation

Although chemical characteristics of catalysts must be uniquely described for each catalyst, certain common physical characteristics are useful in quantifying pressure drop, mass and heat transfer, catalyst strength, and surface accessibility. These characteristics include total surface area, pore size and distribution, pellet diameter and shape, pellet and bed densities, crushing strength, and geometric or aerodynamic area. Definitions of these physical properties along with descriptions of measuring methods and associated principles are given in Table 2.7.

Table 2.7 Physical Characteristics of Catalysts

Property	*Symbol*	*Description*
Bulk Density	ρ_b	• Mass of catalyst per unit volume of bed. • Determined routinely by loading a weighing flask of known volume while tapping to produce a dense-packed bed. Reported by catalyst manufacturer in catalogs or by request.
Particle Density	ρ_p	• Mass of catalyst per unit volume of particles. • For uniform shapes, such as tablets all of the same size, can weigh a known number of particles, each of which would have the same volume. For extrudates that normally have varying lengths, liquid immersion in which the liquid displaced by the catalyst is used, but even mercury penetrates large catalyst pores to a degree.

		• Most accurate procedure is to measure ΔP of a bed packed each time to a different bulk density and calculate the best value of ρ_p using linear relations rationalized from the Egun Equation [Egun, S., Anal. Chem., 23, 151 (1951)].
Solid Density	ρ_s	• Mass per unit volume of solid catalyst (excluding pores).
		• Determined by expanding a nonabsorbing gas (He) into a vessel containing a known mass of catalyst. The difference between the volume of the original system and the final volume, back calculated from the change in pressure, is the volume of the solid.
Geometric Surface Area	a_s	• Exterior surface area per unit volume of catalyst.
	a_m	• Exterior surface area per unit mass of catalyst ($a_m = a_s / \rho_p$)
		• Calculated for cylinders and spheres of uniform dimensions or average dimensions Cylinders: $a_s = \frac{4}{d_c} + \frac{2}{h_c}$ Spheres and cylinders with $h_c = d_c$ $a_s = \frac{6}{d}$

Table 2.7 *(continued)*

Property	*Symbol*	*Description*
		• For accurate transport calculations and for mixed sizes the aerodynamic area determined from ΔP measurements is recommended [see Ergun, S., Anal. Chem., 24, 388 (1952)].
Equivalent Spherical Diameter	D_p	• Used in pressure-drop correlations for any shape catalyst. $D_p = \frac{6}{a_s} = \frac{6\rho_p}{a_m}$ for spheres and cylinders with $h_c = d_c$, $D_p = d$ or d_c, where d = sphere diameter, d_c = cylinder diameter, and h_c = cylinder height.
Total Surface Areas	S_g	• Total surface area per unit mass of catalyst. • Determined by nitrogen adsorption on an evacuated sample at the boiling point of nitrogen, -195.8°C, and over a range of pressures below 1 atmosphere. • The semi-empirical equation of Brunauer-Emmett-Teller (BET) is used to obtain the volume of nitrogen adsorbed as a monolayer, V_m, corrected to O°C and 760 mmHg. $\frac{P}{V^o(P_o - P)} = \frac{1}{V_m c} + \frac{(c-1)P}{cV_m P_o}$

where c is a constant; V^o is the volume of gas adsorbed calculated at O°C and 760 mm Hg, P is the pressure, and P_o is the vapor pressure of N_2 at 195.8°C.

- The value of V_m can be obtained from the slope (S_ι) and intercept (I) of the linear plot of $P/V^o(P_o - P)$ vs P/P_o.

$$V_m = 1/\left(I + S_\iota\right)$$

$$S_g = \frac{V_m N_o}{M_s\, 22400}\,\alpha_a$$

where N_o is Avogadro's number (6.02×10^{23}), in molecules/mole, M_s mass of sample in grams, and α_a is the projected area of an adsorbed N_2 molecule (16.2×10^{-20} m^2).

- Special automated equipment capable of multiple sample analysis is manufactured.

Average Pore Radius $\quad \bar{r}_p$

- A fictitious value based on assuming all pores of the same radius and length.
- $$\frac{\varepsilon_p / \rho_p}{S_g} = \frac{\pi \bar{r}_p^{\,2}}{2\pi \bar{r}_p} = \frac{\bar{r}_p}{2}$$

where $\varepsilon_p = 1 - \dfrac{\rho_p}{\rho_s}$

Pore-Size Distribution

A quantitative measure of the pore volume due to various ranges of pore radii, usually presented as a plot of pore volume versus pore radius.

Table 2.7 *(continued)*

Property	*Symbol*	*Description*
		• Determined in two ways. 1. Gas adsorption Nitrogen adsorbed until saturation of pores is reached. Pressure is then reduced incrementally and the amount desorbed noted. The Kelvin equation relating vapor pressure to radius of curvature in a capillary is used to estimate radii sizes [see Barrett, E. P., Joyner, L. G., and Halenda, P. P., J. Am. Chem. Soc., 73, 373 (1951)]. 2. Porosimetery Mercury is forced into the catalysts at various pressures, and the volume of Hg entering the capillaries is noted. The force on the system is a function of the interfacial tension in the capillaries which varies with the radius of the capillary [see Ritter, H. L., and Drake, L. C., Ind. Eng. Chem., Anal. Ed., 17, 787 (1945)]. • The two methods do not always agree but Method 2 is better for sizes 100 Å and above, and Method 1 for sizes below 100 Å.

Crush Strength	• The crush strength of a catalyst is determined by applying an increasing load on the opposite sides of the curved surface of a catalyst and reporting the load at failure as force per area. •The values can vary widely, depending on sampling and type of instrument. Bulk testing instead of multiple tests of single pellets has been used. Bulk tests are made by applying an increasing force to a bed of catalyst and noting the displacement for each level of force [see AIChE Sym. Series, 70, No. 143 (1974)].
Attrition Characteristics	• Measures tendency of catalyst pellets to produce fines when rolling over each other. Commonly done by gentle turning of a sample of catalyst in a rotating horizontal cylinder with a single baffle. Test length is often set at 1 hour, and the result is reported as weight percent attrited [see AIChE Sym. Series, 70, No. 143 (1974)].

QUANTIFICATION OF RATE PROCESSES INVOLVING SOLID CATALYSTS

The general concept of heterogeneous catalysis and its corollaries provide a general guide for producing operational empiricisms that have been most useful in predicting rates of heterogeneous catalysis.

Though conceptually useful, they offer no guidance to a priori knowledge of the chemical steps that occur between the reactants and catalyst nor any insight into the nature of active surface species of the catalyst. Briefly, reasoning from the general to the particular is essentially impossible. Rarely are elementary steps, as required for Corollary 6, known. Careful experimentation, however, on a specific catalyst system within the framework of the general concept and corollaries can lead to some explicit hypotheses on the chemistry occurring on a particular catalyst. A useful empirical rate form can usually be derived, which, in some cases, may reflect this behavior in its structure. In any event, if any predictive calculations are to be made, it is essential to have a rate equation or rate equations and the necessary data for calculating the value of the temperature-dependent constants appearing therein.

In most cases, rate equations for catalytic reactions are developed from experimental data by using one of the three catalytic rate forms described in the next sections. The third and most detailed form is only occasionally justified, and then only when more insight into catalyst-reactant interaction exists.

Form 1: Power-Law Expressions

Power-law expressions are based on the simple fact that when the rates of a series of explicit chemical steps involving various intermediates are algebraically manipulated to yield a single rate form involving only reactants and products, that rate equation will involve power functions of the reactants and products. When the steps are not known, the exponents can be determined by trial-and-error fitting to experimental data or by a more orderly regression analysis. In such cases it is common practice to restrict the exponents to integral or half integral values:

$$a\mathrm{A} + b\mathrm{B} \rightleftarrows r\mathrm{R}$$

$$\hat{r}_\mathrm{A} = k_p P_\mathrm{A}^e P_\mathrm{B}^f P_\mathrm{R}^g - k_p' P_\mathrm{A}^{e'} P_\mathrm{B}^{f'} P_\mathrm{R}^{g'} \tag{2.1A}$$

where k_p, k_p' are, respectively, apparent rate constants for the forward and reverse reaction in moles (mass cat.)$^{-1}$(time)$^{-1}$(atm)$^{-n}$; P_A, P_B, P_R are partial pressures of the indicated components in atmospheres; and $\hat{r}_\mathrm{A}$ is the rate of disappearance of A in moles (mass cat.)$^{-1}$(time)$^{-1}$.

Eq. 2.1A can be written in concentration units as

$$\hat{r}_\mathrm{A} = \hat{k}_c C_\mathrm{A}^e C_\mathrm{B}^f C_\mathrm{R}^g - \hat{k}_c' C_\mathrm{A}^{e'} C_\mathrm{B}^{f'} C_\mathrm{R}^{g'} \tag{2.1B}$$

To consider every possibility, products are included in the forward step and reactants in the reverse step. Thus, if these entities inhibit the rate, the exponents for them will be negative. Any entity that has no effect will have an exponent of zero. Often, one or more exponents will be zero.

Exponents determined from experimental data must make thermodynamic sense. Criteria can be derived by comparing the equilibrium expression for the overall reaction with that obtained by setting Eq. 2.1A equal to zero [1]. (See Chapter 3 for discussion of equilibrium.)

For the foregoing example ($a\text{A} + b\text{B} \rightarrow r\text{R}$) and an ideal gas forming an ideal solution, the equilibrium constant K is related to the partial pressures at equilibrium ($P_\text{A}^* \cdots P_\text{R}^*$) as follows:

$$K = \frac{(P_\text{R}^*)^r}{(P_\text{A}^*)^a (P_\text{B}^*)^b} \tag{2.2}$$

At equilibrium Eq. 2.1A becomes

$$\frac{k_p}{k_p'} = \frac{(P_\text{R}^*)^{g'-g}}{(P_\text{A})^{e-e'} (P_\text{B})^{f-f'}} \tag{2.3}$$

Since k_p / k_p' and K are functions only of temperature,

$$\frac{k_p}{k_p'} = f(K) \tag{2.4}$$

When the RHS of Eqs. 2.2 and 2.3 are equated through the functional statement of Eq. 2.4, the equality is satisfied if $k_p / k_p' = K^\alpha$. Thus,

$$(P_\text{R}^*)^{g'-g} (P_\text{A}^*)^{e'-e} (P_\text{B}^*)^{f'-f} = [(P_\text{R}^*)^r (P_\text{A}^*)^{-a} (P_\text{B}^*)^{-b}]^\alpha \tag{2.5}$$

Hence, for thermodynamic consistency Eq. 2.5 applies when

$$\frac{g'-g}{r} = \frac{e-e'}{a} = \frac{f-f'}{b} = \alpha \tag{2.6}$$

where $\alpha > 0$.

Example: $N_2 + 3\,H_2 \rightleftarrows 2\,NH_3$

Assume that kinetic data are being gathered for this reaction over promoted iron-oxide catalyst, and data on the forward rate have been analyzed.

$$\hat{r} = k_p P_{N_2}^e P_{H_2}^f P_{NH_3}^g - k_p' P_{N_2}^{e'} P_{H_2}^{f'} P_{NH_3}^{g'}$$

$$K = \frac{P_{NH_3}^2}{P_{N_2} P_{H_3}^3}$$

$$\frac{g'-g}{2} = \frac{e-e'}{1} = \frac{f-f'}{3}$$

Exponents for the forward rate were found to be $g = -1$, $e = 1$, and $f = 1.5$.

Thus

$$\frac{g'+1}{2}=\frac{1-e'}{1}=\frac{1.5-f'}{3}$$

There are several possibilities for the reverse exponents, all of which are thermodynamically consistent. The following list represents three sets of possibilities.

Form	e'	α	f'	g'
A	0	1	−1.5	1
B	−0.5	1.5	−3	2
C	−1.0	2	−4.5	3

The form selected must be the one that best fits the data from experiments in which the reverse reaction plays an important role. For this example, Form A was found to be the best fit. In this case, a mechanistic argument was used, and an equation based on the adsorption of nitrogen being the controlling step [2] was derived and tested. The same result would have been possible by a totally empirical approach of finding the best fit of the forward exponents and then the reverse exponents, with the latter confined to the thermodynamically consistent values. If in the initial stages of an experimental kinetic study, when the constant and exponents of the forward rate have been determined and further investigation reveals that one of the components does not affect the reverse rate and thus would not appear in it, only one value of α is possible. In such instances, further experimental effort can be greatly minimized.

Form 2: Hougen-Watson (H-W) Equations (Modified Langmuir-Hinshelwood) [2]

In Form 2 an attempt is made to separate the three distinct chemical processes described in the general concept of catalysis.

1. Chemisorption of reactants
2. Surface reaction of chemisorbed reactants and reverse reaction of chemisorbed products
3. Chemidesorption of products

The following assumptions and strategies are followed:

1. Simplified Langmuir adsorption theory applies, which means that there is no interaction between adsorbed molecules (the heat of adsorption is constant) and components that adsorb affect each other only by competing for active sites.

2. One of the steps is considered rate-controlling, while the others are assumed at equilibrium. An equation is derived, and its ability to fit experimental data is determined. The process is repeated for all other possible controlling steps, and the one giving the best fit is used for all calculational purposes.

3. Constants are determined from reaction data only, not by independent adsorption measurements.

4. No attempt is made to postulate the details of the chemistry at the molecular level, including steps that might occur during the surface reaction.

5. An ideal surface is envisioned with geometrically arranged sites in a geometric pattern.

Quite simply put, the equation that fits is the result of an idealization that should not be referred to as a mechanism. Even though most evidence suggests that the Langmuir theory does not apply to heterogeneous reactions, these equations are frequently used because they have conceptual value, as shown after the following example derivation, and they often yield good fits to experimental data.

Example Derivation of an H-W Equation

For $A + B \rightleftarrows R + S$, derive an equation for which the surface reaction between adsorbed reactants and products is controlling. The symbol ℓ represents an active site on the catalyst surface.

Chemisorption of reactants:

$$a + \ell \rightleftarrows A\ell$$

$$B + \ell \rightleftarrows B\ell$$

Surface reaction:

$$A\ell + B\ell \rightleftarrows R\ell + S\ell$$

Chemidesorption of products:

$$R\ell \rightleftarrows R + \ell$$

$$S\ell \rightleftarrows S + \ell$$

Express rate in terms of the controlling step, and express all other steps at equilibrium. The rate of the surface step is considered proportional to the concentration of pairs of adjacently adsorbed reactants, which, as shown in Fig. 2.3, is proportional to $sC'_A C'_B / L_R$ in the forward direction and $sC'_R C'_S / L_R$ in the reverse, where C'_A, C'_B, C'_R, C'_S are concentrations (moles/mass of catalyst) of indicated adsorbed species, s is the number of adjacent active sites equidistant to any site, L_R is the total number of active sites (moles/mass of catalyst), and k_s, k'_s are surface reaction-rate constants.

The net rate of consumption of A is

$$\hat{r}_A = k_s s \frac{C'_A C'_B}{L_R} - k'_s s \frac{C'_R C'_S}{L_R}$$

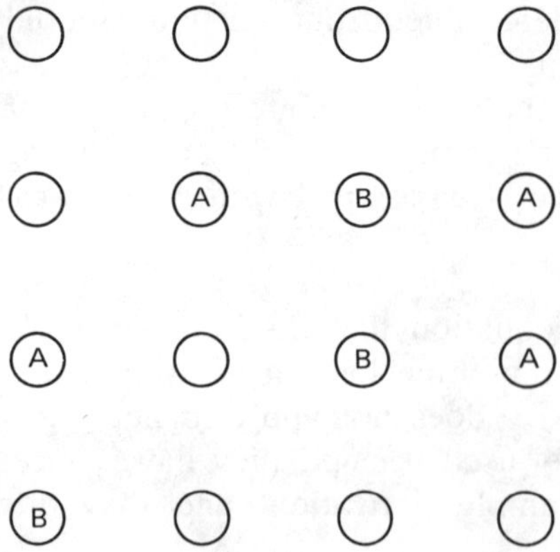

Figure 2.3 Surface reaction proportionalities
Portion of an idealized surface shown: Ⓐ = adsorbed reactant A; Ⓑ = adsorbed reactant B; ◯ = vacant site.

$$P_n = sA\frac{B}{S_T} = (4)(4)\frac{3}{16} = 3$$

More accurately,

$$P_n = 4\,A\frac{B}{16 - A} = (4)(4)\frac{3}{16 - 4} = 4$$

The first solution is usually close enough over a larger area. Nomenclature: $s = 4$, adjacent active sites, equidistant to any site; A = number of adsorbed A; B = number of adsorbed B; S_T = Total number of sites; P_n = number of pairs of A and B.

The number of adjacent sites is simply combined with the rate constants:

$$\hat{r}_A = \frac{k_s}{L_R}\left(C'_A C'_B - \frac{C'_R C'_S}{K'}\right) \tag{2.7}$$

where K' is the equilibrium constant for the surface reaction.

All other steps are represented as at equilibrium, which, in this case, is adsorption equilibrium expressed by the Langmuir form. Hougen and Watson used activities for the bulk-phase components, which for ideal gases become partial pressures [3]:

$$K_A = \frac{C'_A}{a_A C'_\ell}, \quad K_B = \frac{C'_B}{a_B C'_\ell}, \quad K_R = \frac{C'_R}{a_R C'_\ell}, \quad K_S = \frac{C'_S}{a_S C'_\ell}$$

where C'_ℓ is the concentration of vacant sites in moles/mass catalyst. Substituting in Eq. 2.7 gives

$$\hat{r}_{\mathrm{A}} = L_R k_s \left(K_{\mathrm{A}} a_{\mathrm{A}} C'_\ell K_{\mathrm{B}} a_{\mathrm{B}} C'_\ell - \frac{K_{\mathrm{R}} a_{\mathrm{R}} C'_\ell K_{\mathrm{S}} a_{\mathrm{S}} C'_\ell}{K'} \right) \tag{2.8}$$

The s term is omitted since it is an unknown characteristic of the catalyst and can be assumed part of k_s. To eliminate C'_ℓ, write an expression for total number or concentration of sites, L_R:

$$L_R = C'_\ell + C'_{\mathrm{A}} + C'_{\mathrm{B}} + C'_{\mathrm{R}} + C'_{\mathrm{S}} \tag{2.9}$$

From adsorption equilibrium relations,

$$L_R = C'_\ell + C'_\ell K_{\mathrm{A}} a_{\mathrm{A}} + C'_\ell K_{\mathrm{B}} a_{\mathrm{B}} + C'_\ell K_{\mathrm{R}} a_{\mathrm{R}} + C'_\ell K_{\mathrm{S}} a_{\mathrm{S}} \tag{2.9A}$$

or

$$C'_\ell = \frac{L_R}{1 + K_{\mathrm{A}} a_{\mathrm{A}} + K_{\mathrm{B}} a_{\mathrm{B}} + K_{\mathrm{R}} a_{\mathrm{R}} + K_{\mathrm{S}} a_{\mathrm{S}}} \tag{2.10}$$

Substituting Eq. 2.10 in Eq. 2.8 gives

$$\hat{r}_{\mathrm{A}} = \frac{k_s L_R (K_{\mathrm{A}} a_{\mathrm{A}} K_{\mathrm{B}} a_{\mathrm{B}} - K_{\mathrm{R}} a_{\mathrm{R}} K_{\mathrm{S}} a_{\mathrm{S}} / K')}{(1 + K_{\mathrm{A}} a_{\mathrm{A}} + K_{\mathrm{B}} a_{\mathrm{B}} + K_{\mathrm{R}} a_{\mathrm{R}} + K_{\mathrm{S}} a_{\mathrm{S}})^2} \tag{2.11}$$

A relationship between the adsorption equilibrium constants and K' and K, the overall equilibrium constant, can be obtained by setting Eq. 2.11 equal to zero, which, in effect, considers the limit of all steps at equilibrium because the other steps were already at equilibrium:

$$\frac{a_{\mathrm{R}} a_{\mathrm{S}}}{a_{\mathrm{A}} a_{\mathrm{B}}} = \frac{K_{\mathrm{A}} K_{\mathrm{R}} K'}{K_{\mathrm{R}} K_{\mathrm{S}}} = K \tag{2.12}$$

or

$$\frac{K_{\mathrm{R}} K_{\mathrm{S}}}{K'} = \frac{K_{\mathrm{A}} K_{\mathrm{B}}}{K} \tag{2.13}$$

Substituting Eq. 2.13 into Eq. 2.11 yields

$$\hat{r}_{\mathrm{A}} = \frac{k_s L_R K_{\mathrm{A}} K_{\mathrm{B}} (a_{\mathrm{A}} a_{\mathrm{B}} - a_{\mathrm{R}} a_{\mathrm{S}} / K)}{(1 + K_{\mathrm{A}} a_{\mathrm{A}} + K_{\mathrm{B}} a_{\mathrm{B}} + K_{\mathrm{R}} a_{\mathrm{R}} + K_{\mathrm{S}} a_{\mathrm{S}})^2} \tag{2.14}$$

Derivations for all other possibilities follow an identical pattern. The rate is written in terms of the controlling step, and all other steps are expressed at equilibrium and used to eliminate C'_ℓ, the concentration of vacant sites. Such sameness deserves a simple tabular scheme for rapid statement of various possibilities. Table 2.8 A, B reproduces such an aid, written in terms of partial pressures, the more commonly used form.

Table 2.8A Terms for Hougen-Watson-Type Rate Equations

Driving Potential Terms for Hougen-Watson-Type Rate Equations

Reaction	$A \rightleftarrows R$	$A \rightleftarrows R + S$	$A + B \rightleftarrows R$	$A + B \rightleftarrows R + S$
Adsorption of A controlling	$p_A - \frac{p_R}{K}$	$p_A - \frac{p_R p_S}{K}$	$p_A - \frac{p_R}{K\,p_B}$	$p_A - \frac{p_R p_S}{K\,p_B}$
Adsorption of B controlling	0	0	$p_B - \frac{p_R}{K\,p_A}$	$p_B - \frac{p_R p_S}{K\,p_A}$
Desorption of R controlling	$p_A - \frac{p_R}{K}$	$\frac{p_A}{p_S} - \frac{p_R}{K}$	$p_A p_B - \frac{p_R}{K}$	$\frac{p_A p_B}{p_S} - \frac{p_R}{K}$
Surface reaction controlling	$p_A - \frac{p_R}{K}$	$p_A - \frac{p_R p_S}{K}$	$p_A p_B - \frac{p_R}{K}$	$p_A p_B - \frac{p_R p_S}{K}$
Impact of A controlling (A not absorbed)	0	0	$p_A p_B - \frac{p_R}{K}$	$p_A p_B - \frac{p_R p_S}{K}$
Homogeneous reaction controlling	$p_A - \frac{p_R}{K}$	$p_A - \frac{p_R p_S}{K}$	$p_A p_B - \frac{p_R}{K}$	$p_A p_B - \frac{p_R p_S}{K}$

Kinetic Terms for Hougen-Watson Rate Equations

Multiply each of the terms below by $\eta\, L_R$ except for the homogeneous reaction

Adsorption of A controlling	k_A
Adsorption of B controlling	k_B
Desorption of R controlling	$k_R K$
Adsorption of A controlling with dissociation	k_A
Impact of A controlling	$k_A K_B$
Homogeneous reaction controlling	k

Surface reaction controlling

	$A \rightleftarrows R$	$A \rightleftarrows R + S$	$A + B \rightleftarrows R$	$A + B \rightleftarrows R + S$
Without dissociation	$k_S K_A$	$k_S K_A$	$k_S K_A K_B$	$k_S K_A K_B$
With dissociation of A	$k_S K_A$	$k_S K_A$	$k_S K_A K_B$	$k_S K_A K_B$
B not adsorbed	$k_S K_A$	$k_S K_A$	$k_S K_A$	$k_S K_A$
B not adsorbed, A dissociated	$k_S K_A$	$k_S K_A$	$k_S K_A$	$k_S K_A$

Reproduced by permission of the American Institute of Chemical Engineers, "Determination of Mechanism of Catalyzed Gaseous Reactions," K. H. Yang and O. A. Hougen, *Chem. Eng. Prog.*, **46**, 146 (1950).

Conceptual Value of H-W Equations

Hougen-Watson equation forms are more difficult to manipulate and require extra experimental effort to evaluate the multiple constants. They do provide a conceptual framework that is at least qualitatively valuable. The form of the equation makes this point clear, as do the tables.

Table 2.8B Terms for Hougen-Watson-Type Rate Equations
Replacements in the General Adsorption Terms $[1 + K_A p_A + K_B p_B + K_R p_R + K_S p_S + K_I p_I]^n$ *for Hougen-Watson-Type Rate Equations*

Reaction	$A \rightleftarrows R$	$A \rightleftarrows R + S$	$A + B \rightleftarrows R$	$A + B \rightleftarrows R + S$
Where adsorption of A is rate controlling, replace $K_A p_A$ by	$\frac{K_A p_R}{K}$	$\frac{K_A p_R p_S}{K}$	$\frac{K_A p_R}{K p_B}$	$\frac{K_A p_R p_S}{K p_B}$
Where adsorption of B is rate controlling, replace $K_B p_B$ by	0	0	$\frac{K_B p_R}{K p_A}$	$\frac{K_B p_R p_S}{K p_A}$
Where desorption of R is rate controlling, replace $K_R p_R$ by	$K K_R p_A$	$K K_R \frac{p_A}{p_S}$	$K K_R p_A p_B$	$\frac{K K_R p_A p_B}{p_S}$
Where adsorption of A is rate controlling with dissociation of A, replace $K_A p_A$ by	$\sqrt{\frac{K_A p_R}{K}}$	$\sqrt{\frac{K_A p_R p_S}{K}}$	$\sqrt{\frac{K_A p_R}{K p_B}}$	$\sqrt{\frac{K_A p_R p_S}{K p_B}}$
Where equilibrium adsorption of A takes place with dissociation of A, replace $K_A p_A$ by	$\sqrt{K_A p_A}$	$\sqrt{K_A p_A}$	$\sqrt{K_A p_A}$	$\sqrt{K_A p_A}$
	and similarly for other components adsorbed with dissociation.			
Where A is not adsorbed, replace $K_A p_A$ by	0	0	0	0

Exponents of Adsorption Terms for Hougen-Watson-Type Rate Equations

Adsorption of A controlling without dissociation	$n = 1$
Desorption of R controlling	$n = 1$
Adsorption of A controlling with dissociation	$n = 2$
Impact of A without dissociation $A + B \rightleftarrows R$	$n = 1$
Impact of A without dissociation $A + B \rightleftarrows R + S$	$n = 2$
Homogeneous reaction	$n = 0$

	$A \rightleftarrows R$	$A \rightleftarrows R + S$	$A + B \rightleftarrows R$	$A + B \rightleftarrows R + S$
No dissociation of A	1	2	2	2
Dissociation of A	2	2	3	3
Dissociation of A (B not adsorbed)	2	2	2	2
No dissociation of A (B not adsorbed)	1	2	1	2

$$\hat{r}_A = \frac{(\text{Kinetic term})\ (\text{Driving potential term})}{\text{Adsorption term}} \tag{2.15}$$

Kinetic Term: The kinetic term contains L_R, the total number of sites, which clearly focuses on the effect on activity when sites are destroyed by sintering or permanent adsorption of an impurity. If the fresh value of L_R is taken as unity and rates are determined at the same conditions for the fresh and used catalysts, the ratio of these two rates is the fractional value of L_R for the used catalyst, called the *fractional activity* or *fraction of sites remaining active*:

$$\frac{(\hat{r}_A)_{\text{used}}}{(\hat{r}_A)_{\text{fresh}}} = \frac{L'_R}{L_R} \tag{2.16}$$

where L'_R represents the total number of active sites on used catalyst or fractional activity when L_R is called unity.

The product of constants for the case of surface reaction controlling explains in a useful way the rise and fall of the apparent rate constant ($k = k_s K_A K_B$) with increasing temperature. The true rate constant k_s increases continuously with temperature, but adsorption constants decrease. Phenomenologically this observation is explained by realizing that as temperature is raised the amount adsorbed declines, even though the intrinsic reaction rate per reaction event increases. Obviously, at some point, the net rate of disappearance of A will decline as temperature is further increased.

Driving Potential Term: The driving potential term is the conventional term related to the distance from equilibrium. At zero rate, it reduces to the correct overall equilibrium expression.

Adsorption Term: The adsorption term contains all components that are significantly adsorbed and, in the context of this model, inhibit the rate. If an *inert* (a compound not participating in the reaction) is adsorbed strongly enough, it would appear as $C'_\ell K_I a_I$ in Eq. 2.11 and, finally, as an additive term $K_I a_I$ in Eq. 2.14, the final form. Such mathematical behavior fits the actual behavior of what is called a temporary poison. When present, it inhibits the reaction. When removed from the feed, it desorbs; the activity decline that was due to its presence is restored.

Thus the form of these equations is appealing to quick reasoning about reactions. There is continued debate over whether better fits are possible with these equations than with the power-law model. The professional must be flexible and use either form, since experimental data will be correlated according to the subjective leanings of the researcher. There is no reason to redo the correlation unless it is based on insufficient data to justify the more complex case.

Form 3: Mechanistic Model

Although even the mechanism of the simplest catalytic reaction, ortho/para hydrogen conversion, remains a subject of lively debate, researchers often boldly offer hypotheses for the most complex reaction systems and at times derive a single rate equation based on the various postulated steps. Such efforts are commendable, for they provide a useful, though far from perfect, mechanistic view that often increases one's intuitive and predictive sense concerning a reaction system. An equation thus developed should be welcomed, particularly if it fits the data well. Unfortunaltely, such developments are few. Many years of tedious experimentation is necessary for even an approximate approach to a catalytic mechanism, and such effort can seldom be justified in a commerical environment.

Example of a Mechanistic Form

The oxidation of SO_2 over V_2O_5 supported on alumina involves a single reaction with no side reactions. It has been the subject of much study, and mechanistic hypotheses have been numerous.

It is generally agreed that SO_2 is oxidized by oxygen from the V_2O_5 catalyst, and then that oxygen is restored to the catalyst by oxygen from the gas phase. One such scenario suggests the following [4]:

$$SO_2 + 2V^{5+} + O^{2-} \rightleftarrows SO_3 + 2V^{4+} \tag{2.17}$$

$$O_2 + 2V^{4+} \rightleftarrows 2V^{5+} + 2O^{-} \tag{2.18}$$

$$O^{-} + V^{4+} \rightleftarrows V^{5+} + O^{2-} \tag{2.19}$$

Various equations can be obtained, depending on the assumed controlling step. One such equation is based on Eq. 2.18 being controlling and Eqs. 2.17 and 2.19 at equilibrium:

$$(-\hat{r}_{SO_2}) = k_2\left[(C_{V^{4+}})^2(P_{O_2})\ \frac{(C_{V^{5+}})^2(C_{O^-})^2}{K_2}\right] \tag{2.20}$$

$$K_1 = \frac{P_{SO_3}(C_{V^{4+}})^2}{P_{SO_2}(C_{V^{5+}})^2(C_{O^{2-}})} = \frac{K_M}{C_{O^{2-}}} \tag{2.21}$$

$$K_3 = \frac{(C_{V^{5+}})(C_{O^{2-}})}{(C_{O^-})(C_{V^{4+}})} \tag{2.22}$$

By solving for $C_{O^{2-}}$ from Eq. 2.21 and C_{O^-} from Eq. 2.22 and substituting in Eq. 2.20, we get

$$\hat{r}_{SO_2} = k_2 C_{V_t} \frac{K_M P_{SO_2}}{[P_{SO_3}^{1/2} + (K_M P_{SO_2})^{1/2}]^2}\left[P_{O_2} - \left(\frac{P_{SO_3}}{K P_{SO_2}}\right)^2\right] \tag{2.23}$$

where C_{V_t} is the total vanadium concentration of the catalyst that can be combined with k_2 to yield an apparent rate constant for the particular catalyst $k = k_2C_{V_t}$, and

$$K = \frac{P_{SO_3}}{P_{SO_2}P_{O_2}^{1/2}}$$

is the overall equilibrium constant, which is related to the other equilibrium constants by

$$K^2 = K_1^2K_2K_3^2$$

Value of Mechanistic Form

When detailed studies have yielded a rate equation that includes a strong and independently confirmed mechanistic framework, the ability to interpret catalyst behavior is much improved for the user. It must be emphasized, however, that such studies are rare and those that do exist are never completely confirmed. In the seemingly simple and straightforward SO_2 oxidation, the system is known to be quite complex because the higher temperatures that occur in the first bed cause the V_2O_5 to be in the molten state while it is in solid form in subsequent beds. This situation makes it difficult to obtain a single equation applicable to both catalyst forms.

Recommendations on Rate Forms

If involved with the task of developing a rate equation or equations from experimental data, search the literature for articles on similar catalysts. Although the quantitative values may not apply due to differences in catalyst preparation, the form of the rate equation may if the chemical components of the catalysts are essentially the same. If a Form 3 equation has been shown to fit well, use it in preference to any other. Otherwise use Forms 1 or 2, whichever fits better with some attention to custom for the particular system. If it appears that Form 2 equations are commonly used and the fit is good, go along with the custom for that reaction system. It will save time in the long run by avoiding useless explanations.

Pseudo-First-Order Equations

When only a small amount of data have been obtained and one is pressed for predictions, a pseudo-first-order assumption can sometimes prove most helpful in the following situations:

- Excess of one reactant is in feed
- Complex reaction systems sometimes show first-order diappearance of the primary reactant, even though higher-order rates may be involved in side reactions

- Most reactions when plotted as conversion of a limiting reactant versus reciprocal space velocity for a given temperature yield linear portions in both the low and high conversion regions, as shown in Fig. 2.4.

Conversion approaches unity ($X_A = 1$) or equilibrium conversion ($X_A = X_A^*$). For the general reversible case in the low-conversion region [3],

$$\hat{r}_A = \frac{X_A^* - X_A}{\ell} \tag{2.24}$$

Initial rate:

$$\hat{r}_{A_0} = \frac{X_A^*}{\ell} \tag{2.25}$$

$$\hat{r}_A = \frac{\hat{r}_{A_0}}{X_A^*}(X_A^* - X_A) \tag{2.26}$$

$$\frac{W}{F_A} = \int_0^{X_A} \frac{dX_A}{\hat{r}_A} = \frac{X_A^*}{\hat{r}_{A_0}}\left[\ln\left(\frac{1}{1 - X_A/X_A^*}\right)\right] \tag{2.27}$$

where W is the mass of catalyst and F_A is the feed rate of component A. This equation has the familiar logarithmic form for a first-order reaction and produces a linear plot beginning at the origin. The reciprocal of Eq. 2.27 is

$$\frac{F_A}{W} = \frac{\hat{r}_{A_0}}{X_A^*}\left[1/\ln\left(1 - \frac{1}{X_A/X_A^*}\right)\right] \tag{2.28}$$

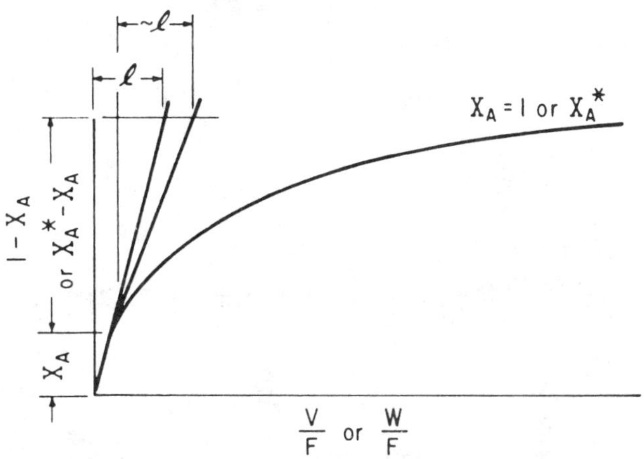

Figure 2.4 Typical conversion versus W/F plot
(Reproduced by permission: H.F. Rase, "Chemical Reactor Design for Process Plants," copyright 1977, John Wiley & Sons, Inc., New York.)

Equation 2.28 yields linear plots in the high conversion region and passes through the new zero coordinates. One good data point on each plot can permit sketching the entire curve by extending the low-range line on the high-range plot, and vice versa. The two do not meet, but a little imagination or, better, one or two intermediate data points will lead to reasonable interpolations. See Fig. 2.5.

Effect of Temperature on Rate Constants

As shown in Eq. 3.28, the reaction equilibrium constant varies with temperature as a function of the heat of reaction ($\Delta H°$ at constant pressure and $\Delta U°$ at constant volume).

$$\left(\frac{\delta K}{\delta T}\right)_P = \frac{\Delta H°}{RT^2} \tag{2.29}$$

In the 1860s, Van't Hoff proposed that, since the equilibrium constant for an elementary reaction was related to the rate constants, the rate constants must exhibit a functional relation to temperature similar to that of the equilibrium constant. Arrhenius [5] added to this concept the following, which he confirmed by experimental data.

From the Guldberg and Waage law of mass action, published in 1867, the rate is zero at equilibrium and, for an elementary reaction of $a\mathrm{A} + b\mathrm{B} \rightleftarrows r\mathrm{R} + s\mathrm{S}$,

$$k_p P_\mathrm{A}^a P_\mathrm{B}^b - k'_\mathrm{P} P_\mathrm{R}^r P_\mathrm{S}^s = 0 \tag{2.30}$$

or

$$\frac{k_p}{k'_p} = \frac{P_\mathrm{R}^r P_\mathrm{S}^s}{P_\mathrm{A}^a P_\mathrm{B}^b} = K \tag{2.31}$$

Hence, $\ln K = \ln k_p - \ln k'_p$. Differentiating this expression with respect to T gives

$$\left(\frac{\partial \ln K}{\delta T}\right)_P = \left(\frac{\partial \ln k_p}{\delta T}\right)_P - \left(\frac{\partial \ln k'_p}{\delta T}\right)_P \tag{2.32}$$

or, since

$$\left(\frac{\partial \ln K}{\delta T}\right)_P = \frac{\Delta H°}{RT^2} \quad \text{(from Eq. 2.29)} \tag{2.33}$$

$$\left(\frac{\partial \ln k_p}{\delta T}\right)_P = \frac{E}{RT^2} + I \tag{2.34}$$

$$\left(\frac{\partial \ln k'_p}{\delta T}\right)_P = \frac{E'}{RT^2} + I \tag{2.35}$$

Experimentally $I = 0$.

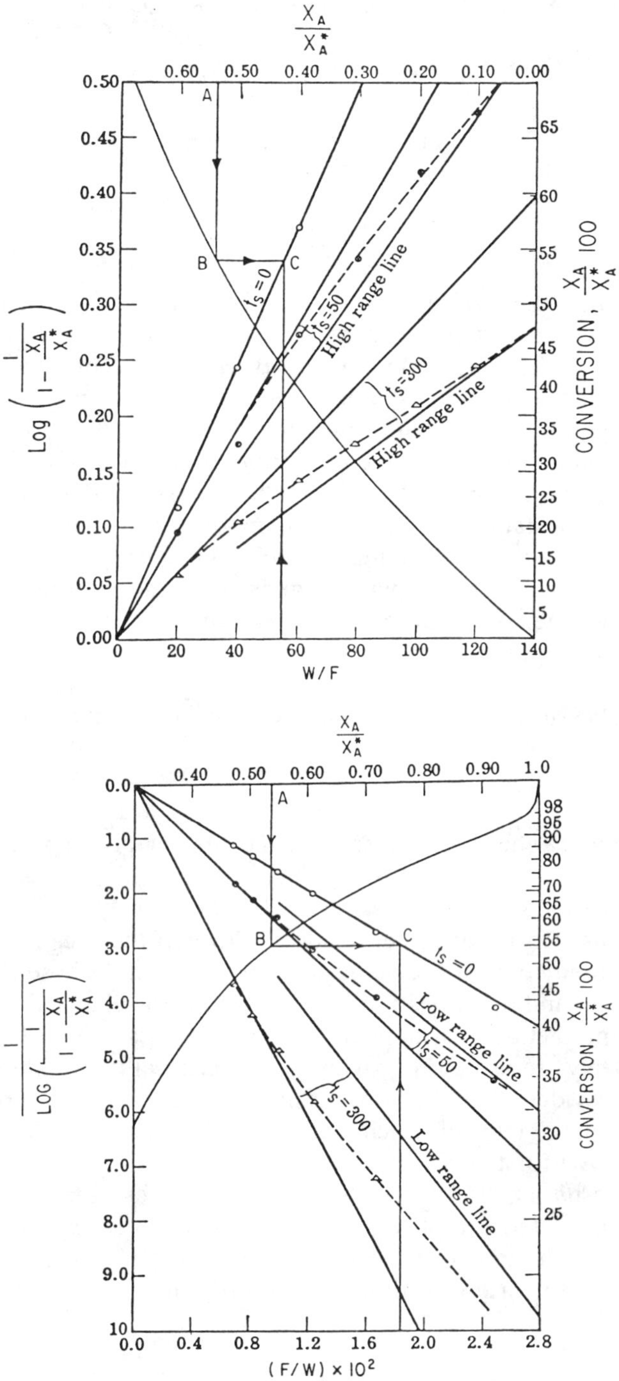

Figure 2.5 Linear pseudo-first-order plots
(Reproduced by permission: O.A. Hougen and K.M. Watson, "Chemical Process Principles, Part 3, copyright 1947, John Wiley & Sons, Inc., New York.)

From Eqs. 2.32 to 2.35, $\Delta H° = E - E'$. Therefore,

$$\ln k_p = \frac{E}{RT} + \text{constant} \tag{2.36}$$

or

$$k_p = Ae^{-E/RT} \tag{2.36A}$$

where A is called the *Arrhenius constant*, *frequency factor*, or *preexponential factor*, and E is the *activation energy*.

Equation 2.36A, called the *Arrhenius equation*, continues to be used. Through countless experiments it has been shown to apply over reasonable ranges of temperature. Since an observed k_p can actually be the product of a rate constant and one or more equilibrium constants, it is possible, if the equilibrium constant represents an exothermic step, that the apparent or observed rate constant will decline or moderate its rate of ascent in a higher-temperature region as temperature is increased. Although the actual rate constant always increases with temperature, the exothermic equilibrium constant will decline. A useful restatement of the Arrhenius law in terms of its general empirical application is

> The natural logarithm of the apparent rate constant for any reaction is a linear function of the reciprocal of the absolute temperature over reasonable ranges of temperature normally considered for practical reaction systems.

The Arrhenius equation has encouraged further probing into the concept of an energy barrier for reactions, which has provided a useful means for visualizing reaction systems. Over the 100-year span of time since its introduction, various efforts have been made to develop theories for reactions that could lead to prediction of the values of rate constants. These include various collision theories and the transition-state theory. If the classical operational test of accurate predictions is applied, these theories must be assessed as inadequate. They are useful in providing a conceptual base, and the transition-state theory provides a general concept involving the formation of an activated complex, the energy level of which constitutes the barrier for forward reaction. A similar analysis for the reverse reaction applies.

In the transition-state theory [6], reactants are assumed to be at equilibrium with an activated complex. The rate of reaction is expressed in terms of the number of such complexes crossing through the transition state, with one half going in the forward direction and the other half in the reverse direction:

$$\text{Rate} = \frac{C^{\dagger}/2}{t^{\dagger}} \tag{2.37}$$

where $C^{\dagger}$ is the concentration of the activated complex and $t^{\dagger}$ is the time required for passage.

For A + B → activated complex at equilibrium,

$$K_c^\dagger = \frac{C^\dagger}{C_A C_B} \tag{2.38}$$

or

$$C^\dagger = C_A C_B K_c^\dagger \tag{2.39}$$

Substituting Eq. 2.39 in Eq. 2.38, expressing $K_c^\dagger$ in terms of partition coefficients, evaluating $t^\dagger$ by using the distribution law, and converting to partial pressure and the thermodynamic equilibrium constant, we obtain

$$\text{Rate} = \kappa \frac{\mathbf{k}}{\mathbf{hR}} K^\dagger P_A P_B \tag{2.40}$$

$$= \kappa \frac{\mathbf{k}}{\mathbf{hR}} e^{-\Delta G^\dagger/RT} P_A P_B \tag{2.41}$$

or

$$k_p = \kappa \frac{\mathbf{k}}{\mathbf{hR}} \exp\left(\frac{\Delta S^\dagger}{R} - \frac{\Delta H^\dagger}{RT}\right) \tag{2.42}$$

where **k**, **h**, **R** are, respectively, the Boltzman constant, the Planck constant, and the gas constant in PVT units; $\Delta G^\dagger$, $\Delta H^\dagger$, $\Delta S^\dagger$ are, respectively, the free energy, enthalpy, and entropy of activation between the reactants and the activated complex; and κ is the *transmission coefficient*. This term was described by Eyring, who introduced the theory, as a "factor for tucking any loose ends created by this nasty problem in mechanics" [7].

When Eq. 2.42 is compared to the Arrhenius expression (Eq. 2.36A), a more conceptually useful idea of the meaning of A becomes apparent:

$$A = \frac{\mathbf{k}}{\mathbf{hR}} e^{\Delta S^\dagger/R} \tag{2.43}$$

We see that A is a function of the entropy of activation and is, therefore, related to the difference in entropy between the activated state and the reactants. Thus, for a catalytic reaction, $\Delta S^\dagger$ would be negative in most cases because of the immobility of the chemisorbed complex. For the reaction to be faster than a comparable homogeneous reaction, the decline in $\Delta S^\dagger$ ($\Delta S^\dagger_{cat} - \Delta S^\dagger_{homo}$) must be such that the change in $T\,\Delta S^\dagger$ is less than the change in $\Delta H^\dagger$ ($\Delta H^\dagger_{cat} - \Delta H^\dagger_{homo}$). By similar reasoning, $\Delta S^\dagger$ for a reaction between two polyatomic molecules is expected to be more negative than for a reaction between two atoms.

Values of A and E or $\Delta S^\dagger$ and $\Delta H^\dagger$ are evaluated from experimentally determined rate constants at different temperatures from the linear form of

$\ln k$ as a function of $1/T$ (Eq. 2.36) or the nonlinear form of the original equation (2.36A). Traditionally, a plot of $\ln k$ versus $1/T$ has been used to determine the slope and thus the activation energy E. Of course, linear least squares on this logarithmic form using conveniently available software is more accurate. As shown by the following reasoning, minimizing the square of the deviation between measured $\ln k_n$ and the true value, $\ln k_n^\circ$, minimizes the fractional error ε_n [8], which is a rational goal since most kinetic data involve analytical error that is a constant percentage error.

$$k_n = k_n^\circ + \varepsilon_n k_n^\circ = k_n^\circ(1 + \varepsilon_n)$$

$$\ln k_n - \ln k_n^\circ = \ln(1 + \varepsilon_n)$$

$$\ln(1 + \varepsilon_n) = \varepsilon_n - \tfrac{1}{2}\varepsilon_n^2 + \tfrac{1}{3}\varepsilon_n^3 \cdots \cong \varepsilon_n \qquad \text{for } \varepsilon_n \ll 1$$

or

$$\ln k_n - \ln k_n^\circ = \varepsilon_n$$

Rates of Deactivation

As can be seen in Table 2.2, catalyst life in successful fixed-bed processes is usually several years, which means that deactivation is relatively slow. The gradual decline in activity is normally offset by gradual increases in operating temperature that, over time, can accelerate the pace of deactivating phenomena, such as sintering and agglomeration. Catalysts with shorter lives often require regeneration, the time for which can vary with the process conditions. In these and other similar situations the ability to predict activity changes with time can be most helpful in planning operating control strategies.

Because deactivation phenomena occur on the catalyst surface and often involve relatively small numbers of molecules or crystallites in complex transformations, the development of deactivation rate forms based on mechanistic arguments is seldom practical. Fortunately, simple empirical forms (Table 2.9) have been proposed [9] and shown through numerous experimental observations to correlate the observed data satisfactorily. Just as with other rate forms, it is suggested that the user try several forms that make sense for the situation at hand and then use the one providing the best fit.

The equations given in Table 2.9 are in the form of the change with time of the activity factor, which is defined as follows:

$$\alpha = \frac{(\hat{r}_A)_{\text{used}}}{(\hat{r}_A)_{\text{fresh}}}$$

These rates are at surface conditions. If significant diffusional resistances occur, gradients of concentration need to be included, or effectiveness factors need to be used for both the main reactions and the deactivating reactions. Alternatively, the effectiveness factors may be considered part of the rate constant and its use strictly limited to the narrow range of interest and a single pellet size.

Table 2.9 Deactivation-Rate Equations: Power-Law Forms

Deactivating Process	*Rate Form*	*Typical Range of Exponent Values*
Sintering	$\frac{d\alpha}{dt} = -k_d \alpha^n$	$n = 2 - 8$ (higher values in reducing atmosphere); most common value: $n = 2$
Poisoning and Fouling*	$\frac{d\alpha}{dt} = -k_d \alpha^n$	$n = 0 - 2$ Most common value: $n = 1$ or 2
Parallel: $A \rightarrow R$, $A \rightarrow P\downarrow$	$\frac{d\alpha}{dt} = -k_d P_A^m \alpha^n$	
Series: $A \rightarrow R \rightarrow P\downarrow$	$\frac{d\alpha}{dt} = -k_d P_R^m \alpha^n$	
Impurity Deactivation: $A \rightarrow R$, $P \rightarrow P\downarrow$	$\frac{d\alpha}{dt} = -k_d P_p^m \alpha^n$	

Nomenclature:
α = activity factor (≤ 1.0)
k_d = deactivation rate constant
P_a = partial pressure of limiting reactant
P_p = partial pressure of poison
P_R = partial pressure of product
A, R, P, P↓ are, respectively, limiting reactant, product, poison and fouling precursor, and adsorbed poison or foulant (coke or inorganic deposit).

*Levenspiel, O., J. Catal., 25, 265 (1972).

It is interesting to consider the effect of diffusional resistance, which becomes larger at higher temperature. Although calculations have been presented, the most convincing evidence is actual experimental data involving coking of pellets at different temperatures, as shown in Fig. 2.6 for parallel and series poisoning [10]. At high temperatures the reactant gradient is steep, and

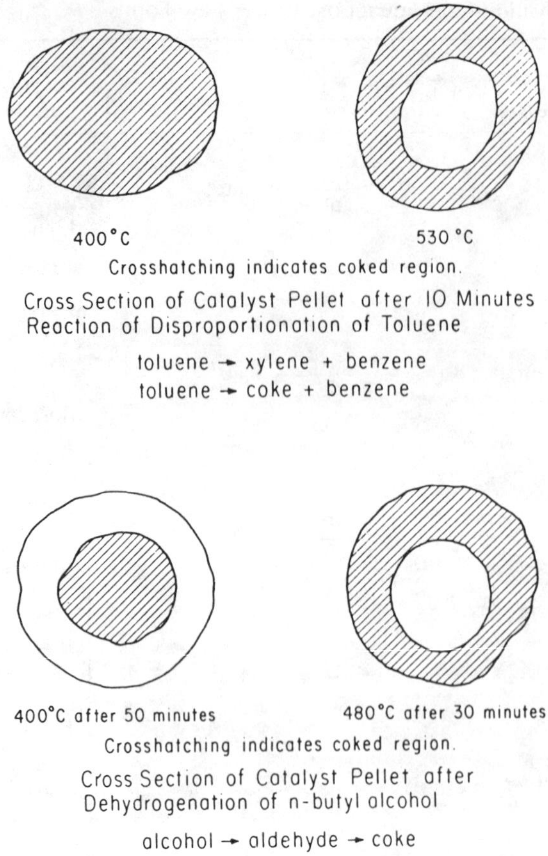

Figure 2.6 Drawings made from photographs comparing bisected coked catalyst particles
(Adapted from photographs in Y. Murakeami et al. *Ind. Eng. Chem. Fundamentals, 7*, 599 (1968), copyright 1968, American Chemical Society; reproduced by permission: H.F. Rase, "Chemical Reactor Design for Process Plants", copyright 1977, John Wiley & Sons, Inc., New York.)

coking, which was readily observed by splitting pellets after reaction and measuring the blackened areas, was confined to the outer layers for parallel fouling. A similar result would be found for parallel poisoning. At lower temperatures, gradients are minimal and coke deposition was uniform. For the series reaction the intermediate and poison or coke precursor R is highest at the inner position at low temperature. At high temperature, R is higher at the outer layers because of the diffusional gradient of A, and thus R. Hence, the deposit of coke is highest in these outer layers.

Avoiding deactivation of a sudden or catastrophic kind is most crucial to successful processes. Understanding the causes is absolutely essential to long-term satisfactory and economical solutions to such problems. The key to each method for preventing or moderating deactivation as summarized in Table 2.3 obviously requires some clear knowledge concerning the causative agent or phenomenon. Such knowledge is also essential in developing improved catalysts that are less prone to rapid deactivation. Modern techniques for studying surfaces make insightful and revealing experiments possible, the results of which can lead to new discoveries or clever innovations. The literature contains much useful experimental background that can be valuable in planning and executing a development program.

LITERATURE CITED

1. Denbigh, K., "The Principles of Chemical Equilibrium," Cambridge University Press, Cambridge, 1964.
2. Temkin, M.H., *Zh. Fiz. Khim.*, *24*, 1372 (1950).
3. Hougen, O.A. and Watson, K.M., "Chemical Process Principles," Part 3, Wiley, New York, 1947.
4. Regner, A. and Simecek, A., *Coll. Czech. Chem. Comm.*, *33*, 2540 (1968).
5. Arrhenius, S., *A. Phys. Chem.*, *4*, 226 (1889).
6. Eyring, H., Gershinowitz, H., and Sun, C.E., *J. Chem. Phys.* *3*, 783 (1935).
7. Eyring, H. and Erying, E.M., "Modern Chemical Kinetics," Van Nostrand Reinhold, New York, 1963.
8. Rawlings, J.R., private communication, May 1988.
9. Levenspiel, O., *J. Catal.*, *25*, 265 (1972).
10. Makurami, Y., *Ind. Eng. Chem. Fundam.*, *1*, 599 (1968).

CHAPTER 3

Chemical Equilibrium

Because of the extensive and accurate thermodynamic data on gases and the highly successful equations of state for quantifying nonideality of gases, precise prediction of equilibrium compositions under a variety of conditions is possible without any laboratory measurements. Thus, without ever leaving desk or workstation, one can determine the range of ideal conditions for a reaction system before making any costly commitments to laboratory or pilot-plant studies.

BASIS FOR EQUILIBRIUM CALCULATIONS

Any professional involved with a reaction system should have knowledge of equilibrium compositions within the anticipated operating range of interest. In the development of such information, it is instructive to recall the basis for equilibrium calculations.

As shown in Fig. 3.1, starting with the first and second laws and then imposing the restraints of a nonflow reversible process, four exact differential equations can be derived that contain only intrinsic point properties and are, therefore, applicable to any process. Of these, the equation for free energy (G) is the most useful when we are interested in applying constraints of constant temperature and pressure. In fact, G was invented or defined to express the common condition of interest, equilibrium at constant temperature and pressure; and it is more useful for chemical reactions than those expressions involving the other state properties.

As shown in Fig. 3.1 a spontaneous process at constant T and P will exhibit a decline in free energy (block 4). The minimum value is reached at equilibrium, at which point an infinitesimal reversible change will produce no change in free energy. Thus the criterion for equilibrium at constant temperature and pressure is $(dG)_{T,P} = 0$.

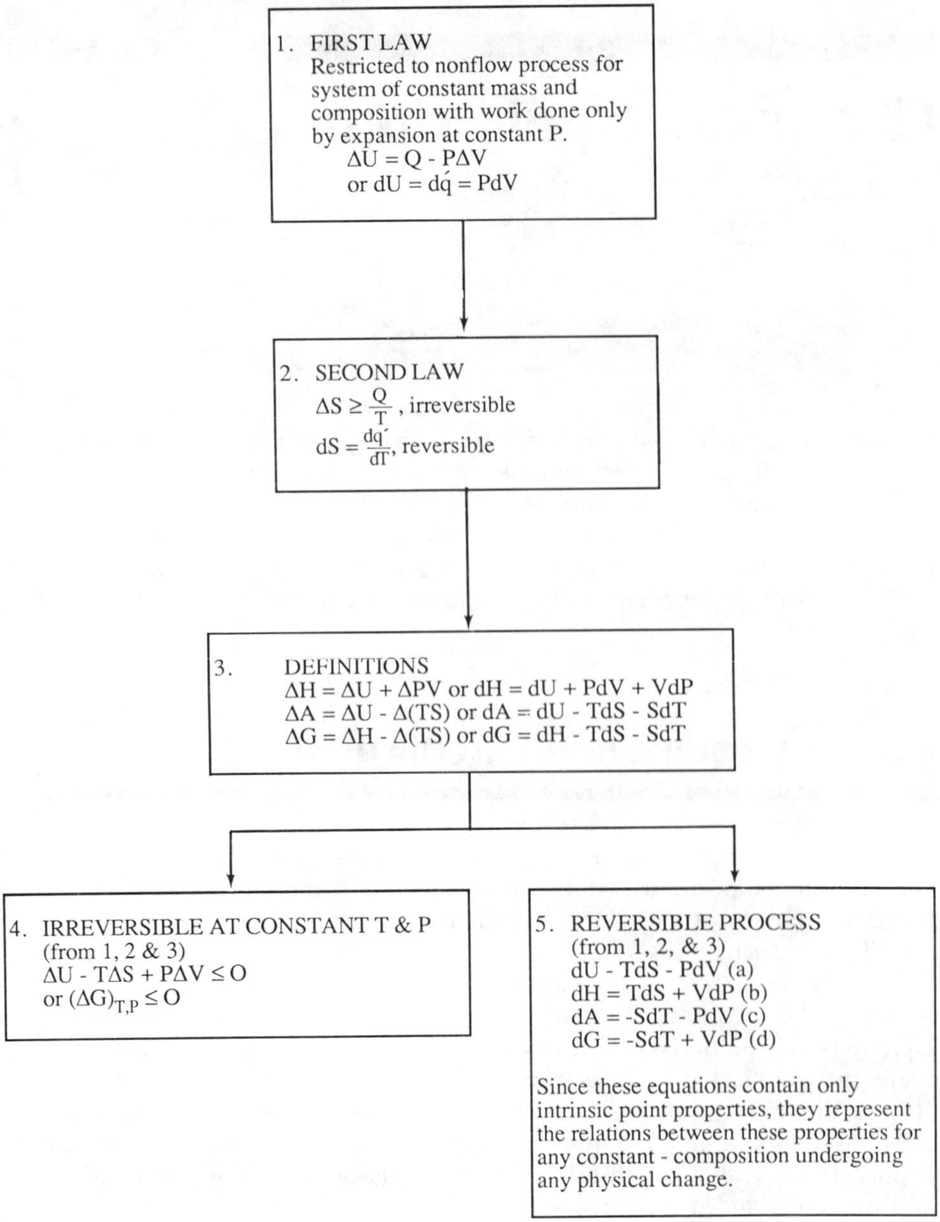

Figure 3.1 Energy functions for systems of constant mass and composition

For chemical reactions systems, composition changes with the extent of the reaction, and the total differential of G must include the effect of compositional changes:

$$dG = \left(\frac{\partial G}{\partial T}\right)_{P,n_1,n_2,\ldots,n_c} dT + \left(\frac{\partial G}{\partial P}\right)_{T,n_1,n_2,\ldots,n_c} dP + \sum_{j=1}^{c}\left(\frac{\partial G}{\partial n_j}\,dn_j\right)_{T,P,n_i \neq n_j} \tag{3.1}$$

where $n_1, n_2, \ldots, n_c$ are moles of indicated components, with c being the last in the series; j is a designation for components; and i refers to components other than j.

Since the first two terms are restricted to constant composition, substitutions can be made from Eq. (d) of block 5 in Fig. 3.1. The partial derivative of G with respect to the composition variable is the *partial molal free energy* $\bar{G}_j$, also called the *chemical potential* μ_j:

$$dG = -S\,dT + V\,dP + \sum_{j=1}^{c} \mu_j\,dn_j \tag{3.2}$$

$$dG = -S\,dT + V\,dP + \sum_{j=1}^{c} \bar{G}_j\,dn_j \tag{3.3}$$

For chemical equilibrium calculations at fixed T and P, Eq. 3.3 becomes

$$dG = \sum_{j=1}^{c} \bar{G}_j\,dn_j = 0 \tag{3.4}$$

FREE ENERGY RELATIONS IN TERMS OF G_f° FOR IDEAL GAS

In order to make use of the large amount of data on standard free energy of formation, it is necessary to define $\bar{G}_j$ in terms of $G^\circ_{f_j}$, the standard free energy of formation at the temperature of interest and a standard state of 1 atmosphere.

$$\bar{G}_j = G^\circ_{f_j} + (\bar{G}_j - G^\circ_{f_j}) \tag{3.5}$$

Although $G^\circ_{f_j}$ is at 1 atm, the actual value of $\bar{G}_j$ is for an ideal gas at the partial pressure of component j. Hence, $\bar{G}_j - G^\circ_f$ is a correction term from 1 atm to a partial pressure P_j. From Eq. 3.3 at constant temperature and composition,

$$dG = V\,dP$$

For an ideal gas

$$V=\frac{n_T RT}{P}=\frac{n_j RT}{\frac{n_j}{n_T}P}=\frac{n_j RT}{P_j}$$

where n_T = total moles

$$dP=\sum dP_j$$

$$\therefore dG=\sum \frac{n_j RT}{P_j}\,dP_j$$

Integrating between 1 atm and P_j gives

$$\sum n_j(\bar{G}_j-G^\circ_{f_j})=\sum_{j=1}^{c} n_j RT\ln\frac{P_j}{1}=\sum_{j=1}^{c} n_j RT\ln P_j \tag{3.6}$$

$$\therefore G=\sum n_j\bar{G}_j=\sum n_j(G^\circ_{f_j}+RT\ln P_j) \tag{3.7}$$

or

$$(dG)_{T,P}=\sum_{j=1}^{c}(G^\circ_{f_j}+RT\ln P_j)\,dn_j \tag{3.8}$$

REACTION EQUILIBRIUM FOR IDEAL GASES

Consider the reaction $aA+bB\rightarrow rR$ at constant T and P:

$$dG=(G^\circ_{f_R}+RT\ln P_R)\,dn_R+(G^\circ_{f_A}+RT\ln P_A)\,dn_A$$
$$+(G^\circ_{f_B}+RT\ln P_B)\,dn_B \tag{3.9}$$

From the reaction stoichiometry,

$$-\frac{1}{a}\,dn_A=\frac{1}{r}\,dn_R=-\frac{1}{b}\,dn_B \tag{3.10}$$

Substituting Eq. 3.10 in Eq. 3.9 gives

$$dG=-\frac{r}{a}(G^\circ_{f_R}+RT\ln P_R)\,dn_A+(G^\circ_{f_A}+RT\ln P_A)\,dn_A$$
$$+\frac{b}{a}(G^\circ_{f_B}+RT\ln P_B)\,dn_A \tag{3.11}$$

or

$$a\left(\frac{\partial G}{\partial n_{\rm A}}\right)_{T,P} = -r(G^\circ_{f_{\rm R}} + RT \ln P_{\rm R}) + a(G^\circ_{f_{\rm A}} + RT \ln P_{\rm A}) + b(G^\circ_{f_{\rm B}} + RT \ln P_{\rm B}) \tag{3.12}$$

At equilibrium $(\partial G/\partial n_{\rm A}) = 0$, as seen from Eq. 3.4. Rearranging gives

$$RT(r \ln P_{\rm R} - a \ln P_{\rm A} - b \ln P_{\rm B}) = -[rG^\circ_{f_{\rm R}} - (aG^\circ_{f_{\rm A}} + bG^\circ_{f_{\rm B}})]$$

or

$$RT \ln \frac{P^r_{\rm R}}{P^a_{\rm A} P^b_{\rm B}} = -\Delta G^\circ, \quad \text{the standard free energy for the reaction at } T \text{ and 1 atm} \tag{3.13}$$

Since the partial-pressure terms are in atmospheres and are actually P_j atm/1 atm, then R should be 1.987 cal/(mole °K) for ΔG° in cal/mole. For the general case,

$$RT \ln \frac{P^r_{\rm R} P^s_{\rm S}}{P^a_{\rm A} P^b_{\rm B}} = -\Delta G^\circ \tag{3.14}$$

where

$$\Delta G^\circ = \sum_{j=1}^{c} \nu_j G^\circ_{f_j}$$

and ν_j is the stoichiometric coefficient for any component j, positive for products and negative for reactants.

The partial-pressure term ratio on the LHS of Eq. 3.14 is called the *equilibrium constant*, K; for ideal gases it has the general form

$$K = \frac{P^r_{\rm R} P^s_{\rm S}}{P^a_{\rm A} P^b_{\rm B}} = \frac{n^r_{\rm R} n^s_{\rm S}}{n^a_{\rm A} n^b_{\rm B}} \left(\frac{P}{n_{\rm T}}\right)^{r+s-(a+b)} \tag{3.15}$$

where $n_{\rm T}$ denotes total moles, R and S refer to products, and A and B to reactants. Since K is dimensionless, the user, when checking, should remember that P_j is actually $P_j/1$ atm and is therefore dimensionless. Perhaps, more simply, always use pressure in atmospheres, and the results will be quantitatively correct.

Substituting K into Eq. 3.14 yields

$$RT \ln K = -\Delta G^\circ$$

or

$$\ln K = \frac{-\Delta G^\circ}{RT} \tag{3.16}$$

Taking the antilog gives

$$K = \exp\left(\frac{-\Delta G^\circ}{RT}\right) \tag{3.17}$$

Again, $G^\circ_{f_j}$ is the standard free energy of formation of a compound j at the temperature T of interest and 1 atm, relative to the elements composing that compound at temperature T and 1 atmosphere. Precise values for gases are tabulated in Stull et al. [1], and other compilations at various convenient temperatures. For some compounds, enthalpies and heats of formation at 25°C (298°K) may be the only data available. In such cases fitted heat-capacity equations for ideal gases [2] can be used to evaluate ΔH° and ΔG° and, thus, K at the desired temperature. See Example 1.

$$c_{p_j} = A_j + B_jT + C_jT^2 + D_jT^3$$

$$(\Delta H^\circ)_T = I_H + \sum_{j=1}^{c} \nu_j \left(A_jT + \frac{B_jT^2}{2} + \frac{C_jT^3}{3} + \frac{D_jT^4}{4}\right) \tag{3.18}$$

$$R \ln K = -\frac{I_H}{T} + I_K + \sum_{j=1}^{c} \nu_j \left(A_j \ln T + \frac{B_jT}{2} + \frac{C_jT^2}{6} + \frac{D_jT^3}{12}\right) \tag{3.19}$$

where $(\Delta H^\circ)_T$ is the standard enthalpy of the reaction at temperature T; I_H is obtained from Eq. 3.18, evaluated at 298°K; and I_K is evaluated from Eq. 3.19 at $T = 298$°K. These two equations are derived in Example 1.

REACTION EQUILIBRIUM FOR NONIDEAL GASES

The convenient relation between the partial pressure and the partial free energy for ideal gases ($PV = n_T\mathrm{RT}$) forming perfect mixtures ($\bar{H}_j = H_j$ and $\bar{V}_j = V_j$), given in Eq. 3.7, can be altered by the fugacity term of G.N. Lewis [3] so that it applies to real gases:

$$G = \sum_{j=1}^{c} n_j G^\circ_{f_j} + \sum_{j=1}^{c} RTn_j \ln \hat{f}_j \tag{3.20}$$

where $\hat{f}_j$ is the partial fugacity of a reacting component j in the mixture. The fugacity of a component of a real gas, like partial pressure for the ideal gas, represents the escaping tendency of that component; and as $P \to 0$, $\hat{f}_j \to P_j$.

Lewis defined the fugacity such that the convenient form for ideal gases applied for nonideal gases. For chemical equilibrium, $\hat{f}_j$ is substituted for P_j and the same derivation as for ideal gases yields the dimensionless expression $\hat{f}_j/1$, since the denominator is really the initial limit or standard state of 1 atm as in Eq. 3.13. Equation 3.20 is often written for any system as

$$G = \sum_{j=1}^{c} n_j G^{\circ}_{f_j} + \sum RTn_j \ln a_j \tag{3.21}$$

where a_j is the activity of component j expressed as $\hat{f}_j/f^{\circ}_j$, where f°_j is the standard-state fugacity of component j at the conditions of interest. Thus, the general equilibrium equation applicable to any system that results is, by analogy with Eq. 3.14,

$$RT \ln \frac{a_{\mathrm{R}}^r}{a_{\mathrm{A}}^a a_{\mathrm{B}}^b} = -\Delta G^{\circ} \tag{3.22}$$

or, conveniently, for gases,

$$RT \ln \frac{\hat{f}_{\mathrm{R}}^r}{\hat{f}_{\mathrm{A}}^a \hat{f}_{\mathrm{B}}^b} = -\Delta G^{\circ} \tag{3.23}$$

where $\hat{f}_j$ is understood to be $\hat{f}_j$ atm/1 atm. Now the equilibrium constant can be defined as previously:

$$RT \ln K = -\Delta G^{\circ}$$

where ΔG° is again calculated from $G^{\circ}_{f_j}$ values at temperature T and 1 atm and is, therefore, not affected by pressure. The value of K as here defined is thus the same in a given system in either the ideal or nonideal condition.

Nonideal Gases Forming Ideal Solutions

When ideal solution behavior applies,

$$\hat{f}_j = Y_j f_j \qquad \text{and} \qquad f_j = \phi_j P$$

or

$$\hat{f}_j = \phi_j Y_j P \tag{3.24}$$

where ϕ_j is the pure-component fugacity coefficient and f_j is the pure-component fugacity for any component j at the temperature and pressure of the system. The working equation for K becomes

$$K = \frac{f_{\mathrm{R}}^r f_{\mathrm{S}}^s}{f_{\mathrm{A}}^a f_{\mathrm{B}}^b} = \frac{n_{\mathrm{R}}^r n_{\mathrm{S}}^s}{n_{\mathrm{A}}^a n_{\mathrm{B}}^b} \left(\frac{P}{n_{\mathrm{T}}}\right)^{r+s-(a+b)} \frac{\phi_{\mathrm{R}}^r \phi_{\mathrm{S}}^s}{\phi_{\mathrm{A}}^a \phi_{\mathrm{B}}^b} \tag{3.25}$$

The pure-component fugacity coefficient is readily calculated by using an appropriate equation of state and existing software, which is available in most commerical simulator programs. A thorough summary and commentary on equations of state is also useful in selecting the ideal one for a particular system (see Ref. 4 for such background). Figure 3.2 presents a means for estimating values of ϕ_j for each component by using the applicable reduced properties. For components that would be liquids if present alone, the gas-phase curve should be extrapolated. It is wise to check a set of values determined by means

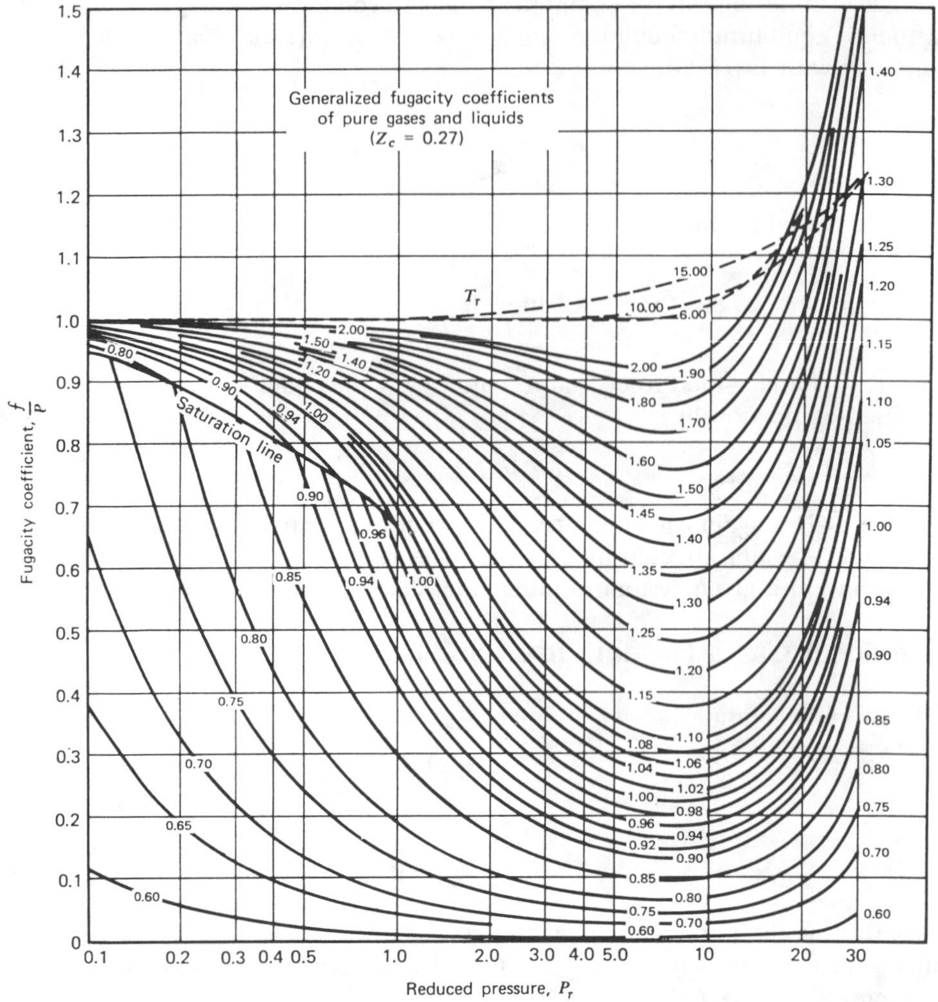

Figure 3.2 Generalized correlation for pure-component fugacity for $z_c = 0.27$, a critical compressibility that is within 4% of the values of over 50% of all compounds, for estimates only. (Reproduced by permission: S.I. Sandler, "Chemical and Engineering Thermodynamics," copyright 1977, John Wiley & Sons, Inc., New York.)

of the normally more accurate equation of state by an independent means such as Fig. 3.2.

Ideal solution behavior is common for gases and occurs under any of the following conditions [5, 6]:

- Low pressure (usually for pseudocritical pressure of 0.8 or less)
- Wide range of pressures and compositions when physical properties of the components are similar
- For a component, under any condition, present in excess

Nonideal Gases Forming Nonideal Solutions

For this case,

$$\hat{f}_j = \hat{\phi}_j Y_j P \tag{3.26}$$

where $\hat{\phi}_j$ is the partial fugacity coefficient for component j: $\hat{f}_j / Y_j P$. See Tables 3.1 and 3.3 for applicable equilibrium formulation. The pure-component fugacity coefficient is composition independent, but the partial fugacity coefficient required for nonideal solutions is not. Although expressions for the partial value based on equations of state can be used, they contain interaction coefficients between the several components that are not always known. See Ref. 4 for an excellent summary on this issue.

EFFECT OF TEMPERATURE ON EQUILIBRIUM

The effect of temperature on equilibrium is readily obtained from the differential of the standard free energy for a reaction with respect to temperature. Since the standard free energy, enthalpy, and entropy for gases are all based on a pressure of 1 atmosphere, the expression may be written as a complete differential:

$$\Delta G^\circ = \Delta H^\circ - T\,\Delta S^\circ$$

or

$$\frac{\Delta G^\circ}{T} = \frac{\Delta H^\circ}{T} - \Delta S^\circ$$

$$\frac{d(\Delta G^\circ / T)}{dT} = \frac{1}{T}\left(\frac{d\,\Delta H^\circ}{dT}\right) - \frac{\Delta H^\circ}{T^2} - \frac{d\,\Delta S^\circ}{dT} \tag{3.27}$$

Since $dH_j = T\,dS_j + V\,dP$,

$$\frac{dH_j}{dT} = T\frac{dS_j}{dT} \qquad \text{for constant } P$$

or

$$\frac{d\,\Delta H^\circ}{dT} = T\,\frac{d\,\Delta S^\circ}{dT}$$

Differentiating the relation between K and ΔG° of Eq. 3.16, and then substituting in Eq. 3.27, gives

$$R\,\frac{d\ln K}{dT} = -\frac{d(\Delta G^\circ/T)}{dT}$$

Therefore

$$\frac{d\ln K}{dT} = \frac{\Delta H^\circ}{RT^2} \tag{3.28}$$

Now using the standard empirical heat capacity equations, we have

$$\frac{d\,\Delta H^\circ}{dT} = \sum_{j=1}^{c} \upsilon_j(A_j + B_jT + C_jT^2 + D_jT^3)$$

Integrating, using the integration constant I_H, gives

$$\Delta H^\circ = I_H + \sum_{j=1}^{c} \upsilon_j\left(A_jT + \frac{B_jT^2}{2} + \frac{C_jT^3}{3} + \frac{D_jT^4}{4}\right) \tag{3.29}$$

Substituting Eq. 3.29 into Eq. 3.28 and integrating gives

$$R\ln K = -\frac{I_H}{T} + I_K + \sum_{j=1}^{c} \upsilon_j\left(A_j\ln T + \frac{B_jT}{2} + \frac{C_jT^2}{6} + \frac{D_jT^3}{12}\right) \tag{3.30}$$

where I_K and I_H are integration constants.

Thus K may be obtained at any temperature from a known value of ΔH° at any temperature, usually 298°K, and a value of K at any temperature, also often known at 298°K. The integration constant I_H can then be evaluated from Eq. 3.29 at the T for which ΔH° is known, and I_K can be evaluated from Eq. 3.30 from the known value of K. The values thus obtained can be used in Eq. 3.30 for calculating K at any temperature within the range of the heat-capacity equations.

For rapid estimates over ranges where ΔH° can be considered constant, the integrated form of Eq. 3.28 at constant ΔH° is useful:

$$\ln \frac{K_2}{K_1} = \frac{\Delta H^\circ}{R}\left(\frac{1}{T_1} - \frac{1}{T_2}\right) \tag{3.31}$$

If the reaction is exothermic, ΔH° is negative and K declines with an increase in temperature, but it increases with a temperature increase for an endothermic reaction for which ΔH° is positive. An increase in temperature corresponds to an energy input to the molecules, which causes the endothermic direction to be accentuated relative to the exothermic. The effects of temperature and other variables on equilibrium are summarized in Table 3.1. The value of the standard free energy for a reaction at the temperature of interest as a convenient criterion for reaction promise is summarized in Table 3.2.

Table 3.1 Effect of Various Conditions on Equilibrium Conversion

Example: $aA + bB \rightleftarrows rR + sS$

$$\frac{n_R^r n_S^s}{n_A^a n_B^b} = K \frac{\phi_A^a \phi_B^b}{\phi_R^n \phi_S^s}\left(\frac{n_t}{P}\right)^{r+s-(a+b)}$$

K is a function of temperature only since standard state is 1 atmosphere.

$$\left(\frac{\partial \ln K}{\partial T}\right)_P = \frac{\Delta H^\circ}{RT^2}$$

ΔH° is negative for exothermic reaction and positive for endothermic reaction

Variable	Effect of Changing Indicated Variable on Equilibrium Conversion: Increase Variable	Decrease Variable
Temperature		
Endothermic	Increase	Decrease
Exothermic	Decrease	Increase
Pressure		
If [r + s - (a + b)] >0, (increase in moles)	Decrease	Increase
If [r + s - (a + b)] <0, (decrease in moles)	Increase	Decrease
Also affects fugacity coefficient ratio, depending on deviations from ideal gas behavior of products relative to reactants.		
Inerts		
If [r + s - (a + b)] >0, (increase in moles)	Increase	Decrease
If [r + s - (a + b)] <0, (decrease in moles)	Decrease	Increase

Table 3.1 (*continued*)

Excess Reactant	Increase	Decrease
Recycled Product(s) in Feed	Decrease (an undesired fast reaction can be prevented from occurring by including its products at equilibrium composition in the feed)	Increase

Table 3.2 Criteria for Reaction Promise Based on ΔG° at Temperature of Interest

	$K = \exp(-\Delta G^\circ/RT)$
$\Delta G^\circ < 0$	Reaction is promising.
$\Delta G^\circ > 0 < 10{,}000$	Warrants study for possible economic viability if products of high value.
$\Delta G^\circ > 10{,}000$	Very unfavorable, but could be feasible for very high valued products.

SOLVING FOR EQUILIBRIUM COMPOSITION

The composition at equilibrium is most easily determined for a single reaction by solving the equilibrium constant expression written in terms of conversion or extent of reaction. These same forms can be used for complex multiple-reaction systems if the various stoichiometric equations are defined or may be postulated. Alternatively, free-energy minimization may be used, which does not require postulating stoichiometric equations. The different procedures are described in the following sections, and examples are given in Part III, Examples 1 through 5.

Compositions in Terms of Conversions or Extents of Reaction

The equilibrium equations summarized in Table 3.3 require expression of the moles of each component in terms of one variable for each reaction considered. This goal can be accomplished by using extent of reaction or conversion. The following procedure, which bases all mole quantities on 1 mole of a limiting reactant fed, is recommended as most convenient for simulation of reactors and associated equipment:

$$aA + bB \rightleftarrows rR + sS$$

$$\text{Conversion} = dX_A = \frac{\text{moles A converted}}{\text{mole of A fed}} = -dn_{Aa}$$

where A is the limiting reactant (i.e., the one of major economic importance) and n_{Aa} is the moles of A at any point per mole of A fed. The differential change of extent of reaction, **e**, is formulated in terms of the stoichiometric numbers:

$$d\mathbf{e} = \frac{-dn_{Aa}}{a} = \frac{-dn_{Ba}}{b} = \frac{dn_{Ra}}{r} = \frac{dn_{Sa}}{s} \qquad (3.32)$$

It is convenient to follow the custom of writing all reactions so that at least one reactant has a stoichiometric number of one.

In the general case, $a = 1$ and dX_A and $d\mathbf{e}$ are equivalent.

$$dX_A = d\mathbf{e} = -dn_{Aa} = \frac{-dn_{Ba}}{b} = \frac{dn_{Ra}}{r} = \frac{dn_{Sa}}{s} \qquad (3.32A)$$

Table 3.3 Summary of Equilibrium Equations

IDEAL GAS, IDEAL SOLUTION (Standard State = 1 atm)

$$K = \frac{P_R^r P_S^s}{P_A^a P_B^b} = \frac{Y_R^r Y_S^s}{Y_A^a Y_B^b} P^{r+s-(a+b)} = \frac{n_R^r n_S^s}{n_A^a n_B^b}\left(\frac{P}{n_T}\right)^{r+s-(a+b)}$$

Example Relations for A:

$$P_A = \frac{n_A}{n_T} P$$

$$\left(d\bar{G}_A = RT\, d\ln P_A\right)_T$$

$$\left(\Delta\bar{G}_A = RT \ln \frac{P_A}{1}\right)_T$$

Table 3.3 *(continued)*

NONIDEAL GAS, IDEAL SOLUTION (Standard State = 1 atm)

$$K = \frac{\hat{f}_R^r \hat{f}_S^s}{\hat{f}_A^a \hat{f}_B^b} = \frac{Y_R^r Y_S^s}{Y_A^a Y_B^b} P^{r+s-(a+b)} K_\phi = \frac{n_R^r n_S^s}{n_A^a n_B^b} \left(\frac{P}{n_T}\right)^{r+s-(a+b)} K_\phi$$

$$K_\phi = \frac{\phi_R^r \phi_S^s}{\phi_A^a \phi_B^b}$$

Example Relations for A:

$$\hat{f}_A = \phi_A Y_A P = \phi_A \frac{n_A}{n_T} P$$

$$\phi_A = \frac{f_A}{P}, \text{ where } f_A = \text{pure component fugacity at T \& P of system}$$

$$\left(d\bar{G}_A = RT\, d\ln \hat{f}_A\right)_T \qquad \left(\Delta\bar{G}_A = RT\ln\frac{\hat{f}_A}{1} = RT\ln a_A\right)_T$$

NONIDEAL GAS, NONIDEAL SOLUTION (Standard State - 1 atm)

$$K = \frac{\hat{f}_R^r \hat{f}_S^s}{\hat{f}_A^a \hat{f}_B^b} = \frac{Y_R^r Y_S^s}{Y_A^a Y_B^b} P^{r+s-(a+b)} K_{\hat{\phi}} = \frac{n_R^r n_S^s}{n_A^a n_B^b} \left(\frac{P}{n_T}\right)^{r+s-(a+b)} K_{\hat{\phi}}$$

$$K_{\hat{\phi}} = \frac{\hat{\phi}_R^r \hat{\phi}_S^s}{\hat{\phi}_A^a \hat{\phi}_B^b}$$

Example Relations for A:

$$\hat{f}_A = \hat{\phi}_A Y_A P = \hat{\phi}_A \frac{n_A}{n_T} P$$

$$\hat{\phi}_A = \frac{\hat{f}_A}{Y_A P} \qquad \left(d\bar{G}_A = RT\,d\ln \hat{f}_A\right)_T \text{ or } \left(\Delta\bar{G}_A = RT\ln\frac{\hat{f}_A}{1}\right)_T$$

K is the same for all three cases.

or the conversion and extent of reaction become equal for those reactions in which the limiting reactant appears:

$$X_A = \mathbf{e} = -\Delta n_{Aa} = \frac{-\Delta n_{Ba}}{b} = \frac{\Delta n_{Ra}}{r} = \frac{\Delta n_{Sa}}{s}$$

Thus for a basis of 1 mole of A and n_{Ba_0} moles of B,

$$n_{Aa} = 1 - X_A$$
$$n_{Ba} = n_{Ba_0} - bX_A$$
$$n_{Ra} = rX_A$$
$$n_{Sa} = sX_A$$
$$n_T = \overline{1 + n_{Ba_0} - X_A - bX_A + rX_A + sX_A}$$

These relations emerge automatically from Eq. 3.32A, and chances of error are minimized by this procedure. The values of n for each component can be substituted in the appropriate equilibrium expression, and the equation solved for X_A. The best procedure is to use a nonlinear algebraic equation solver. See Part III for examples.

When multiple reactions are considered, the variable X_{na} will be used for each equation, with n representing the reaction number and "a" the basis of limiting reactant A. These values will represent extents of reaction per mole of limiting reactant fed.

$$\begin{array}{cccc} \text{M} & \text{O} & \text{F} & \text{W} \end{array}$$

$$CH_3OH + \frac{1}{2}\,O_2 \leftrightarrows CH_2O + H_2O \qquad (1)$$

$$CH_2O + \frac{1}{2}\,O_2 \leftrightarrows CO + H_2O \qquad (2)$$

$$\begin{array}{cccc} \text{F} & \text{O} & \text{CO} & \text{W} \end{array}$$

For simplicity, the second subscript, indicating the basis, is omitted on the mole terms (e.g., n_W for n_{Wa}):

$$X_{1a} = -(\Delta n_M)_1 = -\frac{(\Delta n_O)_1}{0.5} = (\Delta n_F)_1 = (\Delta n_W)_1$$

$$X_{2a} = -(\Delta n_F)_2 = -\frac{(\Delta n_O)_2}{0.5} = (\Delta n_{CO})_2 = (\Delta n_W)_2$$

Basis: 1 mole CH_3OH, n_{O_0} moles O_2, and n_{I_0} moles of inerts in feed

$$\begin{aligned} n_M &= 1 - X_{1a} \\ n_F &= X_{1a} - X_{2a} \\ n_{CO} &= X_{2a} \\ n_W &= X_{1a} + X_{2a} \\ n_O &= n_{O_0} - 0.5(X_{1a} + X_{2a}) \\ n_I &= n_{I_0} \\ \hline n_T &= 1 + n_O + n_I + 0.5(X_{1a} + X_{2a}) \end{aligned}$$

X_{1a} and X_{2a} are extents of reaction per mole of M, methanol, fed for reactions 1 and 2. X_{1a} is also the conversion of methanol. See Part III for example calculations for multiple reactions.

Actual Free-Energy Path of a Reaction

It is intuitively enlightening to consider the source of the actual total free energy of a reaction system. For this purpose, the important industrial catalytic reaction, the oxidization of SO_2 to form SO_3, is chosen. A typical outlet condition from the first bed of a four-bed adiabatic reactor with cooling between stages is around 900°K and 50 in. of H_2O above atmospheric pressure. For illustrative purposes, the calculations were made at 900°K and 1 atm.

From Eq. 3.7, for an ideal gas mixture,

$$G = \sum n_j G^\circ_{f_j} + \sum n_j RT \ln P_j \qquad \text{where } P_j = n_j P$$

Values of $G^\circ_{f_j}$ were obtained from Stull et al.'s compilation [1] at 900°K, and calculations were made at various conversions. The total free-energy values are shown in Fig. 3.3 and are relative to the elements at 900°K and on the basis of 1 mole of SO_2 fed along with the other quantities of feed components shown on the graph. Values of total enthalpy and entropy are also plotted in order to illustrate the change in the two functions that compose free energy. The total enthalpy relative to the elements is readily obtained from the same data compilation.

$$H = \sum_{j=1}^{c} n_j H^\circ_{f_j}$$

The total entropy relative to the elements is then simply given as

$$S = \frac{H - G}{T}$$

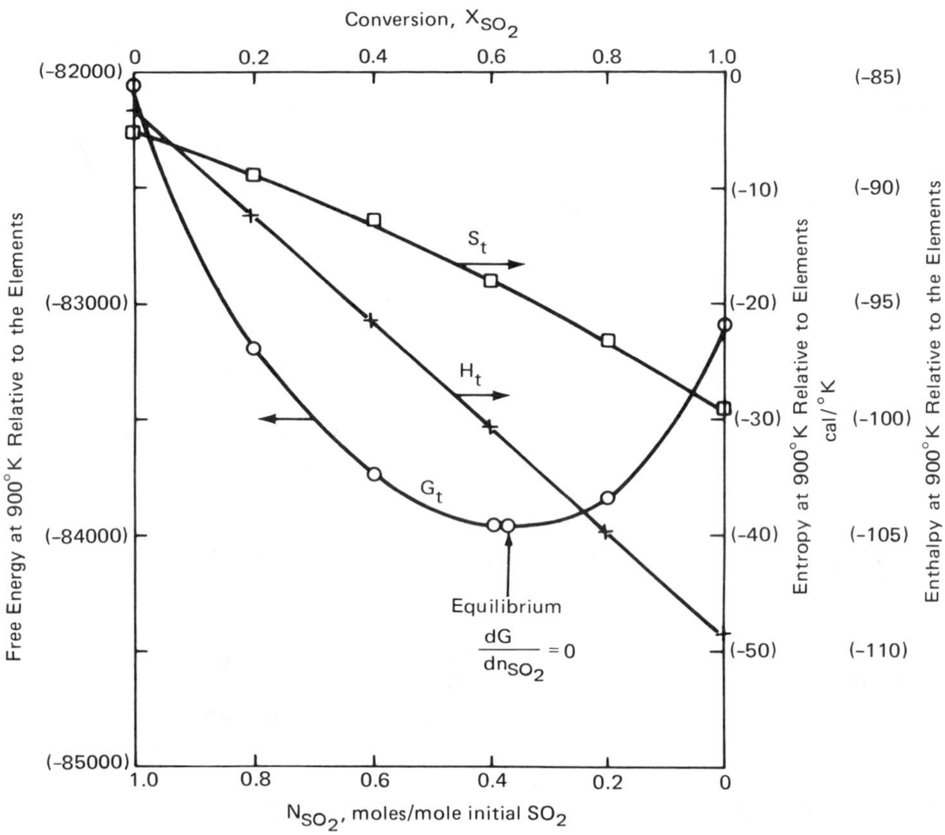

Figure 3.3 Actual free energy versus moles SO_2 remaining or conversion for the reaction $SO_2 + \frac{1}{2}O_2 \rightleftarrows SO_3$; initial conditions: 1 mole SO_2, 1 mole O_2, 7.52 moles N_2; temperature = 900°K, pressure = 1 atm.

It is, of course, possible to calculate S directly from entropies relative to the elements:

$$S = \sum_{j=1}^{c} n_j(S^\circ_{f_j} - R \ln P_j)$$

where $S^\circ_{f_j}$ is the entropy of formation at the temperature of interest. For example,

$$S^\circ_{f_{SO_2}} = S_{SO_2} - S_{O_2} - S_S$$

where the entropy terms on the RHS are absolute values for the indicated components at 1 atmosphere and the temperature of interest.

Note in Fig. 3.3 that the free energy reaches a minimum value. Like the proverbial marble resting in the bottom of a bowl, a system is said to be at equilibrium if no active unbalanced tendency toward change exists. This condition for a reacting system occurs at the minimum of the free energy of that system. The validity of this statement becomes clearer for a chemical reaction if one considers Fig. 3.3. more closely.

In the figure are plotted the entropy and enthalpy relative to the elements along with the free energy. The enthalpy and entropy gauge the two tendencies for reaction, the attractive forces leading to bonding with an attendant release of energy and an entropy increase representing the tendency of systems toward greater randomness. From left to right ($SO_2 + \frac{1}{2} O_2 \rightarrow SO_3$) beginning at 0 conversion, the decline in free energy is caused by the decline in enthalpy due to the release of energy associated with the bonding of oxygen to SO_2, forming SO_3. This attractive force is offset increasingly by a decline in entropy (less randomness), which is counter to a tendency for reaction.

From right to left ($SO_3 \rightarrow SO_2 + \frac{1}{2} O_2$), the effect is opposite. The enthalpy rises since energy must be supplied to overcome the attractive forces so that oxygen may be separated from SO_3. This impediment to reaction is overcome by the simultaneous rise in entropy to produce a net decline in free energy and a favorable potential toward SO_2 and oxygen. These two counteracting tendencies produce a minimum free energy beyond which the reaction cannot proceed from left to right or from right to left.

Although the plots presented are useful for understanding, it is fortunate that they are not needed to determine the equilibrium composition for single reactions. The value of G at the minimum does represent the equilibrium composition point, and this fact was used to develop the expressions for the equilibrium constant by differentiating the equation for G and setting it equal to zero (see Eqs. 3.12 and 3.13). Compositions of complex reaction mixtures can be obtained by simultaneously solving equilibrium constant equations for each reaction in the form of one of the equations in Table 3.3 or by finding the minimum of the total free-energy expression as detailed in the next section.

Calculation of Complex Reaction Equilibria

The literature on techniques for determining equilibrium compositions for multiple-reaction systems is quite large, and much of it predates the wide availability of commercial software applicable to such problems. Busy professionals will find the following two procedures the most efficient. In both instances, the user should confine the analysis to products that, based on prior experience, would appear during the time for a practical reaction to occur.

Simultaneous Equilibrium Expressions

The procedure described here is no different than that for a single reaction except that it requires the simultaneous solution of two or more equations written in terms of the extents of reaction.

Basis: 1 mole of A and n_{B_0} moles of B fed.

$$A + 2B \rightleftarrows 3R + S$$

$$S + B \rightleftarrows R + 3C$$

$$X_{1a} = -(\Delta n_A)_1 = \frac{-(\Delta(n_B)_1}{2} = \frac{(\Delta n_R)_1}{3} = (\Delta n_S)_1$$

$$X_{2a} = -(\Delta n_S)_2 = -(\Delta n_B)_2 = (\Delta n_R)_2 = \frac{(\Delta n_C)_2}{3}$$

Moles at an extent of reaction (X_{1a}, X_{2a}):

$$n_A = 1 - X_{1a}$$

$$n_B = n_{B_0} - 2X_{1a} - X_{2a}$$

$$n_R = 3X_{1a} + X_{2a}$$

$$n_S = X_{1a} - X_{2a}$$

$$n_C = 3X_{2a}$$

$$n_T = 1 + n_{B_0} + X_{1a} + 2X_{2a}$$

$$K_1 = \frac{n_R^3 n_S}{n_A n_B^2} \frac{P}{n_T} K_{\phi_1}$$

or

$$F(1) = \frac{n_R^3 n_S P}{n_A n_B^2 n_T} - \frac{K_1}{K_{\phi_1}} \tag{3.33}$$

$$K_2 = \frac{n_R n_C^3}{n_S n_B} \frac{P^2}{n_T^2} K_{\phi_2}$$

or

$$F(2) = \frac{n_R n_C^3 P^2}{n_S n_B n_T^2} - \frac{K_2}{K_{\phi_2}} \tag{3.34}$$

where K_{ϕ_1} and K_{ϕ_2} are the appropriate fugacity coefficient quotients, which for an ideal solution are

$$K_{\phi_1} = \frac{\phi_R^3 \phi_S}{\phi_A \phi_B} \quad \text{and} \quad K_{\phi_2} = \frac{\phi_R \phi_C^3}{\phi_S \phi_B}$$

These equations can be solved for X_{1a} and X_{2a} with a good nonlinear algebraic equation solver. It is convenient to enter equilibrium constants, fugacity coefficient term, pressure, and initial moles of feed components as adjustable parameters so that studies under various conditions can be rapidly made once the equations are coded. Equations in the form of Eqs. 3.33 and 3.34 are used with values of n_j in terms of X inserted, or a separate subroutine for stoichiometric terms may be used. See Example 3.

This method has the advantage of allowing the use of stoichiometric equations for reactions thought or known to occur, and ties the results to these reactions thereby. All nonlinear algebraic equation solvers require initial guesses, and some systems can be very sensitive to these initial guesses, with convergence highly dependent upon them. The following suggestions will help avoid most difficulties.

1. Select only those components present in the feed and products known to be present at or near equilibrium based on experimental or plant observations.
2. Write stoichiometric equations that yield the products. Use no more than one way to make each product or products [4]. Of course, a given product may appear in several equations when each of them uniquely defines other products.
3. When in doubt, check the independence of the equations. One quick procedure involves combining two or more reactions by products or quotients of the K values. If these result in a value of K equal to a K of one of the other proposed reactions, that reaction is not independent. This procedure is identical to adding and/or subtracting equations to arrive at a new equation that is dependent.

See Example 3.

Free-Energy Minimization

As has been shown, the total free energy for a reacting system reaches a minimum at equilibrium, and the corresponding composition can be found by determining the combination of component concentrations that yields such a minimum. According to Eq. 3.7, the following applies for the ideal gaseous system:

$$\frac{G}{RT} = \sum_{j=1}^{c} n_j\left(\frac{G^\circ_{f_j}}{RT} + \ln P + \ln \frac{n_j}{n_T}\right) \tag{3.35}$$

For nonideal gases, the mole fraction terms for each component must be multiplied by the applicable fugacity coefficient, ϕ_j or $\hat{\phi}_j$.

In order to find the number of moles of each component at equilibrium, the values determined must produce a minimum in the value of G/RT, as given in Eq. 3.35, and simultaneously be constrained by the elemental balance requirement as follows:

$$\sum_{j=1}^{c} \alpha_{ij} n_j - b_i = 0 , \qquad i = 1, \ldots, m \tag{3.36}$$

where α_{ij} is the number of atoms of element i in component j and b_i is the total number of atomic weights of element i in the system.

Lagrange multipliers are introduced in the usual manner to aid in this search for a minimum by adding a term to the free energy to obtain a new function $\mathcal{L}$, known as the *Lagrangian*:

$$\mathcal{L} = \sum_{j=1}^{c} n_j \left(\frac{G^\circ_{f_j}}{RT} + \ln P + \ln \frac{n_j}{n_T} \right) - \sum_{i=1}^{m} \lambda_i \left(\sum_{j=1}^{c} \alpha_{ij} n_j - b_i \right) \tag{3.37}$$

where λ_i is a Lagrange multiplier.

Since the new term from Eq. 3.36 is zero, it will produce an identical minimum to Eq. 3.35. The partial derivatives with respect to compositions set to zero provide the desired criteria for this constrained minimum:

$$\frac{\partial \mathcal{L}}{\partial n_j} = \frac{G^\circ_{f_j}}{RT} + \ln P + \ln \frac{n_j}{n_T} + 1 - \sum_{i=1}^{m} \lambda_i \alpha_{ij} = 0, \qquad j = 1, \ldots, c \tag{3.38}$$

Various methods proceeding from this point have been described [4]. All involve starting estimates of the mole fractions, which modern software usually automatically sets at equal values for all components. The derivatives expressed by Eq. 3.38 are solved for the mole fraction, and further manipulation eliminates n_T, yielding additional equations from which the Lagrange multipliers may be calculated and used for estimating new values of mole fractions at equilibrium for the next iteration. See Example 5 for a demonstration of this method. In addition to the Lagrange multiplier technique, any nonlinear programming procedure for constrained minimizations can be used. The most popular are SQP (successive quadratic programming) and GRG (generalized reduced gradient).

As in the method using equilibrium constants, components should be excluded that could be present at equilibrium but known not to appear during the reaction time being considered. In most cases, because of preliminary experimental data, a reasonable idea of important components that can appear within a practical time period will be known. The unique value of the method is that simultaneous knowledge of the chemical reactions that led to such products is not required. One would hope, however, that if the chemical process is further developed, such knowledge, where feasible, should be sought by other means. Another great advantage of this free-energy minimization method is that the solution method can proceed with arbitrary initial guesses of the equilibrium composition.

THE VALUE OF EQUILIBRIUM CALCULATIONS

Because of the extensive thermodynamic data extant on gases, it is possible to make useful decisions on ideal reaction conditions, reactor type and operating procedure, and even design of reactors in some cases. All this can be done without any laboratory work and should precede any attempts to gather rate data or to use rate data in the decision process. These applications of equilibrium calculations are illustrated on pages 205–240.

LITERATURE CITED

1. Stull, D.R., Westrum, E.F., Jr., and Sinke, G.C., "The Chemical Thermodynamics of Organic Compounds," Wiley, New York, 1969.
2. Reid, R.C., Prausnitz, J.M., and Poling, B.E., "The Properties of Gases and Liquids," 4th Edition, McGraw-Hill, New York, 1987.
3. Lewis, G.N., *Proc. Am. Acad. Sci.*, *37*, 49 (1901).
4. Walas, S.M., "Phase Equilibria in Chemical Engineering," Butterworths, Boston, 1985.
5. Prausnitz, J.M., "Molecular Thermodynamics of Fluid-Phase Equilibria," Prentice-Hall, Englewood Cliffs, N.J., 1969.
6. Hougen, O.A., Watson, K.M., and Ragatz R.A., "Chemical Process Principles," Vol. 2, 2nd Edition, Wiley, New York, 1959.

CHAPTER

4

Transport Processes and the Laws of Conservation

Any attempt to quantify the phenomena occurring in a fixed bed must take account of the transport of heat and mass between the catalyst and the fluid and within the catalyst, the pressure drop across the bed, and the certainty of the conservation of energy and mass.

SIMULTANEOUS TRANSPORT PROCESSES

As the reaction proceeds on the catalyst surfaces, compounds are consumed and others are formed as energy is released or absorbed. Gradients of temperature and concentration (or partial pressure) are thus produced and constitute the driving forces for the transport processes of mass and heat between the interior of the catalyst and the flowing fluid. In addition to these complex transport processes, flow patterns of the bulk fluid as it passes through the bed are tortuous and unpredictable, since a priori knowledge of particle arrangement in a bed is not possible. General flow characteristics and molecular movements, however, can be described in an intuitively satisfying way, as shown in Fig. 4.1.

The various gradients established at the steady state in a fixed-bed reactor can be clearly illustrated by an example of an exothermic reaction, the production of ethylene oxide by the oxidation of ethylene. For simplicity, important side reactions are ignored.

$$C_2H_4 + \tfrac{1}{2}\,O_2 \rightarrow C_2H_4O$$

Figure 4.2 illustrates qualitative gradients of concentration (partial pressure) and temperature for this single-reaction system at an axial position AA′ and radial position BB′.

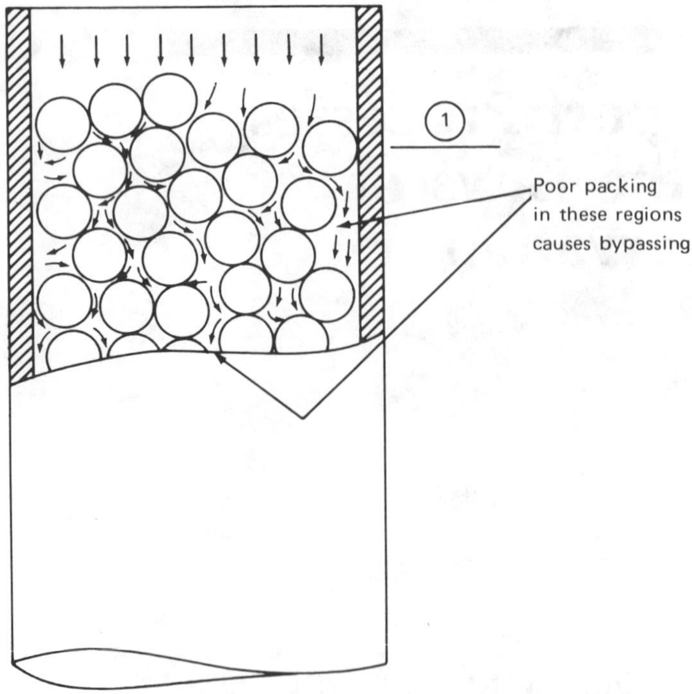

Figure 4.1 General description of fluid and molecular movements in a packed bed

1. Fluid enters the portion of bed illustrated containing product, produced in the bed above position 1, and unreacted reactants.
2. The fluid moves through the gaps between particles where mixing occurs. Elements of fluid are rapidly moved in a random fashion in this region. Immediately adjacent to the catalyst surface the movement is slower, and this slow-moving region is often idealized as a stagnant film. At this film or boundary layer the movement consists of elements of fluid contacting the layer, during which time molecules will be interchanged by molecular diffusion and energy transferred by the exchange of higher-energy molecules from the higher-temperature region with lower-energy molecules from the lower-temperature region, a process called conduction in a fluid.
3. Fluid also penetrates the pores of the catalyst, but at the steady state the predominant motion is Brownian in character, in which the molecules move randomly with the rapidity of movement being proportional to temperature. For very low-density gases and/or very fine pores, gasesous molecules will collide with the pore walls more often than with each other. This movement is described as Knudsen flow or Knudsen diffusion. With larger pores classical molecular diffusion occurs, in which collisions between molecules predominate. In either type of flow, molecules strike the surface. Some are chemisorbed if they strike an active region and are not blocked by other adsorbed molecules or layers of deposits. The adsorbed molecules can react with others and then desorb as products and enter the fluid phase in the pore. Random molecular movement must be reconciled with the well-known experimental observation that molecules are transferred along a concentration

General Defining Relationships for Transport of Mass and Heat

The phenomenological explanations given in Figs. 4.1 and 4.2 are useful intuitively but do not provide the tools for calculating the magnitudes of the variables that must be known for reactor modeling. For such complex situations that defy analysis based on first principles of molecular and atomic processes involved in diffusion and heat transfer, it is necessary to emulate classical thermodynamics and to avoid atomistic description altogether. Instead the problem is simplified and made tractable by expressing rates of heat transfer and mass transfer as proportional to temperature and concentration gradients, respectively, with proportionality constants generally empirically obtained to fit the system being analyzed. The following equations, written in incremental form, were originally proposed over a century ago. They have served well, and the chemical engineer has adapted the same forms to complex systems that are strongly influenced by the equipment used to effect the transfer.

Conduction

$$N_q = -\lambda \frac{\Delta T}{b} \qquad \text{Biot (1804, 1816)}$$

$$= -\frac{\lambda}{\rho c_p} \frac{\rho c_p \Delta T}{b} = -\alpha_t \frac{\rho c_p \Delta T}{b} \qquad \text{Fourier (1822)} \tag{4.1}$$

where N_q is the heat flux expressed as a positive quantity (cal/cm^{2}sec) in the direction of heat flow, λ is the thermal conductivity of the material through which heat is being conducted, b is the thickness of the conducting material, ρ is the fluid density, c_p is the heat capacity, and ΔT is the temperature gradient, which is negative in the direction of flow. The term $\rho c_p \Delta T$ is an energy term

(*Text continues on page 80.*)

Figure 4.1 (*continued*)

gradient from a region of higher to lower concentration. At a given position consider equal elements of volume on each side of the designated position line. In molecular diffusion, on the average equal fractions of the molecules in each element will move across the line in opposite directions. If there are more molecules of A in the element on one side than on the other, a net transfer of A in one direction will occur. In turbulent diffusion caused by high flow rates and tortuous flow path, the elements themselves move with the same result but with greater rapidity. If A is consumed in one region and supplied from another, at the steady state a gradient will be established between the supply region and the region of consumption.

4. The random flow of elements of fluid has a radial component as these elements make their way back and forth along the pockets between catalyst. The bulk of the fluid flows axially toward the exit of the reactor, in this case downward.

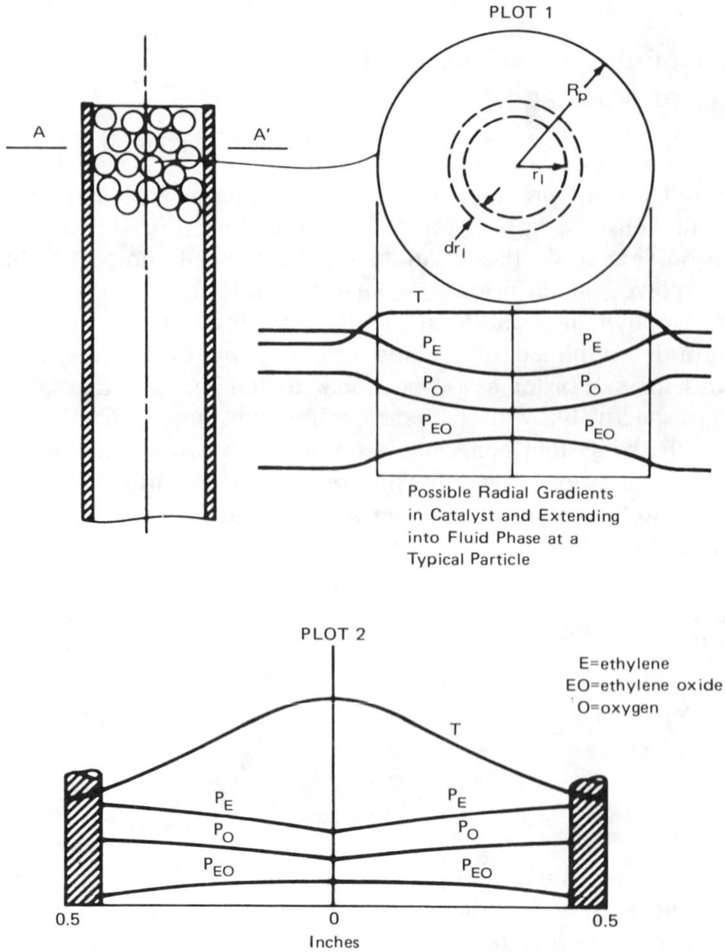

Possible Radial Gradients at Axial Position AA′ in the Fluid Phase (Slopes Exaggerated)

Figure 4.2 Plot 1: Moderate chemical rate
Ethylene and oxygen react as they penetrate the length of a pore. Because the reaction is assumed modest in rate for this plot, all of the reactants are not consumed instantaneously, and the remainder continue moving along the pore in the same manner. The plot represents steady-state average partial pressures at the cross section for all the pores at that position. Reactant molecules that penetrate further into the pores will strike the surface more than those that do not, and chances for random movement back out the pore without reaction are greatly diminished, such that, at the steady state, reactant concentration toward the center is lower than that toward the exterior. Product concentration for similar reasons shows a reverse effect. The total pressure declines toward the center. This gradient aids the diffusion of the reactants and hinders that of the product, which must diffuse in against the gradient. In Knudsen diffusion the total pressure gradient would have no effect, since this type of diffusion is affected primarily by wall collisions, not by pressure-dependent molecular collisions.

The concentration of reactants and the products and temperature within a pocket of fluid between catalyst particles will be constant because of the mixing that results. In essence, these pockets may be thought of as completely mixed elements. The rather stagnant region of these pockets adjacent to the catalyst exterior surfaces, however, present the main resistance to both mass and heat transfer, and gradients across them can be significant. For the gaseous system illustrated at moderate chemical rate, the concentration gradients between the fluid and the exterior of the catalyst are small compared with that within the catalyst. This situation is caused by the easier transport of molecules in the fluid phase, even near the catalyst surface, than the more difficult and constricted region within the catalyst. By contrast, the gradient of temperature is greater exterior to particle. In fact, in most cases that in the interior of the catalyst is essentially flat. Heat is transferred by several means within the catalyst pellet; conduction along with some bulk movement in the fluid, conduction in the solid, and radiation at high temperatures. Although upon initial operation the catalyst may assume rather high temperature differences due to changes in concentration of reacting components with depth of penetration, at the steady state, conduction in the pellet will cause all areas to reach essentially the same temperature. The temperature gradient between the exterior of the particle and the fluid is more significant than the concentration gradient. All the energy generated in the pellet passes through this region adjacent to the particle, as do the reactants and products. The temperature gradient is simply the energy generated divided by the heat capacity of the fluid layer:

$$\Delta T = \frac{(-\Delta H_E)(\Delta C_E)}{\rho c_p}$$

where ΔH_E is heat of reaction per mole ethylene reacted, ΔC_E is concentration gradient of ethylene in moles/volume at catalyst-fluid interface, ρ is fluid density in mass/volume, and c_p is heat capacity of fluid on a mass basis. The term $(-\Delta H_E)/\rho c_p$ is rather large, so even for a small value of ΔC_E the ΔT can be significant.

High chemical rate

In this case, the chemical rate is so fast that, at the steady state, reactants do not reach the central portion of the catalyst. The gradients are not only steep within the particle but now become more significant between the exterior surface and the fluid. Temperature, especially, exhibits a large gradient at the exterior, but the pellet remains isothermal, as explained above. Other conditions can be visualized: a slow chemical step with negligible gradients of concentration and temperature, or a very fast reaction confined to the exterior surface only.

Radial gradients (Plot 2) and axial gradients

In the example, exothermic reaction heat is being removed from the tube at the wall, which means that a radial gradient of energy, and thus temperature, exists in the bed and across the tube wall. Because of this gradient, concentration gradients must also exist, since the reaction will be more rapid in pellets near the center than in those at the wall. The random motions earlier described will cause a net transfer of reactants toward the center and products toward the wall and thereby establish concentration gradients ending at the wall.

Heat is transferred by conduction within and between particles and by turbulence in the pockets and movement between pockets. The events are quite complex, with each pellet being isothermal but of a different temperature than one at an adjacent radial or axial position. The pellets adjacent to the wall are contacting the metal in most cases, and the movement of fluid on that side is quite different than within the bed. More axial flow (by passing) can occur at the wall due to the relatively greater space for flow created by normal loading procedures.

While all this radial bed and internal catalyst activity is taking place, the major flow of fluid and molecules is in the axial direction, and axial movement caused by random motion is usually negligible.

per unit volume and is analogous to concentration in the mass-transport equations that follow. The term α_t is $\lambda/\rho c_p$, which is called the *thermal diffusivity*.

Convective Heat Transfer

$$
\begin{aligned}
N_q &= -h\,\Delta T \\
&= -\frac{h}{\rho c_p}\,\rho c_p\,\Delta T \qquad \text{Newton (1701)}
\end{aligned}
\tag{4.2}
$$

where h is the heat-transfer coefficient, which can be thought of as λ/b, in which b is the unknown thickness of an assumed film of fluid along the transfer surface.

Diffusion (Molecular)

$$
N_A = -\mathscr{D}_A\,\frac{\Delta C_A}{b} = -\mathscr{D}_A\,\frac{\Delta P_A}{\mathbf{R}Tb} \qquad \text{Fick (1855)} \tag{4.3}
$$

where $\mathscr{D}_A$ is the molecular diffusivity of A in the mixture; ΔC_A and ΔP_A are the concentration and partial-pressure gradients of A across the distance b; N_A is the moles of A transferred/(time)(area of transfer), and is a positive quantity in the direction of transport; and $\mathbf{R}$ is the gas-law constant in PVT units.

Convective Mass Transport

$$
N_A = -k_a\,\Delta C_A = -\frac{k_a}{\mathbf{R}T}\,\Delta P_A \tag{4.4}
$$

where k_a is a mass-transfer coefficient with concentration gradient. As with the heat-transfer coefficient, k_a may be compared to $\mathscr{D}_A/b$.

Assessment of Transport Flux Equations

The transport flux equations are all similar, and simply say that the rate of the indicated transfer is proportional to a gradient, and the proportionality constants are defined by the equation. The hope, as always, was to be able to determine the constants by calculation from theory. This dream has only been realized for diffusivities of ideal gases and thermal conductivity for monatomic gases. The most accurate equations for low-pressure gases evolve from the Chapman-Enskog kinetic theory, and can be used when appropriate values of collision integrals and characteristic lengths are available or estimable. Semiempirical equations of more general utility, as assembled in Table 4.1, are suggested for quick estimates when data cannot be found. See Reference 1 for more detail on various estimating procedures. Data sources are given in Table 4.1.

(*Text continues on page 86.*)

Table 4.1 Transport Properties of Gases

DIFFUSIVITY

Binary Gas-Low Pressure [2]

$$\mathcal{D}_{AB} = \frac{0.00143T^{1.75}}{P\,M_{AB}^{0.5}\left[\left(\Sigma v_i\right)_A^{1/3} + \left(\Sigma v_i\right)_B^{1/3}\right]^2}$$

where $\mathcal{D}_{AB}$ is the diffusivity for A diffusing into B in cm^2/sec; P is pressure in atm; T is temperature in °K; $M_{AB} = 2\,[\,(1/M_A) + (1/M_B)\,]^{-1}$; M_A and M_B are molecular weights of A and B, respectively; v_i is atomic diffusion volume to be summed over atoms, groups, and structural features of diffusing compound values for which are listed below (Example for CH_4 15.9 + 4(2.31) = 25.14).

A. Atomic and structural diffusion volume increments, v_i's

C	15.9	(F)	14.7
H	2.31	(Cl)	21.0
O	6.11	(Br)	21.9
(N)	4.54	(I)	20.8
(S)	22.9	Aromatic or Heterocyclic Ring	−18.3

B. Diffusion volumes of atom and simple molecules, Σv_i's

He	2.67	CO	18.0
Ne	5.98	CO_2	26.7
Ar	16.2	N_2O	35.9
Kr	24.5	NH_3	20.7
Xe	32.7	H_2O	13.1
H_2	6.12	SF_6	71.3
D_2	6.84	(Cl_2)	38.4
N_2	18.5	(Br_2)	69.0
O_2	16.3	(SO_2)	41.8
Air	19.7		

Effect of Pressure on Binary Diffusivity [3]

Use empirical correlation as shown on the following graph, based on reduced pressure and temperature. Pseudocritical temperatures and pressures are used with a mole fraction weighting.

$$T_c = Y_A(T_c)_A + Y_B(T_c)_B$$

$$P_c = Y_A(P_c)_A + Y_B(P_c)_B$$

$$T_r = \frac{T}{T_c} \quad P_r = \frac{P}{P_c}$$

Table 4.1 *(continued)*

where T_c and P_c are critical temperature and pressure; T_c' and P_c' are pseudocritical properties of the binary mixture; T_r and P_r are reduced temperature and pressure, and Y is the mole fraction of the indicated component.

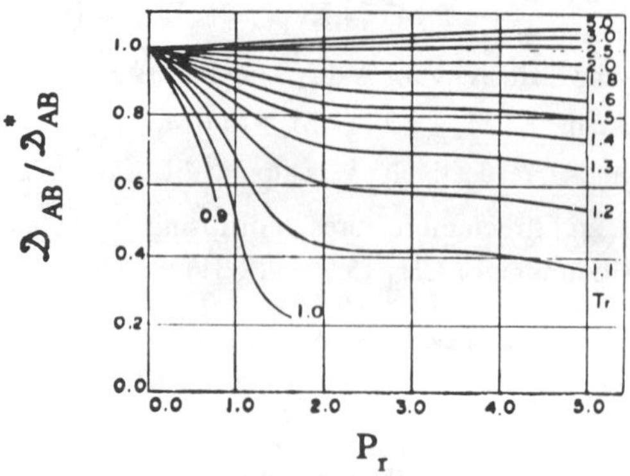

$\mathcal{D}_{AB}$ is the binary diffusivity at the desired pressure in cm²/sec, and $\mathcal{D}^*_{AB}$ is the low-pressure or ideal gas value.

Graph reproduced by permission, Takahashi, S., J. Chem. Eng. Japan, 7, 417 (1974).

Multicomponent Gas Mixture-Approximation [1]

$$\mathcal{D}_{im} = \left(\sum_{j=1}^{n} \frac{Y_j}{\mathcal{D}_{ij}} \right)^{-1}$$

where $\mathcal{D}_{im}$ is the diffusivity of i diffusing into a homogeneous mixture, m; $\mathcal{D}_{ij}$ is the binary diffusivity of i diffusing into j present as Y_j mole fraction.

THERMAL CONDUCTIVITY

Since thermal conductivity is a significant factor in estimating heat-transfer coefficients, actual values at the conditions of interest are preferred. References 3 to 9 are useful for this purpose. Estimating procedures are complex because of the effect of molecular velocities and translational, rotational, and vibrational degrees of freedom on thermal conductivity. The procedures have been thoroughly assessed [1] and some are included in commercial simulator programs. If one value of thermal conductivity is known, simpler procedures are possible.

LOW PRESSURE (Up to 10 atm)

Rough Estimate:

Monatomic Gases:
$$\frac{\lambda_{f2}}{\lambda_{f1}} = \frac{T_2}{T_1}$$

Organic Vapors [8]
$$\frac{\lambda_{f2}}{\lambda_{f1}} = \left(\frac{T_2}{T_1}\right)^{1.786}$$

Improved Estimate: Atmospheric Pressure [10, 11]

$$\lambda_f = \frac{9.96 \times 10^{-5}\left[\exp.\left(0.0464\,T_r\right) - \exp.\left(-0.2412\,T_r\right)\right] + \overset{\circ}{\lambda}_f\, f\left(T_r\right)}{M^{1/2}\, T_c^{1/6}\, P_c^{2/3}}$$

where λ_f is thermal conductivity in cal/cm sec °K; $\overset{\circ}{\lambda}_f$ is Thodos constant unique to each compound and obtainable from group contributions [10,11]; T_r is the reduced temperature; T_c and P_c are critical temperature and pressure in °K and atm, respectively; $f(T_R)$ values are given below.

If a value of λ_f is known at one temperature, one can solve for $\overset{\circ}{\lambda}_f$, and then use the value calculated to determine values of λ_f at other temperatures.

$f(T_r)$ Relations

Saturated hydrocarbons (except methane)-	$-0.152T_r +$	$1.191T_r^2 -$	$0.039T_r^3$
Olefins	$-0.255T_r +$	$1.065T_r^2 +$	$0.190T_r^3$
Acetylenes	$-0.068T_r +$	$1.251T_r^2 +$	$0.183T_r^3$
Naphthalenes and aromatics	$-0.354T_r +$	$1.501T_r^2 +$	$0.147T_r^3$
Alcohols		$1.000T_r^2$	
Aldehydes, ketones, ethers, esters	$-0.082T_r +$	$1.045T_r^2 +$	$0.037T_r^3$
Amines and nitriles		$0.633T_r^2 +$	$0.367T_r^3$
Halides	$-0.107T_r +$	$1.330T_r^2 -$	$0.223T_r^3$
Cyclic compounds	$-0.354T_r +$	$1.501T_r^2 -$	$0.147T_r^3$

Table 4.1 *(continued)*

Effect of Pressure

Nonpolar Gases [2]:

$$\left(\lambda_f - \lambda_f^*\right)\left(M^{1/2} T_c^{\;1/6} P_c^{\;2/3}\right) z_c^5 = a\left(e^{-b\rho_r} - c\right) \times 10^{-8}$$

where λ_f^* is the thermal conductivity at low pressure in cal/cm sec°K, ρ_r is the reduced density (ρ/ρ_c); z_c is the critical compressibility; and ρ_c is the critical density.

Condition	a	b	c
$P_r < 0.5$	14.00	0.535	1.0
$0.5 < P_r < 2.0$	13.10	1.069	1.069
$2.0 < P_r < 2.8$	2.976	2.016	2.016

Diatomic Gases [13]:

$$\lambda_c = 9.415 \times 10^{-5}\left[\frac{P_c^{\;9/16}}{M^{1/2} T_c^{\;1/5}}\right]$$

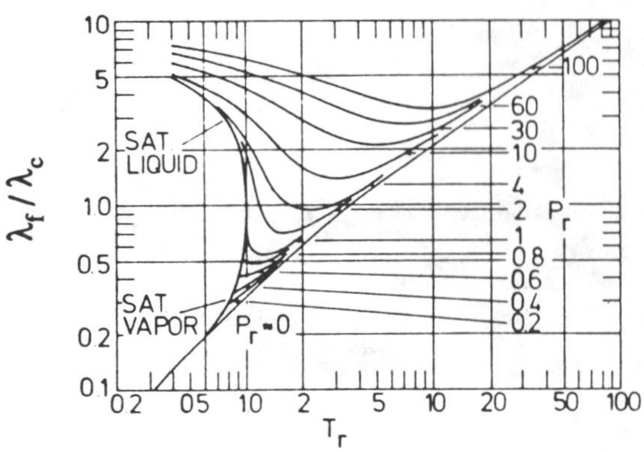

Reproduced by permission: Schaeter, C. A. and Thodos, G. AIChE J., 5, 367 (1959), as adapted by Schunck, M., "Heat Exchanger Design Handbook," Vol. 5, p. 5.1.4.6, Hemisphere Publ, Washington, 1983.

Gas Mixture Thermal Conductivity

$$\lambda_{fm} = \frac{\sum_{j=1}^{n} Y_j \lambda_j M_j^{1/3}}{\sum_{j=1}^{n} Y_j M_j^{1/3}}$$

where λ_{fm} and λj are the thermal conductivities of the mixture and component j, respectively, in cal/cm sec °K ; Y_j is the mole fraction of j; M_j is the molecular weight of j.

Viscosity

See References 3 to 9 for compilation of data and Reference 1 for detailed discussion of estimating methods and recommendations. Where possible, use actual data.

General Estimate at All Pressures (20)

Mean deviation ± 6% except higher for H_2, H_e, paraffins above C_7, and strongly polar molecules.

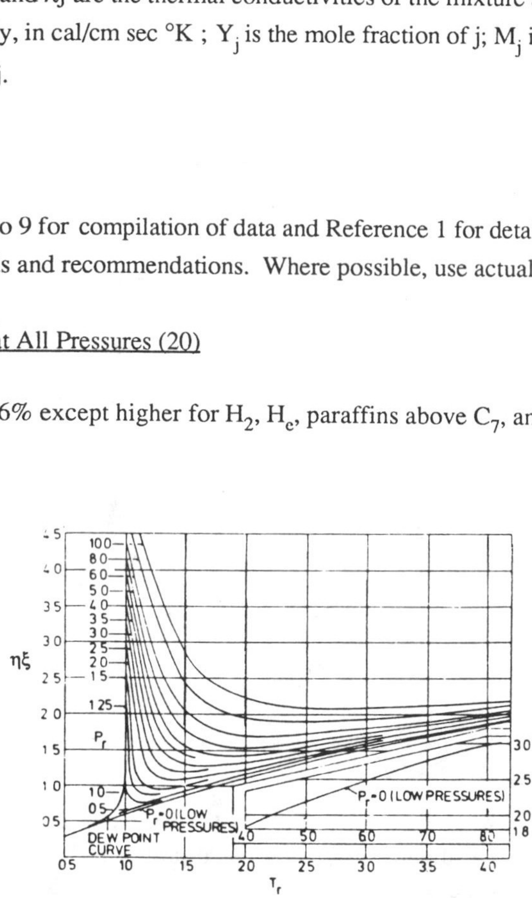

Reproduced by permission: Lucas, K. Chem. Eng. Tech., 46 (4), 157 (1974), as adapted by Shunck, M., "Heat Exchanger Design Handbook," Vol. 5, p. 5.1.4.1, Hemisphere Publ, Washington, 1988.

$$\xi = 3.8 \times 10^9 \quad \frac{T_c^{1/6}}{M^{1/2} P_c^{2/3}}$$

Table 4.1 *(continued)*

The Y axis not dimensionless.

P_c is in N/m^2 (1 atm = 1.01325×10^5 N/m^2);
T_c is in °K
η is viscosity in N sec/m^2 (1 N sec/m^2 = 1000 cp)

Gas Mixture Viscosity [15]

$$\mu_m = \frac{\sum_{j=1}^{n} Y_j \mu_j M_j^{0.5}}{\sum_{j=1}^{n} Y_j M_j^{0.5}}$$

Prandtl Number, $Pr = c_p \mu / \lambda_f$

Typical Approximate Values [16]
Temperature range of 200 - 1000°K and at low pressures

Gas	Pr
Monatomic	0.67
Diatomic	0.73
Triatomic	0.76
Polyatomic	1.0

As is well known, the mass- and heat-transfer coefficients for convection are correlated by generalized dimensionless equations that yield predictions of h and k_a with average deviations seldom less than 20%. What is called convection is really quite complex, even for turbulent flow in a cylindrical tube. Heat energy and mass are carried out by turbulent movement of elements of fluid, the character of which is strongly influenced by the type of equipment used. At the same time conduction and molecular diffusion occurs, which are independent of mass flow. It is clear that in assigning coefficients for mass and heat transfer for a flow system as complex as a fixed-bed reactor we must be content with these rather nebulous proportionality constants, h and k_a, defined as simply as possible.

Defining Relationships for Mass and Heat Transfer for Fixed-Bed Reactors

As illustrated in Figs. 4.1 and 4.2, gradients in concentration and temperature occur within the catalyst pellet, between the pellet and the fluid phase, and along the radial coordinate. The major gradients are in the direction of flow, which is, in all but radial reactors, the axial direction. To express the radial and interphase gradients, it was necessary to borrow the Fickian, Fourier, and Newtonian transport forms and simply define new empirical proportionality factors—effective pellet diffusivities and conductivities, and interphase mass- and heat-transfer coefficients. These empirical quantities are best illustrated by considering the fluxes at the several interfaces of interest (see Figs. 4.1 and 4.2).

Boundary Between Spherical Catalyst and Fluid (Steady State)

Component Transport

$$(\hat{r}_j)_e = -\mathscr{D}_{I_j} a_m \left(\frac{dC_j}{dr_I}\right)_{r_I = R_p} = k_{s_j} a_m (C_{j_s} - C_{j_b}) = k^s_{g_j} a_m (P_{j_s} - P_{j_b}) \qquad (4.5)$$

$$\begin{bmatrix}\text{Observed or}\\ \text{effective rate}\\ \text{of reaction}\\ \text{of } j \text{ (plus for}\\ \text{products, minus}\\ \text{for reactants)}\end{bmatrix} = \begin{bmatrix}\text{Component flux}\\ \text{at catalyst}\\ \text{exterior surface}\end{bmatrix} = \begin{bmatrix}\text{Component transport}\\ \text{between catalyst}\\ \text{exterior surface}\\ \text{and the fluid}\end{bmatrix}$$

where $(\hat{r}_j)_e$ is the effective or global (observed) rate of production (+) or consumption (−) of j; $\mathscr{D}_{I_j}$ is the effective diffusivity of j in the catalyst pellet; a_m is the exterior surface area per unit mass of catalyst; C_{j_s}, C_{j_b} and P_{j_s}, P_{j_b} are the concentrations and partial pressures of j at any defined position, at the exterior catalyst surface, and in the bulk fluid, respectively; r_I is the radial distance in the catalyst, R_p is the radius of the catalyst; and $k^s_{g_j}$ and k_{s_j} are the transfer coefficients between the catalyst exterior and the bulk fluid ($k^s_{g_j} = k_{s_j}/\mathbf{R}T$).

Heat Transfer

$$(-\Delta H_1)(\hat{r}_1)_e = -\lambda_I a_m \left(\frac{dT}{dr_I}\right)_{r_I = R_p} = h_s a_m (T_s - T_b) \qquad (4.6)$$

$$\begin{bmatrix}\text{Heat generation}\\ \text{within particle}\\ \text{(plus for exothermic,}\\ \text{minus for endothermic)}\end{bmatrix} = \begin{bmatrix}\text{Heat flux at}\\ \text{catalyst exterior}\\ \text{surface}\end{bmatrix} = \begin{bmatrix}\text{Heat transfer}\\ \text{between catalyst}\\ \text{surface and}\\ \text{bulk fluid}\end{bmatrix}$$

where $(\hat{r}_1)_e$ is the observed rate of reaction 1 (in this case the only reaction occurring) and a positive quantity, as are all rates of individual reactions; λ_l is the effective radial thermal conductivity in the catalyst particle; h_s is the heat-transfer coefficient between the bulk fluid and the catalyst surface; and T, T_s, T_b are temperatures at any defined point, the surface, and the bulk fluid.

Boundary Between the Cylindrically Shaped Bed and Reactor Wall

Component Transport

$$\mathscr{D}_r a_m \left(\frac{dC_j}{dr_t}\right)_{r_t=R_t} = 0 \tag{4.7}$$

where $\mathscr{D}_r$ is the radial diffusivity or radial dispersion coefficent, r_t is the radial distance in the bed, and R_t is the radius of the bed. The radial flow of components, unlike the heat transfer, stops at the reactor wall, and thus the flux at that point is zero. The general definition of the *effective radial dispersion coefficient* is

$$N_j = -\mathscr{D}_r \frac{dC_j}{dr_t} \tag{4.8}$$

where N_j is the flux of j, positive for product and negative for reactant.

Heat Transfer

$$N_q = \lambda_r \left(\frac{dT}{dr_t}\right)_{r_t=R_b} = h_w(T_w - T_{R_t}) = h_T(T_h - T_{R_t}) \tag{4.9}$$

where N_q is the heat flux, which is positive for heat added to reactor and negative for heat removed; λ_r is the radial thermal conductivity of the combination of catalyst bed and flowing fluid; h_w is the heat-transfer coefficient between the edge of the bed and the wall where a discontinuity occurs, and a slower-moving film of resistance is envisioned at the wall; T_w, T_{R_t}, T_h are, respectively, temperatures at the wall, at the radius of the bed, and of the heat-transfer medium in the jacket. The reciprocal of h_T encompasses several resistances:

$$\frac{1}{h_T} = \frac{D}{h_o D_o} + \frac{1}{h_w} + \frac{b_w D}{\lambda_w D_m} + r_f \tag{4.10}$$

where h_o is the outside jacket film coefficient; b_w is the wall thickness; λ_w is the wall thermal conductivity; r_f is the fouling resistance; D, D_o, D_m are, respectively, inside, outside, and mean diameters. If the gradient within the core of the bed is modest or constant, an average bed temperature T is appropriate at each axial position such that

$$N_q = U_Z(T_h - T) \tag{4.11}$$

$$\frac{1}{U_Z} = \frac{D}{h_o D_o} + \frac{1}{h_i} + \frac{b_w D}{\lambda_w D_m} + r_f \tag{4.12}$$

By using a parabolic radial temperature distribution, the following relation between h_T and U_Z results [16, 17]:

$$U_Z = \frac{h_T}{1 + h_T D/8\lambda_r} \tag{4.13}$$

where U_Z is the overall coefficient based on the inside tube surface area, and h_i is the inside film coefficient defined in terms of a constant or mean radial bed temperature T:

$$N_q = h_i(T_w - T) \tag{4.14}$$

These various transport equations will be used in Chapter 5 for developing fixed-bed reactor models. Empirical equations for predicting the heat- and mass-transfer coefficients and effective conductivity and diffusivity are summarized in Table 4.2, along with the average deviations of predicted from observed values. It is clear that prediction of nonadiabatic reactor performance is very uncertain unless transport coefficients are known for similar reactors from actual tests. Errors in heat-transfer data are the most damaging, since a small

Table 4.2 Recommended Heat and Mass Transfer Correlations for Packed Beds

Type	*Correlation*	*Deviation %*	*Range*
Particle-to-Fluid (a)	Heat Transfer:		
$h_s = \frac{\Sigma(-\Delta H_n)(\hat{r}_n)}{a_m(T_s - T)}$	$\frac{h_s D_p}{\lambda_f} = 2 + 1.1\ Pr^{1/3}\ Re^{0.6}$	25%	
	Mass Transfer:		
$k_{s_j} = \frac{\Sigma \nu_{jn} \hat{r}_n}{a_m(C_{js} - C_j)}$	$\frac{\mu}{\rho \mathcal{D}_{jm}} = 2 + 1.1\ Sc^{1/3} Re^{0.6}$	25%	
Heat-Transfer Coefficient (b) For one-dimensional model $h_i = \frac{q}{A_i(T - T_w)}$	Spherical Particle* $\frac{h_i D}{\lambda_f} = 2.26\ Re^{0.8} Pr^{0.33} \exp\left(-\frac{6D_p}{D}\right)$	21%	15-75 $\frac{kcal}{m^2 hr\ °C}$
where T_w is the wall temp of the fluid at axial position of interest (for homogeneous model, T of fluid and of bed are assumed idential)	$20 \leqq Re \leqq 7600$ and $0.05 \leqq D_p/D \leqq 0.3$		
	Cylindrical Particle* $\frac{h_i D}{\lambda_f} = 1.40\ Re^{0.95} Pr^{0.33} \exp\left(-\frac{6D_p}{D}\right)$	27%	
	$20 \leqq Re \leqq 800$ and $0.03 \leqq D_p/D \leqq 0.2$		

Table 4.2 *(continued)*

Type	*Correlation*	*Deviation %*	*Range*
Wall Heat-Transfer Coefficient (b) For two-dimensional model $h_w = \frac{q}{A_i(T_{R_t} - T_w)}$ where T_{R_t} = Temp at $r_t = R_t$	Spherical Particle* $\frac{h_w D_p}{\lambda_f} = 0.19\ Re^{0.79} Pr^{0.33}$ $20 \leqq Re \leqq 7600$ and $0.05 \leqq D_p/D \leqq 0.3$	14%	100-250 $\frac{kcal}{m^2 hr\ °C}$
	Cylindrical Particle* $\frac{h_w D_p}{\lambda_f} = 0.18\ Re^{0.93} Pr^{0.33}$ $20 \leqq Re \leqq 800$ and $0.03 \leqq D_p/D \leqq 0.2$	33%	
Effective Radial Thermal Conductivity (c) $-\lambda_r \left(\frac{\partial T}{\partial r_t}\right)_{r_t = R_t} = \frac{q}{A_i}$	$\frac{h_w D_p}{\lambda_r} \frac{\varepsilon}{1 - \varepsilon} = 0.27$ Use value of h_w to calculate λ_r Applies: $500 < \frac{D_p G}{\mu(1 - \varepsilon)} < 6000$ $0.05 < D_p/D < 0.15$	approx.	1-10 $\frac{kcal}{m^2 hr\ °C}$ $\frac{\lambda_r}{\lambda_f} = 1 - 12$
Effective Radial Diffusivity (c) $N_{jr} = -\mathcal{D}_r \left(\frac{\partial C_j}{\partial r}\right)_{z,r}$	Typical commercial reactors with Sc≈0.7 $\frac{\varepsilon \mathcal{D}_r}{u_s D_p} = \frac{1}{m} + \frac{0.38}{Re}$ $D_p/D > 0.1$ $m = 11$ for $Re > 400$ $m = 57.85 - 35.36 \log Re + 6.68 (\log Re)^2$ for $20 < Re < 400$ For $D_p/D < 0.1$ divide $\mathcal{D}_r$ calculated from above by $\left[1 + 19.4 (D_p/D)^2\right]$ For more general equation in terms of Re, tortuosity, and ε, see Ref. (d).	very approx.	$\frac{u_s D_p}{E \mathcal{D}_r} = 6\text{-}20$

Notes: $Re = \frac{D_p u_s \rho_f}{\mu}$. D_p is diameter of a sphere or an equivalent sphere, $6/a_v$.

* Original data based on air correlation generalized by multiplying by $(Pr / 0.72)^{0.33}$ to give equation noted.

(a) Wakao, N. and Kaguei, S. "Heat and Mass Transfer in Packed Beds," Gordon and Breach Science Publishers, New York, 1982.

(b) Li, C. and Finlayson, B. A. Chem. Engr. Sci., 32, 1055 (1977).

(c) Kulkarni, B. D. and Doraiswamy, L. K., Catal. Rev. - Sci. Eng., 22 431 (1980).

(d) C. Y. Wen and L. T. Fan, "Models for Flow Systems and Chemical Reactors," Dekker, N.Y., 1975.

error in a predicted temperature will cause a large error in rate. Similarly, adiabatic reactor calculations become dubious for the same reason if a temperature gradient between the fluid and the particle is of such magnitude that it must be calculated in the design model. Adiabatic reactors without such gradients can be designed with assurance, especially for gases for which excellent thermodynamic data are available.

Pressure Drop in Packed Beds

Pressure drop through a reactor, though seldom more than 10% of the total pressure and thus not a major factor in changing the chemical rate in a gaseous reaction, can be a very important calculation for assessing energy costs, especially when recycle is required. Redution of ΔP in such situations can produce astounding savings on a yearly basis, especially for large units. In studies of this nature, the reactor ΔP should include not only the bed ΔP, as discussed here, but also entrance and exit losses and losses through distributors and collectors.

- Entrance and exit losses: evaluate as the kinetic energy of the upstream fluid
- Distributors and collectors: use orifice equation for perforated trays

The path of fluid flow in a packed bed is highly complex, as shown in Fig. 4.1. Since pressure-drop calculations for fluid flowing in a pipe with no packing require an empirical equation, it is reasonable that the more complex packed bed will necessitate a similar approach. The Ergun equation [18, 19] has been preferred for years. It is rationalized on the basis that pressure losses are caused by simultaneous kinetic (inertial) and viscous losses:

$$-C_1 \frac{dP}{dZ} = \left[\alpha \frac{(1-\varepsilon)^2}{\varepsilon^3} \frac{\mu}{D_p^2}\right] u_s + \left[\beta \frac{(1-\varepsilon)\rho_f}{\varepsilon^3 D_p}\right] u_s^2 \tag{4.15}$$

where ε is the void fraction of the bed; μ is the viscosity; D_p is the equivalent spherical diameter $a_s/6$; a_s is the geometric surface area/volume; u_s is the superficial velocity based on the empty reactor cross section; ρ_f is the fluid density; C_1 is a conversion factor for yielding the desired pressure units as follows:

P	C_1
lb_f/ft^2	$32.17(lb_m)(ft/sec^2)/lb_f$
g_f/cm^2	$980.67(g_m)(cm/sec^2)/g_f$
atm	$1.01325 \times 10^6\ (g_m)(cm/sec^2)/atm$

Consistent units must be employed in each case.

For a given catalyst and fluid Eq. 4.15 can be reduced to

$$\frac{-C_1 \operatorname{grad} P}{u_s} = \mathrm{a} + \mathrm{b}\mathbf{G} \tag{4.16}$$

where **G** is the mass velocity. In this linear form it proves most valuable for obtaining precise values of the constants α and β for a given bed and fluid. Good values of these constants will result if a full range of mass velocities is studied covering both the viscous and inertial dominated regions.

From Eq. 4.15,

$$\mathrm{a} = \alpha \frac{(1-\varepsilon)^2}{\varepsilon^3} \frac{\mu}{D_p^2} \tag{4.17}$$

$$\mathrm{b} = \beta \frac{(1-\varepsilon)\rho_f}{\varepsilon^3 D_p} \tag{4.18}$$

Equation 4.15 can be arranged to give

$$\frac{dP}{dZ} = -(A\mathbf{G} + B\mathbf{G}^2) \tag{4.19}$$

where

$$A = \alpha \frac{(1-\varepsilon)^2}{\varepsilon^3} \frac{\mu}{D_p^2 \rho_f C_1}$$

$$B = \beta \frac{1-\varepsilon}{\varepsilon^3} \frac{1}{D_p \rho_f C_1}$$

or

$$\frac{dP}{dZ} = -\frac{f_k \mathbf{G}^2}{C_1 D_p \rho_f} \left(\frac{1-\varepsilon}{\varepsilon^3}\right) \tag{4.20}$$

where

$$f_k = \beta + \alpha \left(\frac{1-\varepsilon}{D_p \mathbf{G}/\mu}\right)$$

Values of $\beta = 1.75$ and $\alpha = 150$ have been used for many years, but, recently, separate values for smooth and rough pellets have been suggested [20]:

	α	β
Smooth	180	1.8
Rough	180	4.0

Recommendations for ΔP Estimates

- For accurate predictions where ΔP plays an important role in economic analysis, use Eq. 4.16 with constants determined by actual pressure-drop tests. Changes in a for other fluids or conditions may be predicted by a viscosity ratio, as suggested by Eq. 4.17.
- For predictions based on estimates of void fraction and D_p, use Eq. 4.19 or 4.20 with $\beta = 1.8$ and $\alpha = 180$ for most regular pellet shapes, such as spheres and cylinders, and $\alpha = 3.0$ for irregular pieces. Determine D_p as suggested in Table 4.3.

Table 4.3 Equivalent Diameters* $D_P = (6/a_s)$

Shape	D_p
Sphere	d_s, the sphere diameter
Cylinder with length (l_c) equal to diameter (d_c)	d_c
Cylinder, $d_c \neq l_c$	$\dfrac{6d_c}{4 + 2\,d_c/l_c}$
Ring with outside diameter of d_o and inside diameter of d_i	$1.5\,(d_o - d_i)$
Mixed Sizes	$\dfrac{1}{\Sigma Y_i/D_{p_i}}$ where Y_i is volume fraction of size i
Irregular Shapes	$S_p \overline{D}_v$ $\overline{D}_v$ = diameter of sphere having same volume as average particle S_p = sphericity = $\dfrac{\text{surface area of sphere}}{\text{average surface area of particle}}$ Approximate $\overline{D}_v$ as average diameter from screen analysis S_p = 0.5 to 0.7 with lower value being for rough surface.

*a_s is exterior surface area/volume of catalyst pellet.

- Unless more definitive information for the catalyst in question is available, use Table 4.4 for estimating the bed void fraction.

The void fraction is a very sensitive parameter, since it appears raised to the second and third powers. It is strongly influenced by the change in environment at the wall of the tube, where a major discontinuity occurs, causing increases in void space in that region. As the tube diameter becomes larger, this effect on the average porosity of the bed declines and all large beds

Table 4.4 Rough Estimate of Bed Void Fraction

Particle of Pellet	*Normal Charge*	*Dense* Packed*	*Multiplier for Small Tubes*
Tablets	0.36	0.31	$1 + 0.43 \frac{D_p}{D}$
Extrudates			
Short	0.40	0.33	$1 + 0.46 \frac{D_p}{D}$
Long	0.46	0.40	$1 + 0.46 \frac{D_p}{D}$
Spheres			
Uniform	0.40	0.36	$1 + 0.42 \frac{D_p}{D}$
Mixed Sizes	0.36	0.32	
Irregular	0.42 (average)		$1 + 0.3 \frac{D_p}{D}$

* The dense packed condition is often reached a short time after start up due to vibrations and fluid movement during downflow operation.

Based on data from Harshaw Chemical Company booklet, "Harshaw Catalysts," by permission. Multipliers for irregular data approximated from Leva, M. and Grummer, M., Chem. Eng. Prog., 43, 713 (1947).

D_p is diameter of an equivalent sphere with same surface area per unit volume, and D is the internal diameter of the tube.

of a particular catalyst of any size will have the same void fraction. This is a reasonable conclusion, since the size of void spaces is a relatively constant fraction of the catalyst size. Particle shape, roughness, and size distribution, as well as the mode of charging, also affect the degree of void space. Roughness impedes settling, and mixed sizes produce lower-porosity beds because the smaller pieces tend to fill spaces between larger pieces.

LAWS OF CONSERVATION AND THE QUANTITATIVE APPROACH TO FIXED-BED REACTOR ANALYSIS

In all phases of engineering design, analysis, and diagnostics, the professional must obtain some quantitative results of known accuracy in order to make intelligent decisions. It is both interesting and somewhat humbling to realize that invariably our reasoning and calculations are based on the laws of conservation (e.g., mass, momentum, and energy). These are laws of nature that serve to unify all science and engineering. They have no formal proofs, but have attained the status of laws by always being found true under certain prescribed conditions. Useful equations can be developed by beginning with one or more of these laws and realistic assumptions for making mathematical analysis of a particular system both tractable and sensible. Such equations permit all sorts of useful predictive calculations, provided the required data are at hand or obtainable by experimental study in a reasonable time. The professional would do well to keep this essential and basic foundation of the conservation laws clearly in mind, for it is not only unifying but greatly simplifying.

Conservation of Mass

Although the conservation of mass was proposed as early as 100 B.C., its final formulation had to await the work of Lavoisier (1770), who showed that, even for reactions in which new substances appear and others disappear, an equal quantity of matter exists before and after the experiment. His work demonstrated that the mass (or moles) of each element remains the same. We now know that in nuclear reactions mass conservation does not apply and only energy is conserved, making the conservation of mass, in effect, a special case of the Law of Conservation of Energy.

General Mass Balance

The general statement of the Law of Conservation of Mass for any open system, except one involving nuclear reactions, is as follows.:

[Total mass entering] = [total mass leaving] + [accumulation of mass] (4.21)

This equation becomes much more valuable for so-called overall material balances on operating units if it is supplanted by a number of component equations, often conveniently written in terms of moles.

Nonreacting Component.

[Moles j entering] = [moles j leaving] + [accumulation of j] (4.22)

Reacting Components. Use elemental balances, since elements do not change.

[Moles element i entering] = [moles element i leaving] + [accumulation of element i] (4.23)

Although Eqs. 4.22 and 4.23 are simply statements of the obvious, much like cash-flow balances, the summation for all i's and j's will conform to Eq. 4.21 if, first, the equation for each i and j is multiplied by the applicable elemental or molecular weights.

Rules for Application of Material Balance Equations

Equations 4.21–4.23 are very useful in determining unknown streams or components from overall balances. Familiar rules for such effort include the following:

- As many independent algebraic balances must be written as there are unknowns.
- When two components bear a fixed relationship to each other in a system in all streams in which they appear, only one independent equation can be written for these substances.
- If the fractional quantities of a nonreacting component (inert) appearing in one inlet and one outlet stream are known in each stream, the relationship between masses or moles of the two streams can be readily calculated.

$$W_{A_1} = \frac{m_{A_1}}{m_{T_1}} \qquad W_{A_2} = \frac{m_{A_2}}{m_{T_2}}$$

where W_{A_1}, W_{A_2} are fractional masses of A in streams 1 and 2, respectively; m_{A_1}, m_{A_2} are masses of A in streams 1 and 2, respectively; m_{T_1}, m_{T_2} are total masses of streams 1 and 2, respectively.

For one inlet and outlet,

$$m_{A_1} = m_{A_2}$$

$$\frac{W_{A_1}}{W_{A_2}} = \frac{m_{A_1}/m_{T_1}}{m_{A_2}/m_{T_2}} = \frac{m_{T_2}}{m_{T_1}}$$

- One independent equation can be written for each compound for which no reaction occurs, or for each element for which reaction occurs.
- Although the total mass balance is the sum of all component and/or atomic balances, it is not independent, but may be substituted for any of the other balances.
- The fact that summation of the mole fractions for any given stream is unity provides another useful independemt equation.
- There are often many more independent equations possible than required. One selects, in such cases, those easiest to manipulate and/or those thought to be more accurately known.
- For systems requiring three or more equations, library matrix-solving routines are recommended. In such cases, concern about extra independent equations is pointless, since they add no perceptible time to the solution effort.

Component Balances Including Reaction Rates

The balance equations presented thus far are useful in accounting for all streams and components around a reactor system and are crucial in establishing recycle and purge flows and compositions. These equations, however, are not applicable to the important task of predicting and analyzing reactor behavior. Formulations for such purposes must include reaction-rate expressions, our only link to the path-dependent phenomena occurring within the reactor itself. For this purpose, another simple statement accounting for a component is expressed conveniently in terms of rates:

$$\overset{\text{I}}{\begin{bmatrix}\text{Molar rate of component } j \\ \text{flowing out of volume element}\end{bmatrix}}$$

$$-\overset{\text{II}}{\begin{bmatrix}\text{molar rate of component } j \\ \text{flowing into volume element}\end{bmatrix}} + \overset{\text{III}}{\begin{bmatrix}\text{molar rate of accumulation} \\ \text{of } j \text{ in volume element}\end{bmatrix}}$$

$$= \overset{\text{IV}}{\begin{bmatrix}\text{molar rate of production} \\ \text{of } j \text{ in volume element}\end{bmatrix}} \quad \text{or} \quad -\begin{bmatrix}\text{molar rate of consumption} \\ \text{of } j \text{ in volume element}\end{bmatrix} \tag{4.24}$$

By multiplying the equation for each component by the applicable molecular weight and summing all such equations, a form of the general conservation of mass results (Eq. 4.21). The summation of all consumption and production terms expressed as mass is zero, as would be expected from the conservation principle. As demonstrated by Lavoisier, the mass that disappears as a compound is consumed reappears in the products:

$$\sum_{j=1}^{c} \hat{r}_{j_m} = 0 \tag{4.25}$$

where $\Sigma\, r_{j_m}$ is the mass rate of production (+) or consumption (−) of all j's.

The compositions and temperature for fixed-bed reactors, just as for all tubular reactors, varies with position, and application of Eq. 4.24 requires statements in terms of point values. Thus both items I and II must be combined in a single differential expressing the change over a differential distance, and the accumulation is also a differential expressing the change of composition with time. Item IV is the point value of the reaction rate or summation of rates for a given component. Fortunately, as will be shown in Chapter 5, many commerical-scale reactors are amenable to such idealizations; and the probable errors are acceptable, provided the rate equations and data are of good quality.

Conservation of Energy

The formulation and acceptace of the Law of Conservation of Energy was preceded almost 100 years by a more restricted principle, the conservation of mechanical energy, largely attributed to Newton's brilliant contributions.

$$\begin{aligned}\text{(Work done by an exterior force)} = {} & \text{(work done against friction)} \\ & + \text{(change in potential energy)} \\ & + \text{(change in kinetic energy)}\end{aligned} \tag{4.26}$$

The incorporation of heat into this equation had to await the development of reliable thermometers and the subsequent formulation by Black (1759–1799) of the principle of the conservation of heat energy when no work is involved:

$$\Delta U = Q \tag{4.27}$$

where Q is the heat added to the system and U is the internal energy.

Thus, for a system involving no work of any kind and no changes in potential energy or kinetic energy, a property depending on the state of the system, called *internal energy* U, is altered by an amount equivalent to the net heat added to or removed from the system. The absolute value of U is not known; only changes between different conditions or a value relative to a base condition may be known. This property U is proportional to the change in temperature of the system and its mass, and the proportionality constant is the heat capacity per unit mass at constant volume for the particular material:

$$\Delta U = c_v m \, \Delta T \tag{4.28}$$

If the system is adiabatic, meaning no heat is added or removed, then

$$\Delta U = 0 \tag{4.29}$$

Following these pronouncements, the brilliant work of Joule established the mechanical equivalent of heat by a number of ingenious experiments. The most accurate series involved stirring water by a paddle stirrer in an insulated tank. The work was obtained by measuring the potential energy dissipated by falling weights that, through a pulley, turned the paddle. The change in internal energy produced by the friction of stirring was calculated from the change in temperature and the mass of the water. After much refinement of procedure, Joule reported a rather consistent value of 4.2 N-m/cal or J/cal, which is now known more accurately to be 4.184. Experiments involving work other than frictional work were studied as well as the conversion of internal energy to work; so the following equation could be written with confidence for systems in which no heat is added or removed:

$$\Delta U = W_a \tag{4.30}$$

where W_a is the work done on the system such that a negative value is assigned to work done by the system. An expression for the First Law of Thermodynamics is obtained by combining Eqs. 4.27 and 4.30:

$$\Delta U = Q + W_a \tag{4.31}$$

General Energy Equation

Equation 4.31 can be expanded to include potential energy, kinetic energy, and friction losses (lost work) by using the mechanical energy equation (Eq. 4.26):

$$\Delta U = \Delta(mU_m) = Q + W_a - W_f - \Delta\left(m \frac{u^2}{2}\right) - \Delta(mgZ) \tag{4.32}$$

where U_m is internal energy per unit mass and m is the mass flowing. For an open (flow) system there must be added the difference between the work done by the fluid within the defined flowing mass on the fluid downstream and the upstream fluid on the defined flowing mass:

$$\Delta U = \Delta(mU_{\mathrm{m}}) = Q + W_{\mathrm{a}} - W_{\mathrm{f}} - \Delta\left(m\,\frac{u^2}{2}\right) - \Delta(mgZ) - \Delta(mPv) \tag{4.33}$$

where v is the volume per unit mass. The last term on the right side may also be written as $\Delta(PV)$.

By moving the PV or mPv term to the LHS and using the definition of enthalpy ($H = U + PV$), Eq. 4.33 can be written as,

$$\Delta H = \Delta(mH_{\mathrm{m}}) = Q + W_{\mathrm{a}} - W_{\mathrm{f}} - \Delta\left(m\frac{u^2}{2}\right) - \Delta(mgZ) \tag{4.34}$$

Differential Energy Equations Based on Various Restrictions

Equations 4.33 and 4.34 are not useful for analyzing the changes occurring as a function of position or time. For such purposes differential equations are needed. The internal energy is a state function, and ΔU depends only on the initial and final state. But Q and W_a are path-dependent properties. Their values can change with the kind of operation, and only their algebraic sum remains constant if the initial and final states are also unchanged. Part of the genius in the energy concept of thermodynamics is the application of appropriate restrictions so that the heat or work term can be written as a proper path-independent differential; and, since U may be so written, the third term must also be independent of path. The following restrictions are illustrative (for brevity, kinetic, and potential energy terms as well as lost work are neglected).

Processes with No Work or No Heat Transfer

$$\Delta U = Q \qquad \text{(no work)} \tag{4.35}$$

The heat term Q in this case is equal to a path-independent change and is therefore path-independent itself, and an appropriate differential equation is possible:

$$dU = dQ \tag{4.36}$$

Adiabatic Process (No Heat Transfer)

$$\Delta U = W_{\mathrm{a}} \tag{4.37}$$

Similar reasoning as above permits,

$$dU = dW_a \tag{4.38}$$

Processes Involving Both Heat and Work. In this case a restriction on the work term, such that it is represented by state properties, again permits appropriate differential equations of great value in dealing logically with practical systems and equipment.

The Differential Energy Functions. A series of highly useful path-independent equations can be derived by starting with a nonflow system restricted to reversible work of expansion only ($W_a = P\,dV$). Based on these restrictions the First Law differential equation becomes

$$dU = dQ - P\,dV \tag{4.39}$$

Another useful form may be obtained from Eq. 4.39, since, from the Second Law, $dQ = T\,dS$ for a reversible process, where S is the entropy:

$$dU = T\,dS - P\,dV \tag{4.40}$$

This equation now contains only path-independent properties and is thus generally applicable to any system of constant composition in the absence of external forces. It forms the basis for the four energy-function equations shown in Fig. 3.1.

The Differential Energy Equation Used for Fixed-Bed Reactors. For flow work (ΔPV) only, the First Law differential equation becomes

$$dU = dQ - d(PV) \tag{4.41}$$

or

$$dU + d(PV) = dQ \tag{4.42}$$

but $dU + d(PV) = dH$ by definition. Hence,

$$dH = dQ \tag{4.43}$$

Equation 4.43 forms the basis for analysis of fixed-bed reactors. Pressure drop occurs due to friction, but the energy effect is small and may be neglected. The change of pressure, however, is important and must be followed by an empirical equation (see Eq. 4.20), especially for gaseous reactions where changes in total pressure affect reaction rates. Further, pressure drop is also an important energy factor over an extended operating period, and the energy required to overcome it using a gas compressor may not be ignored.

LITERATURE CITED

1. Reid, R.C., Prausnitz, J.M., and Poling, B.E., "The Properties of Gases and Liquids," 3rd Edition, McGraw-Hill, New York, 1987.
2. Fuller, E.N., Ensley, K., and Giddings, J.C., *J. Phys. Chem.*, *73*, 3679 (1969).
3. Takaheshi, S., *J. Chem. Eng. Japan*, *1*, 417 (1974).
4. Touloukian, Y.S., Liley, P.E., and Saxens, S.C., "Thermophysical Properties of Matter," Vol. 3, IFI/Plenum, New York, 1970.
5. Vargaftik, N.B., "Handbook of Physical Properties of Liquids and Gases," 2nd Edition, Hemisphere, Washington, D.C., 1979.
6. Gallant, R.W., "Physical Properties of Hydrocarbons," Vols. 1, 2, 2nd Edition, Gulf Pub., Houston, 1984 (contains graphs of properties).
7. Yaws, Carl L., "Physical Properties," *Chemical Engineering*, McGraw-Hill, New York, 1977.
8. Owens, E.J. and Thodas, G., *AIChE J.*, *6*, 676 (1960).
9. API, "Technical Data Book," Amer. Pet. Inst., New York, 1986.
10. Roy, D. and Thodas, G., *Ind. Eng. Chem. Fundam.*, *7*, 529 (1968)
11. Roy, D. and Thodas, G., *Ind. Eng. Chem. Fundam.*, *9*, 71 (1970).
12. Stiel, L.I. and Thodas, G., *AIChE J.*, *10*, 26 (1964).
13. Schaefer, C.A. and Thodas, G., *AIChE J.*, *5*, 367 (1959).
14. Lucas, K., *Chem. Eng. Tech.*, *46*(4), 157 (1974).
15. Horning, F. and Zipperer, L., *Gas-Wasserfach*, *79*, 49 (1936).
16. Beck, J., Jr., and Singer, E., *Chem. Eng. Prog.*, *47*, 534 (1951).
17. Quinton, J.H. and Storrow, J.A., *Chem. Eng. Sci.*, 5, 245 (1956).
18. Ergun, S. and Orning, *Fnd. Eng. Chem.*, *41*, 1179 (1949).
19. Ergun, S., *Chem. Eng. Prog.*, *48*, 89 (1952).
20. MacDonald, I.F., El-Sayed, M.S., Mow, K., and Dullien, F.A.L., *Ind. Eng. Chem. Fundam.*, *18*, 199 (1979).

PART

II

DESIGN AND OPERATION

Part II provides a practical approach to the use of mathematical models and scale-up in designing reactors and mathematical models and diagnostic procedures in resolving operating options and problems related to existing reactors. This purpose is served by recommending suitable models, summarizing data gathering procedures for establishing model parameters, and recommending design strategies, shortcut estimating methods, and diagnostic techniques.

Well-tested reactor models with valid parameter values can be invaluable tools in decision making associated with design and operation. As in all dealings with real-world problems, however, the constraints of time and economics often force some compromise with an ideal data gathering program. Therefore great care must be exercised in the use of predictions from model calculations. For this reason, various recommendations and tabular summaries are presented that are designed to guide the reader in rational decision making based on the existing realities at a particular time.

In doing design or solving operating problems, the engineer should always remember the overriding importance of chemistry in reactor performance. No model has yet incorporated all the details of the chemical phenomena involved. For this reason alone, a thorough understanding of the chemistry of the reactions that occur in a reactor is absolutely essential as a healthy constraint on the breadth of interpretation of calculated outcomes. By knowing the chemical details, the engineer can readily recognize unrealistic conditions and outcomes.

CHAPTER 5

Design and Operating Models

Several types of models have been developed to satisfy particular conditions of operation. As is always the case when dealing with reality, assumptions are made to render the analysis as tractable as possible, consistent with the range of uncertainty introduced by the kinetic data to be used. When viewed from the molecular standpoint, the complex fluid motion and molecular interactions (see Fig. 4.1) defy rational analysis based on any reasonable microscopic approach. Instead, as is common in engineering, macroscopic analyses produce rather useful models that can track the temperature and concentration profiles with acceptable accuracy in many cases.

In reality a fixed-bed reactor is heterogeneous in character. Fluid flows between solid catalyst particles and is transported to the catalyst surface and within its pores where reaction takes place. The major flow of fluid is in the axial direction. The energy flow may be similarly described (see Fig. 4.2). When heat is transferred at the wall, as in multitubular reactors, radial gradients of heat and mass can become significant, especially for tubes exceeding a diameter of 1 inch. Fortunately, these various complexities are often minimized in practical reactors, and simplified models may be used. To develop a rational basis for simplification, one should determine pertinent goals or objectives and should understand conditions at the catalyst and in the bed.

OBJECTIVES OF SIMPLIFICATION

An intelligent set of objectives for simplification in reactor modeling has been presented [1].

1. The model should not be more detailed than absolutely necessary for the purpose.
2. The model should contain as few parameters as possible.
3. Reliable correlations should exist for the parameters of the selected model.
4. As little work as possible should be involved in the mathematical solution of the model equations.

One should always resist the temptation of unnecessary complexities. They inevitably introduce additional parameters for which little reliable data are available and require extensive experimental measurements to evaluate.

CONDITIONS AT THE CATALYST

By focusing attention on the catalyst and the immediate surrounding fluid, some useful insights and reasonable simplifications can be set forth.

Since components and heat must move to and from the catalyst interior, the exterior catalyst surface, and the bulk fluid phase, it is essential that the following equalities apply at the steady state.

Component Transfer

[Transport of component j between bulk fluid phase and catalyst exterior]
= [catalyst interior pore-diffusion flux of j at catalyst exterior]
= [observed rate of consumption or production of j]

For a single reaction, called reaction 1, a limiting reactant A as an example component, and a spherical catalyst with an exterior surface area per unit volume, $a_s = 3/R_p$:

$$
\begin{aligned}
k_{s_A}(C_{A_s} - C_{A_b}) &= \mathscr{D}_{I_A}\left(\frac{dC_A}{dr_I}\right)_{r_I=R_p} = (\hat{r}_A)_e \rho_p \frac{R_p}{3} \\
&= \eta(\hat{r}_A)_s \rho_p \frac{R_p}{3} = \eta_o(\hat{r}_A)_b \rho_p \frac{R_p}{3}
\end{aligned}
\tag{5.1}
$$

See below for nomenclature.

Heat Transfer

[Transport of heat between bulk fluid phase and catalyst exterior]
= [catalyst interior heat flux at catalyst exterior]
= [observed rate of consumption or production of heat]

$$
\begin{aligned}
h_s(T_s - T_b) &= -\lambda_I\left(\frac{dT}{dr_I}\right)_{r_I=R_p} = (-\Delta H_1)(\hat{r}_1)_e \rho_p \frac{R_p}{3} \\
&= (-\Delta H_1)\eta(\hat{r}_1)_s \rho_p \frac{R_p}{3} = (-\Delta H_1)\eta_o(\hat{r}_1)_b \rho_p \frac{R_p}{3}
\end{aligned}
\tag{5.2}
$$

where C_{A_s}, T_s are the catalyst exterior surface concentration and temperature, respectively; C_{A_b}, T_b are the bulk fluid concentation of A and temperature, respectively (often b is omitted); r_I is the radial distance along the spherical catalyst; $\hat{r}_1$ is the rate of reaction 1 (the only reaction in this case) per unit mass of catalyst; $\hat{r}_A$ is the rate of consumption of component A per unit mass of catalyst ($\hat{r}_A = \nu_A \hat{r}_1$). Subscripts e, s, b refer, respectively, to the effective or, often-called, global rate, the intrinsic rate (unaffected by diffusional gradients) determined at the catalyst exterior conditions, and the intrinsic rate determined at the bulk fluid conditions. The rate $\hat{r}_A$ is negative since ν_A, the stoichiometric coefficient, is negative for a reactant. The terms η and η_o are effectiveness factors that allow evaluation of the rate from intrinsic rate expressions that are based on powdered catalyst for which gradients within the pellets would be negligible in most cases. From Eqs. 5.1 and 5.2, it is seen that these effectiveness factors are defined as follows:

$$\eta = \frac{(\hat{r}_j)_e}{(\hat{r}_j)_s} = \frac{\text{effective rate}}{\text{intrinsic rate at exterior surface conditions}} \tag{5.3}$$

$$\eta_o = \frac{(\hat{r}_j)_e}{(\hat{r}_j)_b} = \frac{\text{effective rate}}{\text{intrinsic rate at bulk-fluid conditions}} \tag{5.4}$$

In the context of a competitive business climate requiring rapid reduction to practice or quick analysis and implementation of process improvements, the time required to gather the necessary data for reliable evaluation of effectiveness factors or for simultaneous solution of intraparticle-flux and axial-flow component balances (see Table 5.1) is seldom justifiable. Instead, effective (global) rates are used with data determined in the range of commercial interest, and rate equations are evolved from such data. The concept of intraparticle gradients and effectiveness factors, however, is quite valuable in reasoning and decision making.

Concentration and temperature gradients between the catalyst exterior surface and the bulk fluid are usually negligible in practical reactors. Again, however, the concept is most useful along with the appropriate equations in setting operating conditions so as to avoid such gradients. Both intraparticle and interphase gradients are discussed in the following sections.

Gradients between Catalyst Exterior and Bulk Fluid (Interphase Gradients)

Since both heat- and mass-transfer coefficients are increased by increasing mass velocities, it is possible and practical in most reactors to avoid significant gradients except for some reactors operating near atmospheric pressure, where the allowable pressure loss is very small. Estimated heat- and mass-transfer coefficients are not very reliable. When used, they produce models that are much less reliable than those for reactors without significant gradients. This fact is reason enough to avoid conditions leading to significant gradients, but another important reason is to improve operability of the reactor. Gradients introduce lags in the system that make the reactor system less responsive to necessary changes.

A good sense of the relative importance of interphase temperature and concentration or partial pressure gradients can be obtained from Eqs. 5.1 and 5.2. For simplicity, since the bulk temperature and concentration are the conditions of major interest, the subscript b will, henceforth, not be used unless necessary to avoid ambiguity. For limiting reactant A, the heat balance at the surface is

$$h_s(T_s - T) = k_{s_A}(C_A - C_{A_s})(-\Delta H_A)$$
$$T_s - T = \frac{k_{s_A}}{h_s}(C_A - C_{A_s})(-\Delta H_A) \tag{5.5}$$

The definitions of J_d and J_h are

$$J_d = \frac{k_{s_A}\rho_f Y_{A_f}}{G} \mathrm{Sc}^{2/3}$$
$$J_h = \frac{h_s}{c_p G} \mathrm{Pr}^{2/3}$$

Substituting these definitions into Eq. 5.5 gives

$$T_s - T = \left(\frac{J_d}{J_h}\right)\left(\frac{\mathrm{Pr}}{\mathrm{Sc}}\right)^{2/3}\left(\frac{1}{Y_{A_f}}\right)(C_A - C_{A_s})\frac{-\Delta H_A}{c_p \rho_f}$$

or

$$T_s - T = \left(\frac{J_d}{J_h}\right)\left(\frac{\mathrm{Pr}}{\mathrm{Sc}}\right)^{2/3}\left(\frac{1}{Y_{A_f}}\right)\left(\frac{Y_A - Y_{A_s}}{Y_A}\right)\frac{(-\Delta H_A)Y_A}{c_{p_m}} \tag{5.5A}$$

where c_p is heat capacity per unit mass; c_{p_m} is heat capacity per total moles of mixture; Y, Y_{A_s} are the mole fractions of A in the bulk and at the surface, respectively; Y_{A_f} is the film factor; and ρ_f is the fluid density of the mixture in mass units.

Since $J_d \cong J_h$, the J-factors for mass and heat transfer, and Pr and Sc are each close to unity for gases, as is Y_{A_f}. A simplified expression is

$$T_s - T = \left(\frac{Y_A - Y_{A_s}}{Y_A}\right)\frac{(-\Delta H_A)Y_A}{c_{p_m}} \tag{5.5B}$$

The second term on the RHS has been called the *adiabatic factor* [3], since it represents the adiabatic temperature change if all the reactant is consumed without regard to equilibrium limitations. For an assortment of gaseous fixed-bed reactors, absolute values range from 100 to 700°K. The first term on the RHS represents the fraction of reaction controlled by mass transfer or the fractional error in using bulk values instead of surface values. As long as A is reacting, $Y_A - Y_{A_s}$ will have a finite value; but for practical purposes when the fraction is 0.10 or less, the error in the rate expression is less than errors introduced by inevitable inaccuracies in other parameters. Thus for the range of values of adiabatic factors cited and a fraction of 0.10, corresponding values for $T_s - T$ are 10 to 70°, which are much higher than usually encountered under practical commercial conditions at the steady state. Hence, concentration gradients are not significant in commercial gas-phase fixed beds. This fact has been emphasized by various authors [4, 5].

The fractional error in the rate caused by a temperature error, ΔT_e, can be obtained by ratioing rates at the same composition by using single apparent rate constants:

$$\frac{(\hat{r}_A)_s}{(\hat{r}_A)_b} = \frac{k_{T_s}}{k_T} = \frac{\exp[-E/R(T+\Delta T_e)]}{\exp(-E/RT)} = \exp\left[\frac{E}{R}\left(\frac{1}{T} - \frac{1}{T+\Delta T_e}\right)\right]$$
$$= \exp\left[\frac{E}{R}\left(\frac{\Delta T_e}{T^2 + T\,\Delta T_e}\right)\right]$$

Since ΔT_e is small, a simple solution for ΔT_e is possible.

$$\Delta T_e \cong \frac{RT^2}{E} \ln \frac{(\hat{r}_A)_s}{(\hat{r}_A)_b}$$

where $\Delta T_e = T_s - T$. For 10% error in $\hat{r}_A$,

$$\Delta T_e \cong 0.2 \frac{T^2}{E} \tag{5.6}$$

where E is in cal/g mole and T is the bulk fluid temperature.

Thus for a reaction with an E of 10,000 and a bulk temperature of 500°K, the allowable ΔT_e that will still allow using bulk temperatures for the rate equations is 5°. Apparent activation energies for complex rate forms may be determined from rates calculated at two different temperatures and the same composition:

$$\ln \frac{(\hat{r}_A)_1}{(\hat{r}_A)_2} = \frac{E}{R}\left(\frac{1}{T_1} - \frac{1}{T_2}\right) \tag{5.7}$$

In most commercial reactors in which mass velocities are 500 lb/hr ft^2 (0.07g/cm^2 sec) or above for adiabatic and 1000 lb/hr ft^2 (0.14g/cm^2 sec) or above for nonadiabatic heat-exchange reactors, it is seldom a problem to maintain acceptably low ΔT_e gradients so that bulk conditions can be used. Even when lower mass velocities must be used, as in the case of reactors operating near atmospheric pressure, such as SO_2 oxidation, when the ΔT_e

exceeds a reasonable allowable level, the concentration gradients will not be significant and bulk values of composition may be used.

To determine the required **G** necessary for minimizing ΔT_e, solve the following single-reaction equation for **G**:

$$h_s a_m \Delta T_e = (\hat{r}_1)_b(-\Delta H_1) \tag{5.8}$$

where $(\hat{r}_1)_b$ is evaluated at bulk conditions for the estimate. The heat-transfer coefficient h_s can be evaluated from the equation given in Table 4.2. For multiple reactions use $\Sigma_{n=1}^{n_r} \hat{r}_n(-\Delta H)_n$ and obtain estimated rates from a program for which it is assumed that $T = T_s$.

The mass velocity at which a given ΔT_e is satisfied can be determined by using a heat-transfer correlation that is easily solvable for **G** [6] and substituting it into Eq. 5.8:

$$h_s = \frac{0.357}{\left(\dfrac{d_p \mathbf{G}}{\mu}\right)^{0.359} c_p \mathbf{G}}$$

$$\mathbf{G} = \left[\frac{0.357\left(\dfrac{\mu}{d_p}\right)^{0.359} a_m \Delta T_e}{c_p \sum_{r=1}^{n_r} \hat{r}_n(-\Delta H_n)}\right]^{0.736} \tag{5.9}$$

where d_p is the diameter of the sphere having the same exterior area as the particle in question.

A single reaction criterion based on a 5% error has been given [7]:

$$\frac{|\Delta H_1| \hat{r}_1 D_p \rho_p}{h_s T} < 0.3 \frac{RT}{E} \tag{5.10}$$

Conclusions about Interphase Gradients

In most cases, bulk conditions may be used in fixed-bed reactor design and analysis. When preliminary calculations indicate that ΔT_e values exceed the allowable for desired rate accuracy, include an algebraic equation for the heat transfer between the catalyst and the fluid. It is not necessary to include mass-transfer gradient equations, since these gradients will be acceptably small at the steady state.

Small-scale experimental equipment such as pilot plants will normally have such low flow rates that both mass- and heat-transfer gradients can be quite high unless special arrangements of catalyst and inerts or special recirculating reactors are employed. Discussions of this subject have been presented [3, 7], and criteria given for determining special conditions for avoiding unacceptably high gradients.

Intraparticle Transport

At a given operating condition it is intuitively reasonable to conclude that, as the catalyst pellet size is increased, a significant concentration gradient between the surface and the interior can develop for all components. Conversely, if the catalyst size is held constant and the temperature increased, there may occur a similar gradient due to more rapid consumption of the reactants, because of the exponential rate increase caused by increased temperatures. In both cases, the rate varies with position in the catalyst.

As is most often the case, development work on a process usually leads to emphasis on a single catalyst size. Then one needs to contend with changes in catalyst concentration profiles only as a function of temperature. Over a wide temperature range, a plot of the apparent rate constant obtained directly from the observed effective rate will exhibit varying apparent activation energies, as shown in Fig. 5.1. Fortunately, the range of temperature of interest for an economic process is restricted, and the observed or apparent activation energy will be constant in the range. As will be shown below, this apparent or effective activation energy in the range of substantial gradients is half the intrinsic or true value. But using the apparent value obtained in the operating region of interest is no detriment to the design or analysis of a fixed-bed

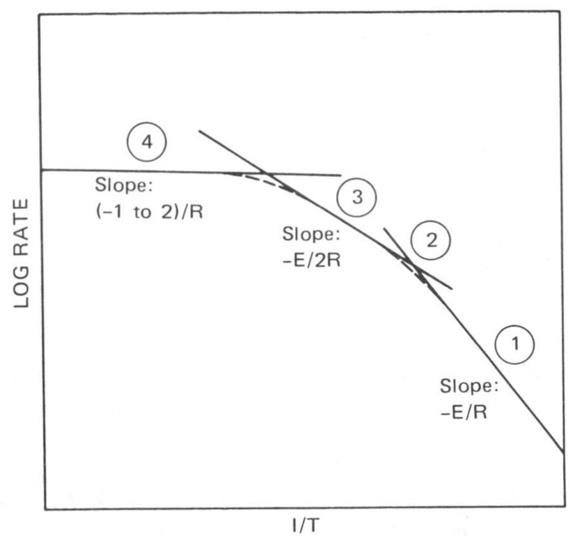

Figure 5.1 Plot of log of rate versus $1/T$ over a wide temperature range
Range 1: true activation energy with chemical rate controlling; Range 2: transition region—intraparticle resistance begins to become important; Range 3: intraparticle resistance controlling; Range 4: mass transfer between fluid and exterior of particle becomes controlling. The practical range of interest normally spans a relatively narrow temperature range, rarely more than two of the indicated ranges.

reactor. The effect of changing catalyst size or the effect of a radical temperature change can be readily seen from the mathematical formulation of the effectiveness factor.

First-Order Effectiveness Factor

Analytical solutions for the effectiveness factor have been obtained for a number of rate forms, but apply only to single reactions. The simplest derivation is that for a single first-order reaction and an isothermal catalyst pellet:

$$\mathscr{D}_{I_A}\left(-\frac{dC_A}{dr_I}\right)_{r_I=R_p} = \eta(\hat{r}_A)_s\rho_p\frac{R_p}{3}$$

$$\eta = \frac{\mathscr{D}_{I_A}(-dC_A/dr_I)_{r_I=R_p}}{(\hat{r}_A)_s\rho_pR_p/3} \tag{5.11}$$

where r_I is the radial distance and R_p is the equivalent spherical particle radius; $(\hat{r}_A)_s$ is the rate of reaction of A based on exterior surface conditions.

The differential of C_A with respect to r_I can be evaluated from a continuity equation representing transport within the spherical particle. As the simplest example, consider a first-order irreversible reaction, designated reaction 1, with A reacting to R and the rate expressed as $\hat{r}_1 = \hat{k}_cC_A$ or $\hat{r}_A = -\hat{k}_cC_A$, where $\hat{k}_c$ is a concentration rate constant based on a unit mass of catalyst, volume/(mass cat.)(time), for first order.

For the general case;

$$\hat{k}_c/(\mathbf{R}T)^n = k_p$$

where n is the order of the reaction.

For a reacting component A diffusing inward, as shown in Fig. 5.2, the following component balance can be performed for a first-order reaction:

Moles of A out of element	Moles of A into element	Moles of A reacted

$$(N_A|_{r_I})4\pi r_I^2 \quad - \quad (N_A|_{r_I+\Delta r_I})4\pi(r_I+\Delta r_I)^2 \quad + \quad \rho_p\hat{k}_cC_A\cdot 4\pi r_I^2\,\Delta r_I = 0$$

By dividing through by $4\pi\,\Delta r_I$ and taking the limit as Δr_I approaches 0, an expression for the change in flux with radial position results.

$$\lim_{\Delta r_I\to 0}\frac{(r_I^2N_A)_{r_I+\Delta r_I}-(r_I^2N_A)_{r_I}}{\Delta r} = \frac{d}{dr_I}\,r_I^2N_A = r_I^2\rho_p\hat{k}_cC_A \tag{5.12}$$

For equimolar counter diffusion N_A is defined as $N_A = \mathscr{D}_{I_A}dC_A/dr_I$ where $\mathscr{D}_{I_A}$ is the "effective" diffusivity of A through the porous structure of the catalyst in

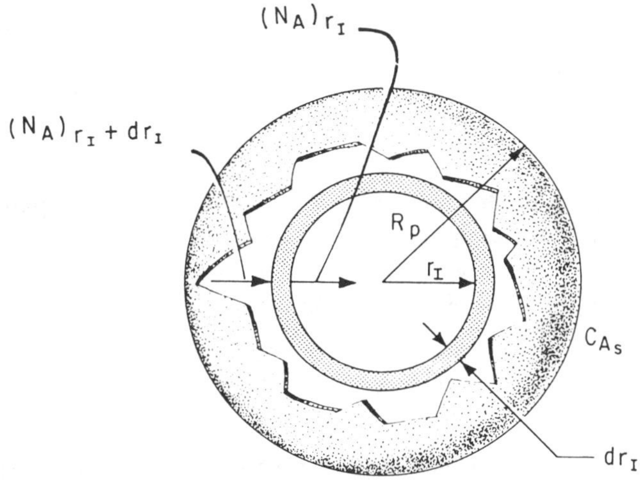

Figure 5.2 Intraparticle diffusion: shell balance on spherical pellet
(Reproduced by permission: H.F. Rase, "Chemical Reactor Design for Process Plants," copyright 1977, John Wiley & Sons, Inc., New York.)

a solution of A and R. Substituting this expression in Eq. 5.12 leads to the differential equation

$$\frac{d^2C_A}{dr_I^2} + \frac{2}{r_I}\frac{dC_A}{dr_I} = \frac{\rho_p \hat{k}_c}{\mathscr{D}_{I_A}} C_A = \left(\frac{3\phi_s^2}{R_p}\right) C_A \tag{5.13}$$

where

$$\phi_s = \frac{R_p}{3}\left(\frac{\rho_p \hat{k}_c}{\mathscr{D}_{I_A}}\right)^{1/2}$$

is the Thiele modulus. For the general case of simple power-law kinetics and where n is the order, the Thiele modulus is defined for a spherical particle as

$$\phi_s = \frac{R_p}{3}\left(\rho_p \hat{k}_c C_{A_S}^{n-1}/\mathscr{D}_{I_A}\right)^{1/2}$$

For the boundary conditions of $C_A = C_{A_S}$ at $r_I = R_p$ and $dC_A/dr_I = 0$ at $r_I = 0$, Eq. 5.12 can be solved analytically to give

$$\frac{C_A}{C_{A_S}} = \frac{1}{(r_I/R_p)}\,\frac{\text{Sinh}[3\phi_s(r_I/R_p)]}{\text{Sinh}(3\phi_s)} \tag{5.14}$$

Now Eq. 5.11 can be solved for the effectiveness factor by substituting the value of the differential of C_A with respect to r_I at $r_I = R_p$ from Eq. 5.14. The result for first-order is

$$\eta = \frac{1}{\phi_s}\left(\frac{1}{\tanh 3\phi_s} - \frac{1}{3\phi_s}\right) \tag{5.15}$$

If the modulus for other shapes, such as cylinders and flat plates, is defined in the same manner as for spheres, with the multiplier being the catalyst pellet volume per unit exterior surface area (the reciprocal of a_s), the resulting equations will fall on essentially the same curve [8]. The resulting modulus is

$$\phi = \frac{1}{a_s}\left(\frac{\rho_p \hat{k}_c C_{A_s}^{n-1}}{\mathcal{D}_{I_A}}\right)^{1/2} \tag{5.16}$$

where a_s is the exterior surface area of a pellet per unit volume; $1/a_s = R_p/3$ for spheres, L_s for a slab (one-half the thickness if exposed on both sides), and $R_p/2$ for cylinders.

The corresponding equations in addition to Eq. 5.15 are

$$\eta = \frac{\tanh \phi}{\phi} \qquad \text{(slab)} \tag{5.17}$$

$$\eta = \frac{1}{\theta}\frac{I_1(2\phi)}{I_0(2\phi)} \qquad \text{(cylinder)} \tag{5.18}$$

where $I_1(2\phi)$ and $I_0(2\phi)$ are two modified Bessel functions.

Since Eq. 5.17 for a slab is the simplest form mathematically and all equations fall on the same line when the modulus is defined as in Eq. 5.16, it is convenient to use it for further mathematical manipulation involving any shape.

Effectiveness for Other Rate Forms

There is a copious literature on effectiveness factors involving various rate forms. Perhaps the most useful result has been the development of a general modulus for arbitrary kinetics [9, 10], which, when applied to power-law kinetics, results in a convenient plot (Fig. 5.3) that gives approximately the same values of effectiveness factors for a given value of the modulus over a rather large range of orders of reaction:

$$\phi_g = \frac{1}{a_s}\left[\frac{n+1}{2}\frac{\rho_p \hat{k}_c C_{A_s}^{n-1}}{\mathcal{D}_{I_A}}\right]^{0.5} \tag{5.19}$$

Many reactions with more complex kinetics fall on the same curve for $\phi_g > 3$, where intraparticle diffusion gradients are high.

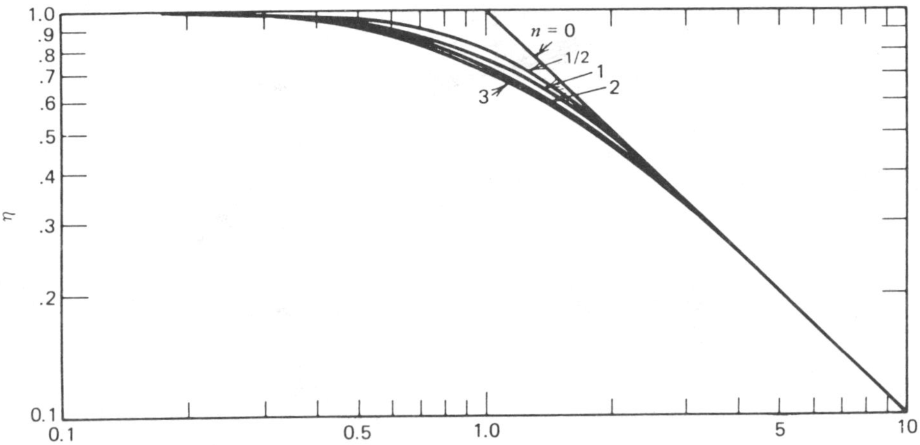

$$\phi_g = \frac{1}{a_s}\left[\frac{n+1}{2}\,\frac{\rho_p \hat{k}_c C_{A_s}^{n-1}}{\mathcal{D}_{I_A}}\right]^{0.5}$$

Figure 5.3 Generalized plot of effectiveness factor for simple-order reactions (Reproduced by permission: G.F. Froment and K.B. Bischoff, "Chemical Reactor Design," copyright 1979, John Wiley & Sons, Inc., New York.)

The expression for observed rate $(\hat{r}_A)_e$ can be solved for $\hat{k}_c$:

$$\hat{k}_c = \frac{(\hat{r}_A)_e}{\eta C_{A_s}^n} \tag{5.20}$$

Substituting in Eq. 5.19 gives

$$\Phi = \phi_g^2 \eta = \frac{(n+1)\rho_p(\hat{r}_A)_e}{2a_s^2 \mathcal{D}_{I_A} C_{A_s}} \tag{5.21}$$

Replotting Fig. 5.3, using this new parameter [10], produces Fig. 5.4, which is quite useful in speculating about preliminary laboratory observations on catalysts of varying physical properties but the same chemical compositions.

Can an Isothermal Catalyst Particle Be Assumed?

At the steady state, the assumption of an isothermal particle makes sense intuitively, and has been confirmed by both theoretical arguments and experimental evidence [3, 11–13]. If the reaction is very fast, it will be confined to the outer layers of the catalyst. At the steady state, the entire pellet will essentially reach a constant temperture, and the main ΔT, if any, will occur

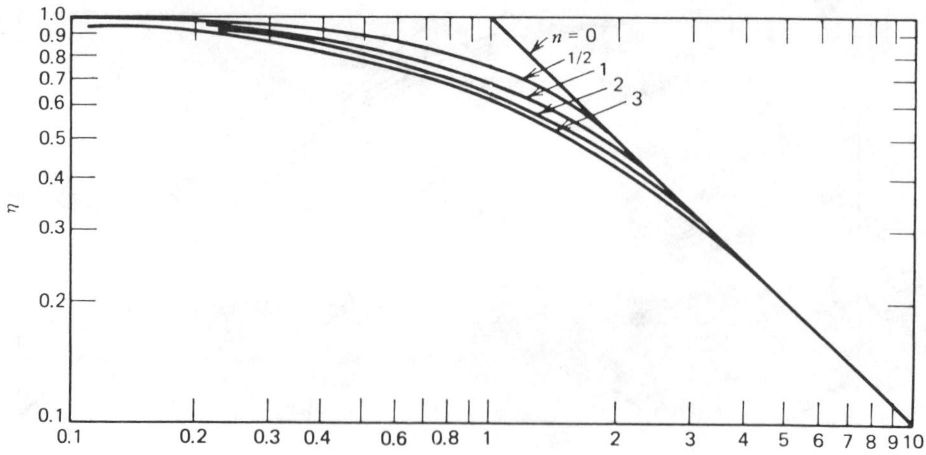

$$\Phi = \phi_g^2 \eta = \frac{(n+1)(\hat{r}_A)_e}{2a_s^2 \mathcal{D}_{I_A} C_{A_s}}$$

Figure 5.4 Effectiveness factor in terms of observables for simple-order reactions (Reproduced by permission: G.F. Froment and K.B. Bischoff, "Chemical Reactor Design," copyright 1979, John Wiley & Sons, Inc., New York.)

across the fluid film between the exterior catalyst surface and the bulk fluid. For a more modest reaction rate, penetration by reactants will be more uniform, and the temperature will again be uniform. Although the steady-state catalyst temperature is dictated by the operating condition and the heat-transfer rate between the bulk fluid and the catalyst, it is essentially constant throughout the pellet. At the same time, it is possible for the ΔT between the fluid and catalyst to be significant when flow rates are low, as discussed previously. In order to demonstrate even small temperature gradients in the particle, it is necessary to use a larger and more active catalyst than would normally be selected for a given reaction system.

Even though temperature gradients in the pellet may be considered negligible, concentration gradients can be significant, which is the reason for the effectiveness-factor concept.

Falsification of Activation Energy and Order

For most purposes, effective rates based on catalyst exterior conditions, or bulk conditions, are recommended, and the use of effectiveness factors avoided when the catalyst size is not changed. Since the apparent activation energies and orders may be different than the true values obtained using powdered catalyst, or smaller pellets and lower temperatures, it is well to understand the reason when data are questioned or challenged.

If diffusional gradients in a pellet are large, $\phi_s > 5$ and Eq. 5.15 reduces to

$$\eta = 1/\phi_s \tag{5.22}$$

Expressing the intrinsic rate as $\hat{k}_c C_A^n$, gives

$$\hat{r}_1 = \eta \hat{k}_c C_{A_s}^n = \frac{\hat{k}_c C_{A_s}^n}{\phi_s}$$

Then substituting Eq. 5.13 for ϕ_s gives

$$\frac{3\mathscr{D}_{I_A}^{1/2} \hat{k}_c C_{A_s}^n}{R_p (\rho_p \hat{k}_c C_{A_s}^{n-1})^{1/2}} = \hat{k}_c^{1/2} \left(\frac{\mathscr{D}_{I_A}}{\rho_p} \right)^{1/2} C_{A_s}^{(n+1)/2}$$

or

$$\hat{r}_1 = \frac{3A^{1/2}}{R_p} \frac{\mathscr{D}_{I_A}^{1/2}}{\rho_p} \exp\left(\frac{-E}{2RT} \right) C_{A_s}^{(n+1)/2} = A_o \exp\left(\frac{-E_o}{RT} \right) C_{A_s}^{n_o} \tag{5.23}$$

where E is the true intrinsic activation energy, n is the intrinsic order, E_o is the observed activation energy, and n_o is the observed order

Since diffusivity for gases is not a strong function of temperature, the relation between observed and intrinsic activation energy for a catalyst of low effectiveness factor is

$$E_o = \frac{E}{2} \tag{5.24}$$

The relation between the observed order and the intrinsic order n is

$$n_o = (n+1)/2 \tag{5.25}$$

Therefore, for intrinsic orders other than $n = 1$, the observed order will be different. It must quickly be reiterated that such falsification of activation and order is certainly no impediment for design or analysis of reactors using the observed or effective rate as long as the catalyst size is the same as that for which the rate equation was originally developed.

Effect of Particle Size on Rate ($\eta \leq 0.2$)

Another interesting and useful conclusion emerges from Eq. 5.23, for catalysts with low effectiveness factors ($\eta \leq 0.2$). At equal conditions of concentration and temperature, the effectiveness factor is inversely proportional to the pellet radius or exterior surface area:

$$\eta \propto \frac{1}{R_p} \propto a_s \tag{5.26}$$

Multiple-Reaction Systems

Composition gradients within a catalyst pellet involved in a multiple-reaction system can be solved simultaneously along with the axial fluid composition and temperature gradients by using the differential equations given at the end of Table 5.1. When it is necessary to determine such gradients, for purposes of catalyst "design", the increase in numbers of parameters, such as the effective diffusivities, often requires excessive experimental effort. A useful and reliable modeling procedure for design and control purposes is to use power-law rate equations, in which every component is considered as a possible parameter in both forward and reverse reactions. Parameter optimization will yield the proper exponents, and these must be restricted to only those that yield thermodynamically consistent expressions (see Eq. 2.6). The final exponents and corresponding activation energies will be unique for a particular type and size of catalyst.

CONDITIONS IN THE BED

Precise knowledge of the fluid pathways and gradients in the bed is perhaps even more difficult to realize than behavior in the pellet. Obviously, empiricisms must be invoked, for it is not ever possible to pack a bed exactly the same in two successive trials.

Radial Velocity Gradients

Although the general smoothed shape of a velocity profile through a packed bed is well known on a macroscopic scale (see Fig. 5.5), it is not possible to predict the magnitudes with any accuracy with position. Fortunately, for larger beds, the deviation from a constant radial velocity at a given axial position is confined to only a small fraction of the cross section adjacent to the wall. Tubes of small diameter, such as 1-in. diameter, used so often in heat-exchange reactors, have a substantial portion of the cross section affected by the wall; and assumption of a flat velocity profile is certainly tenuous. On a smaller scale of observation, the smooth curve in Fig. 5.5 does not exist because of the presence of the packing, which causes mixing and even fluctuations in velocity, but quantitative prediction of such behavior is not possible. Unusual velocity profiles can result from faulty catalyst-loading procedures and inadequate attention to inlet distributor and outlet collectors, as shown in Fig. 5.6. Such situations must be avoided in order to ensure maximum yield of quality product.

The only reasonable solution to this dilemma is to assume a flat velocity profile, design a good flow distribution system, and monitor carefully the loading of catalyst into the bed to prevent gross bypassing of flow. In using a

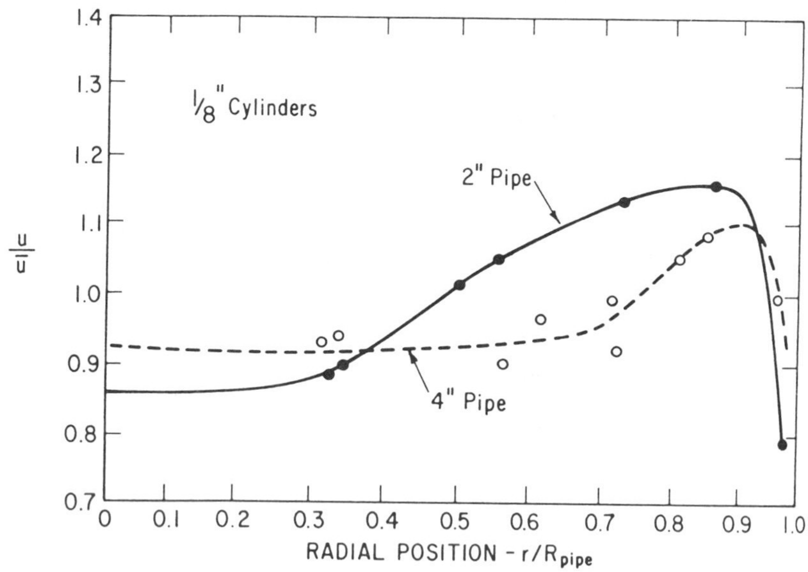

Figure 5.5 Velocity profiles in packed beds
Large beds exhibit flat profiles over most of the cross section except near the walls. Symbols u and $\bar{u}$ represent point velocity and average velocity, respectively. (Reproduced by permission: C.E. Schwarz and J.M. Smith, *Ind. Eng. Chem. 45*, 1209 (1953) copyright 1953, American Chemical Society, Washington, D.C., as adapted by H.F. Rase, "Chemical Reactor Design for Process Plants," copyright 1977, John Wiley & Sons, Inc., New York.)

flat profile, one is not ignoring influences in the radial direction, since they are lumped into parameters applicable to heat and mass transfer and pressure drop if these parameters are determined for similar or analogous conditions.

Axial Gradients

Axial dispersion and axial heat conduction effects in a packed bed for gases generally have no effect on performance for the following easily achieved criteria [3, 7, 13]:

Axial Dispersion

$$\frac{L}{D_p} \geq 50 \tag{5.27}$$

or

$$\frac{L}{D_p} > \frac{20n}{(\mathrm{Pe})_a} \ln \frac{C_{A_0}}{C_{A_e}} \tag{5.28}$$

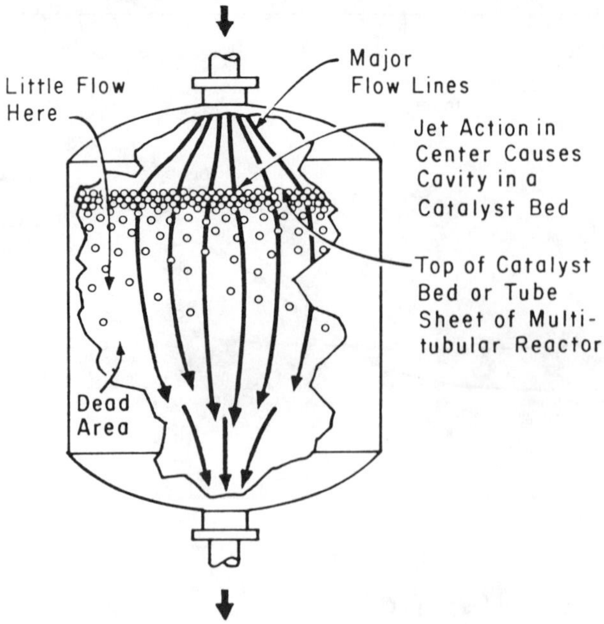

Figure 5.6 Improperly designed distribution produces significant bed bypassing (Reproduced by permission: H.F. Rase, "Chemical Reactor Design for Process Plants," copyright 1977, John Wiley & Sons, Inc., New York.)

Equation 5.28 is a general criterion [7] in terms of a simple power-law rate form with a maximum 5% error in rate. The term $(\mathrm{Pe})_a$ is the axial peclet number (2 for gases when $\mathrm{Re} > 2$, n is the reaction order, and C_{A_0} and C_{A_e} are inlet and exit concentrations of the limiting reactant).

Axial Heat Conduction [14] For a nonadiabatic reactor with cooling or heating at the wall,

$$\frac{\lambda_a(-\Delta H)\rho_b}{(\mathbf{G}c_p)^2(T_0 - T_w)}\hat{r}_0 < 1 \tag{5.29}$$

where λ_a is the axial thermal conductivity, and T_0 and $\hat{r}_0$ refer to inlet values of temperature and rate.

The inequalities represented by Eqs. 5.27–5.29 are invariably satisfied in practical reactors, which, for economic reasons, are usually operated at velocities high enough to require a length-to-diameter ratio far exceeding the minimum that would satisfy the criterion. Short laboratory reactors will not meet the criteria and often fail to mimic commercial performance.

Radial Gradients

Adiabatic commerical reactors exhibit no significant radial gradients since no heat transfer is involved; because of the larger-diameter vessels, wall effects cover only a very small part of the total cross section. Nonadiabatic reactors, however, can develop significant radial temperature gradients that, because of the differing rates generated at any axial position, will produce radial concentration gradients as well. These gradients of temperature and composition occur when large maxima (hot spots in exothermic reactions) or large minima (low-temperature excursions in endothermic reactions) occur. Since a large deviation from an optimum temperature causes undesired side reactions and even safety problems in exothermic reactions, every effort should be made to design and operate in a manner that will minimize radial gradients. Translated into adjustable parameters, this goal requires small-diameter tubes (1″ OD is common), high mass velocities to improve heat transfer ($\mathbf{G} > 1000\,\text{lb/hr ft}^2$), and careful selection of a catalyst to avoid excessively high activity.

Again, the radial temperature gradient is the dominating effect. If it is minimized to a point that ignoring it will cause an error in the rate no more than 5–10%, both radial temperature and concentration gradients may usually be ignored. Further, minimizing the radial gradient, by sufficiently high velocity, also minimizes the interphase (between fluid and catalyst surface) temperature gradient. In fact, the radial gradient becomes limiting first, as one can see by comparing at the same temperature Eq. 5.28 with the following criterion for minimizing that gradient to cause a rate error of no more than 5–10% [7]:

$$\frac{|\Delta H_1|\hat{r}_1 \rho_b D^2}{\lambda_r T_w} < \frac{1.6RT_W}{E} \tag{5.30}$$

(For a 5% error in rate. For a 10% error in rate double RHS.)

Equation 5.30 was derived by solving the two-dimensional heat balance equation (Eq. 5.56). It applies for a single reaction, but a rough idea can be obtained for a multiple reaction by using the E of the most crucial reaction and the summation of heats of reaction contributions:

$$\frac{\Sigma(|\Delta H_n|\hat{r}_n)\rho_b D^2}{\lambda_r T_w} < \frac{1.6RT_w}{E} \tag{5.31}$$

For most nonadiabatic reactors with the commonly used 1″ OD tubes, it is possible to minimize radial gradients satisfactorily, provided a highly active catalyst is not mandated and/or a rapid highly exothermic reaction is not involved. For larger-diameter tubes, such as are common in fired reactors as used in steam reforming (3–6″), gradients cannot be ignored if design or operation calculations based on rate equations are required.

GENERAL TUBULAR REACTOR MODEL

The traditional and most familiar tubular reactor component balance is written in terms of concentration as follows for any component j. See Appendix C-1 for complete development and assumptions.

$$\varepsilon\left(\frac{\partial C_{\mathrm{j}}}{\partial t}\right)_{Z,r_t} + \left(\frac{\partial u_{\mathrm{s}} C_{\mathrm{j}}}{\partial Z}\right)_{t,r_t} - \varepsilon \mathscr{D}_{\mathrm{a}}\left(\frac{\partial^2 C_{\mathrm{j}}}{\partial Z^2}\right)_{t,r_t} - \varepsilon \mathscr{D}_{\mathrm{r}}\left(\frac{\partial^2 C_{\mathrm{j}}}{\partial r_t^2} + \frac{1}{r_t}\frac{\partial C_{\mathrm{j}}}{\partial r_t}\right)_{t,Z}$$

$$= \rho_{\mathrm{b}} \sum_{n=1}^{n_{\mathrm{r}}} \nu_{jn} \hat{r}_n \qquad j = 1, \ldots, c \tag{5.32A}$$

Similarly the heat balance takes the form

$$\left[\varepsilon(\overline{c_p \rho})_{\mathrm{f}} + (1-\varepsilon)(\overline{c_p \rho})_{\mathrm{s}}\right]\left(\frac{\partial T}{\partial t}\right)_Z + \left[\frac{\partial u_{\mathrm{s}}(c_p \rho)_{\mathrm{f}} T}{\partial Z}\right]_t - \lambda_{\mathrm{a}}\left(\frac{\partial^2 T}{\partial Z^2}\right)_t$$

$$- \lambda_r\left(\frac{\partial^2 T}{\partial r_t^2} + \frac{1}{r_t}\frac{\partial T}{\partial r_t}\right)_{t,Z} = \rho_{\mathrm{b}} \sum_{n=1}^{n_{\mathrm{r}}} (-\Delta H_{\mathrm{n}})\, \hat{r}_n \tag{5.32B}$$

where ε is the void fraction of the bed; C_{j} is the bulk concentration of a component designated by numbers 1 through c; c is the number of components; t is the time; Z is the length along the bed (axial distance); r_t is the radial distance; $\mathscr{D}_{\mathrm{a}}$ is the effective axial diffusivity; $\mathscr{D}_{\mathrm{r}}$ is the effective radial diffusivity; ρ_{b} is the bulk density of the bed; ν_{jn} is the stoichiometric coefficient for any component j in reaction n; n_{r} is the total number of reactions; u_{s} is the superficial velocity based on the empty cross-sectional area; c_p is the heat capacity per unit mass; ρ is the density; f and s refer, respectively, to fluid and solid, T is the bulk temperature; λ_{a} and λ_{r} are, respectively, axial and radial thermal conductivities; ΔH_n is the heat of reaction for Reaction n.

From Eqs. 5.32A and 5.32B, the first term at the left is the accumulation term necessary for unsteady-state operation. In the heat balance this term contains two parts: the heat capacity of the fluid and that of the solid. The former is generally negligible. The second term is the flow or convection term. Since u_{s} will vary, it should remain inside the differential for this form of the equation. In the heat-balance second term the $(c_p \rho)_{\mathrm{f}}$ term is usually extracted from the differential, since it does not vary strongly through the bed. The third and fourth terms are axial and radial mass and heat flow related to gradients in the bed. For commercial reactors the axial component may be neglected, for it is insignificant in properly designed units. Finally the RHS of both equations contains the reaction term.

Special-purpose equations in more convenient forms may be derived directly, or, when convenient, they may be derived by proper substitutions in the two previous equations, with nonapplicable terms being eliminated. The recommended equations that follow will be written in terms of moles per mole of limiting reactant, n_{ja}, which creates equation forms most amenable to incorporation into process simulators.

IDEAL PLUG-FLOW (ONE-DIMENSIONAL) PSEUDOHOMOGENEOUS MODEL

Based on the foregoing discussion on gradients, it is clear that for commercial adiabatic reactors and a large percentage of nonadiabatic reactors, only axial gradients due to the progressive flow through the system need be considered. Such a conceptual model is called a *plug-flow pseudohomogeneous model*, the latter term meaning that conditions on the catalyst surface and in the bulk are essentially identical and mathematically viewed as a continuum. The fluid is assumed to move through the bed in a plug, such that each differential element passes through the system without back-mixing (axial dispersion) and has a single temperature and composition at a given axial position throughout its cross section. The criteria for this model can best be summarized as follows:

- *Gradients* Mass flow, temperature, pressure, and composition are uniform over the flow cross section at any axial position; gradients occur, therefore, only in the axial direction.
- *Flat velocity profile* A reasonably flat velocity profile is assured for D/D_p values of 30 or more [15] with a maximum deviation of 20%. This criterion is readily met in most reactors, except small-diameter heat-exchange reactors. The current practice is to use the flat profile for these multitubular reactors with the realization that *a priori* calculations will be less reliable than those for larger diameter beds.
- *Evenly distributed flow over the bed cross section* Load catalyst to ensure even distribution, a result that is more readily realized by carefully distributing small amounts rather than resorting to rapid dumping. In multitubular reactors, the ΔP for all tubes should be as close to the same value as possible (5–10% variation). In all cases, great care must be exercised to avoid attrition, which produces fines that cause increased pressure drop and localized bypassing. Provide low-ΔP distributors and collecting devices, as shown in Fig. 1.1 to prevent bypassing.
- *Negligible gradients of temperature and composition between catalyst and fluid* Use Eqs. 5.8 and 5.9 or the single-reaction criterion given by Eq. 5.10 to evaluate conditions for negligible temperature gradient. If the ΔT gradient is negligible, the concentration will also be essentially zero (see Eqn 5.5).
- *Negligible axial dispersion effects*
  1. $L/D_p > 50$ is normally easily met.
  2. See also Eq. 5.28.

- *Rate equations based on a single catalyst size and shape* Although obtaining the data and employing the equations is simpler and the predictions more accurate, it is essential that they be used only for the catalyst size and over the temperature range on which they are based. The same equation may be used on another catalyst size if independent tests show that such a change does not affect the rate.

DERIVATION OF IDEAL PLUG-FLOW (ONE-DIMENSIONAL), PSEUDOHOMOGENEOUS MODEL ($T_s = T$)

With the wide availability of ODE solving routines, it is most advantageous to write the model equations in a general form so that a program can be written that is easily adapted to any reaction system. Balance equations using well-known terms such as mass velocity, W/F, and moles of a compound per mole of limiting reactant introduce useful simplifications and are preferred in practice. These equations can be developed as in the following sections, or they can be rationalized from Eqs. 5.32A and 5.32B, eliminating, where necessary, nonapplicable terms.

Mole Balance

From Eq. 4.24, the steady-state mole balance for any component j may be written as follows for an increment of catalyst ΔW corresponding to a volume $\Delta W/\rho_b$ with a cross section the same as the reactor cross section:

$$\overset{\text{I}}{\begin{bmatrix}\text{Rate of } j \text{ flowing out}\\ \text{of the volume element}\end{bmatrix}} - \overset{\text{II}}{\begin{bmatrix}\text{rate of } j \text{ flowing into}\\ \text{the volume element}\end{bmatrix}}$$

$$= \overset{\text{IV}}{\begin{bmatrix}\text{rate of production of } j\\ \text{in the volume element}\end{bmatrix}} \quad \text{or} \quad - \begin{bmatrix}\text{rate of consumption of } j\\ \text{in the volume element}\end{bmatrix}$$

$$\Delta \mathscr{F}_j = \Delta W \sum_{n=1}^{n_r} \nu_{jn} \hat{r}_n \qquad j = 1, \ldots, c$$

or, as $\Delta W \to 0$

$$\rho_b \frac{d\mathscr{F}_j}{dW} = \rho_b F_A \frac{dn_{ja}}{dW} = \rho_b \frac{dn_{ja}}{d(W/F_A)} = \frac{\mathbf{G} Y_{A_0}}{M_F} \frac{dn_{ja}}{dZ} = \rho_b \sum_{n=1}^{n_r} \nu_{jn} \hat{r}_n \qquad j = 1, \ldots, c \tag{5.33}$$

where $\mathscr{F}_j$ is the moles of j flowing per unit time; W is the mass of catalyst; F_A is the moles of limiting reactant (or any reactant to be used as the basis for calculation) flowing per unit time; n_{ja} is the moles of j per mole of A fed; **G** is the superficial mass velocity; Y_{A_0} is the mole fraction of A in the feed; ρ_b is the bulk density of the catalyst bed; M_F is the molecular weight of the feed; Z is the axial length parameter along the bed; r_n is the rate of reaction n; n_r is the total number of reactions; ν_{jn} is the stoichiometric coefficient for component j in reaction n with appropriate sign: (+) for production and (−) for consumption.

Since use of the calculus in this manner involves a continuum encompassing the entire cross section of the element, the mass velocity appearing in Eq. 5.33 must be that based on the cross section of the tube. The compositions and temperatures are assumed constant throughout the element with catalyst and fluid conditions identical.

The most generally useful form of Eq. 5.33 is that involving **G** since the mass velocity is an important value in technical discussions because of its frequent association with transport equations and the ease of transferring experimental knowledge related to **G** from one situation to another. Unlike the superficial velocity, **G** is constant throughout the reactor length when, as in these equations, the steady state is assumed.

For each component,

$$\frac{\mathbf{G}Y_{A_0}}{M_F}\frac{dn_{ja}}{dZ} = \rho_b \sum_{n=1}^{n_r} \nu_{jn}\hat{r}_n \qquad j = 1, \ldots, c \tag{5.34}$$

or, in DE form,

$$\frac{dn_{ja}}{dZ} = \frac{M_F\rho_b}{\mathbf{G}Y_{A_0}} \sum_{n=1}^{n_r} \nu_{jn}\hat{r}_n \qquad j = 1, \ldots, c \tag{5.35}$$

The number of these component balance equations required for solution must be at least equal to the number of reactions, since all unrepresented components would be related by algebraic expressions through stoichiometry to those used in the several equations. But, when a library ODE solver is used, it is much simpler and more efficient to write an equation for each component and to let the solver keep track of all values.

Another familiar form of Eq. 5.33 can be written for a limiting reactant A and a single reaction:

$$\frac{\mathbf{G}Y_{A_0}}{\rho_b M_F}\frac{dn_{Aa}}{dZ} = \frac{dn_{Aa}}{\mathrm{d}(W/F_A)} = -\hat{r}_A \quad \text{or} \quad -\frac{dn_{Aa}}{\mathrm{d}(W/F_A)} = \hat{r}_A \tag{5.36}$$

Since $-dn_{Aa} = dX_A$, then

$$\frac{\mathbf{G}Y_{A_0}}{\rho_b M_F}\frac{dX_A}{dZ} = \frac{dX_A}{d(W/F_A)} = \hat{r}_A = (-\hat{r}_A) \tag{5.37}$$

where X_A is called the conversion (moles A converted per mole of A fed), and $(-\hat{r}_A)$ is a symbolic representation of the rate of consumption of A as a positive number.

In addition to the continuity equations, an energy conservation equation and a pressure drop equation (if rate is affected by pressure) are required.

Conservation of Energy Equation

It proves most convenient when there are accurate thermodynamic data for a reaction system, as is the case for gases, to apply the First Law, using as a basis the elements at 25°C. For gases, accurate enthalpies of formation at 25°C are recorded [16].

From Eq. 4.43,

$$dH = dQ$$

Basis: Elements at 25°C (298°K); molar flow rate F_j and differential heat flux dq'; assume approximate ideal gas behavior.

$$\sum d\mathcal{F}_j H_j = dq'$$

$$\sum \mathcal{F}_j dH_j + \sum H_j \, d\mathcal{F}_j = dq'$$

But

$$dH_j = c_{p_j} \, dT \quad \text{and} \quad H_j = (H^\circ_{f_j})_{298} + \int_{298}^{T} c_{p_j} \, dT$$

Substituting and differentiating with respect to V, and assuming* that $(d \ln(F_j c_{p_j})/dV)/(d \ln T/dV) \approx 0$ and $(d \ln c_{p_j}/dV)/(d \ln T/dV) \approx 0$ for all V, gives

---

*

$$\frac{dAB}{dV} = A\frac{dB}{dV} + \frac{B\,dA}{dV} \approx A\frac{dB}{dV}$$

or

$$1 + \left(B\frac{dA}{dB}\right) / \left(A\frac{dB}{dV}\right) \approx 1$$

or

$$\left(B\frac{dA}{dV}\right) / \left(A\frac{dB}{dV}\right) = \frac{d \ln A}{dV} / \frac{d \ln B}{dV} \approx 0 \qquad \text{for all } V$$

where A and B represent $\mathcal{F}_j c_{p_j}$ and T, respectively.

$$\sum \mathscr{F}_j c_{p_j} \frac{dT}{dV} + \sum \left[(H^\circ_{f_j})_{298} + \int_{298}^{T} c_{p_j}\, dT \right] \frac{d\mathscr{F}_j}{dV}$$

$$= \frac{dq'}{dV} = U_Z \frac{4}{D} (T_h - T) \tag{5.38}$$

where $\mathscr{F}_j$ is molar flow rate of any component j; c_{p_j} is heat capacity of any component j; $(H^\circ_{f_j})_{298}$ is standard heat of formation of any component at 298°K and 1 atm; q' is the heat added per unit time; V is the reactor volume; U_Z is the point value of the overall heat-transfer coefficient based on the inside tube surface area; D is the inside tube diameter; $4/D$ is the inside surface area of the tube per unit volume; T_h is the temperature of the heat-transfer fluid in the jacket.

The use of an overall coefficient U_Z, as defined in Eq. 4.13, implies a flat radial temperature profile with all the temperature drop occurring at the bed perimeter. Thus, the inside coefficient in Eq. 4.14 is defined as

$$h_i = \frac{N_q}{T_w - T}$$

where T represents the bulk mean average temperature at any particular axial position, and T_w is the wall temperature.

The terms in the brackets in Eq. 5.38 constitute the ideal-gas enthalpy for component j. At high pressure, the actual enthalpy departure from the ideal-gas enthalpy (H^*) may be significant. For such cases, enthalpy corrections $(H^* - H)$ should be employed based on an appropriate equation of state (see Reference 4, Chapter 3). A quick way to assess the need for a correction is to calculate compressibility factors for the reaction mixture at the inlet and outlet of the reactor. If both are close to unity, the enthalpy correction will be negligible.

A more convenient form than Eq. 5.38 is obtained by the following transformations:

$$dV = A_c\, dZ \tag{5.39}$$

where A_c is the cross-sectional area of reactor.

$$\mathscr{F}_j = n_{ja} F_A = \frac{n_{ja} \mathbf{G} A_c Y_{A_o}}{M_F} \quad \text{and} \quad \frac{d\mathscr{F}_j}{dV} = \rho_b \sum_{n=1}^{n_r} \nu_{jn} \hat{r}_n \tag{5.40}$$

Substituting Eqs. 5.39 and 5.40 into Eq. 5.38, gives

$$\frac{dT}{dZ} = \frac{-\sum_{j=1}^{c} \left\{ \left[(H^\circ_{f_j})_{298} + \int_{298}^{T} c_{p_j}\, dT \right] \rho_b \sum_{n=1}^{n_r} \nu_{jn} \hat{r}_n \right\} - U_Z \frac{4}{D} (T - T_h)}{\frac{\mathbf{G} Y_{A_0}}{M_F} \sum_{j=1}^{c} n_{ja} c_{p_j}} \tag{5.41}$$

Equation 5.41 can also be written in terms of heats of reaction:

$$\frac{dT}{dZ} = \frac{\sum_{n=1}^{n_r} (-\Delta H_n) \hat{r}_n \rho_b - U_Z \frac{4}{D} (T - T_h)}{\frac{\mathbf{G} Y_{A_0}}{M_F} \sum_{j=1}^{c} n_{ja} c_{p_j}} \tag{5.42}$$

where the subscript n refers to a reaction number, and the rate is based on the same component as the heat of reaction and is a positive number.

Rationalization of Heat-of-Reaction Form

A brief development for a single reaction will demonstrate the relationship between Eqs. 5.41 and 5.42:

$$A \rightarrow 2R \qquad \text{Rate} = r_1 \; \frac{\text{moles A converted}}{\text{(time)(mass cat.)}}$$

The first term of the numerator of Eq. 5.41 in this case is:

$$\sum \left[(H^\circ_f)_{298} + \int_{298}^{T} c_{p_j}\, dT \right] \rho_b \sum_{n=1}^{n_r} \nu_{jn} \hat{r}_n = \left[(H^\circ_{f_R})_{298} + \int_{298}^{T} c_{p_R}\, dT \right] 2\hat{r}_1 \rho_b - \left[(H^\circ_{f_A})_{298} + \int_{298}^{T} c_{p_A}\, dT \right] \hat{r}_1 \rho_b \tag{5.43}$$

But any $(H^\circ_{f_j})_{298} + \int_{298}^{T} c_{p_j}\, dT = H_j$, the enthalpy relative to the elements at T, and the numerator of Eq. 5.41 for the adiabatic case becomes

$$2H_R \hat{r}_1 \rho_b - H_A \hat{r}_1 \rho_b = \Delta H_1 \hat{r}_1 \rho_b \tag{5.44}$$

where ΔH_1 is the heat of reaction 1 at temperature T. Substituting in Eq. 5.41 with the heat-transfer term included, gives

$$\frac{dT}{dZ} = \frac{(-\Delta H_1) \hat{r}_1 \rho_b - U_Z (4/D)(T - T_h)}{(\mathbf{G} Y_{A_0} / M_F) \sum n_{ja} c_{p_j}} \tag{5.45}$$

For multiple reactions, Eq. 5.42 results. If the heat of reaction remains essentially constant, Eqs. 5.42 or 5.45 (for a single reaction) are most convenient. When an excess of one component, often an inert, is fed, the heat capacity may also be constant. For that case, substitute $\mathbf{G}c_p$ for the denominator on the RHS of Eqs. 5.42 and 5.45, where c_p is the mean heat capacity on a mass basis. When the heat of reaction and heat capacity vary, Eq. 5.41 is preferred. For gases, the convenient heat-capacity equations as functions of temperature may be used. Even if all the rate equations for a system are not known, the rate of production or consumption of a component can sometimes be empirically described as a function of some known rate.

Algebraic Heat-Balance Equation

For the adiabatic case, the heat-transfer term is omitted for all equations, and it is possible to write a path-independent algebraic equation, as would be expected for the adiabatic case.

Substituting Eq. 5.37, written in terms of reaction n, into Eq. 5.42, using X_{na} as a general designation for conversion based on the same reactant as the particular ΔH_n in each case and restricting to the adiabatic case (omit heat-transfer term), gives

$$\frac{dT}{dZ} = \frac{\sum_{n=1}^{n_r} (-\Delta H_n)\, dX_{na}}{\sum n_{ja} c_{p_j}\, dZ} \tag{5.46}$$

or

$$\sum n_{ja} c_{p_j}\, dT = \sum_{n=1}^{n_r} (-\Delta H_n)\, dX_{na} \tag{5.47}$$

where X_{na} is the moles of designated reactant converted in reaction n per mole of A fed

Integrate by one of two paths:

1. React at T_1 and heat mixture to T_2:

$$\left[\sum n_{ja} \bar{c}_{p_j}(T_2 - T_1)\right]_{\text{OUT}} = \sum_{n=1}^{n_r} (-\Delta H_n)_{T_1} X_{na} \tag{5.48}$$

where OUT refers to final or exit mixture of products, unreacted reactants, and inerts, and $\bar{c}_{p_j}$ is the average heat capacity per mole of component j. Equation 5.49 can also be written in terms of mass of fluid flowing.

$$[m_a\bar{c}_p(T_2 - T_1)]_{\text{OUT}} = \sum_{n=1}^{n_r} (-\Delta H_n)_{T_1} X_{na} \tag{5.49}$$

where m_a is the mass flowing per mole of A fed, and $\bar{c}_p$ is average heat capacity of the outgoing mixture per unit mass.

2. Heat initial feed to T_2 and react at T_1:

$$\left[\sum n_{ja}\bar{c}_{p_j}(T_2 - T_1)\right]_{\text{IN}} = \sum_{n=1}^{n_r} (-\Delta H_n)_{T_2} X_{na} \tag{5.50}$$

or

$$[m_a\bar{c}_p(T_2 - T_1)]_{\text{IN}} = \sum_{n=1}^{n_r} (-\Delta H_n)_{T_2} X_{na} \tag{5.51}$$

The last equation is usually the most convenient.

These equations are useful for quick calculations of adiabatic temperature, using constant or average values of heats of reaction and heat capacity. For a design model using a differential equation solver, it is better to leave the heat balance in differential equation form to avoid writing a separate subroutine for the algebraic form used with the DE solver.

ONE-DIMENSIONAL HETEROGENEOUS MODEL

$$T_s \neq T \quad \text{but} \quad (n_{ja})_s = n_{ja}$$

This model differs from the ideal plug-flow case only in that the catalyst surface temperature (T_s) differs significantly from that of the bulk fluid (T). As discussed earlier, the concentration differences will be quite small for most realistic conditions. Again, any concentration gradient in the catalyst pellet is accounted for in the lumped constant for the effective rate. Thus, it is possible to convert the ideal plug-flow case into the heterogeneous case by simply adding a heat balance between the catalyst and the bulk fluid:

$$\begin{aligned} h_s a_m(T_s - T) &= -\sum_{j=1}^{c} \left\{\left[H_{f_j}^\circ + \int_{298}^{T} c_{p_j}\, dT\right] \sum_{n=1}^{n_r} \nu_{jn}(\hat{r}_n)_s\right\} \\ &= \sum_{n=1}^{n_r} (-\Delta H_n)(\hat{r}_n)_s \end{aligned} \tag{5.52}$$

where a_m is the exterior area of catalyst/mass of catalyst, and s refers to the catalyst exterior surface condition, with $(\hat{r}_n)_s$ meaning the rate evaluated at T_s.

The value of T_s as required by the DE solver can be calculated in a separate subroutine or by software that solves DEs together with algebraic equations (LSODI).

This model is recommended for adiabatic reactors operating at low velocities where substantial differences between T_s and T occur.

RADIAL-FLOW ADIABATIC REACTOR MODEL

Radial-flow adiabatic reactors have been used successfully for reactions that must be operated at or near atmospheric pressure and require, therefore, very low pressure drops across the bed. The model equations for a standard fixed-bed reactor (pseudohomogeneous), such as Eqs. 5.35 and 5.41, may be used by simply replacing the parameter Z by the radial distance r_t. The inlet position is the outside radius of the distributor, and the outlet position is the inside radius of the catalyst basket (see Fig. 1.2). In addition, G should be replaced by $m_f/2\pi r_t L$, where m_f is the total mass flowing per unit time, and L is the length of the catalyst bed in the axial direction. The heat-transfer term is omitted. This recommendation assumes plug flow and no radial dispersion, which is a good assumption for commercial reactors using the radial design, such as the ethylbenzene dehydrogenation reactor for producing styrene. See Example 8 for use of the radial model.

TWO-DIMENSIONAL PSEUDOHOMOGENEOUS MODEL

This model is recommended for heat-exchange (nonadiabatic) reactors in which substantial minimum or maximum temperature excursions occur that cause significant radial gradients of temperature and concentration (criteria of Eqs. 5.30 and 5.31 are violated). For best results, one should use values of effective rates and other parameters obtained directly from single-reactor-tube studies conducted in the narrow range already known to be needed for the particular reaction system of interest. The most reliable design involves the simple scale-up from one tube to many tubes of the same size (see Table 5.2), but a model with reliable values for the parameters provides flexibility in examining operational options. A particular concern is the extent of the hot spot in exothermic systems.

The most widely accepted and rational approach for the complex two-dimensional problem is to apply the familiar continuity equations for both heat and component transport, with the catalyst and fluid considered as a homogeneous entity with equal compositions and temperatures at any point. Because of the high mass velocities required for heat transfer through the tubes, the gradient of temperature between the catalyst and fluid is generally negligible. The terms of Fick's and Fourier's laws are invoked for transport in the radial direction by defining an effective radial diffusivity and an effective radial conductivity. The actual mechanism for transport is quite complex, involving, for heat transfer, conduction in the solid, convection in the fluid, and radiation between phases.

It is convenient and reasonable, because of the diminished accuracy of this model, to invoke the following simplifications.

- Use constant heats of reaction and heat capacity per unit mass evaluated in the temperature range of interest.
- Use single average values of the heat-transfer parameters.
- Use a single constant effective radial diffusivity, even though modest differences exist between the several components.
- Reduce the number of differential equations to the minimum required for solution, which is the same as the number of reactions plus the heat balance. In this way, the complexity of the model is reduced by limiting the number of variables solved directly from the DEs to designated key components or extents of reactions. Linear algebraic stoichiometric expressions can then be used to determine the composition.

With reference to Fig. 5.7 the change in both the axial and radial directions of moles of key component k per mole of limiting reactant and temperature can

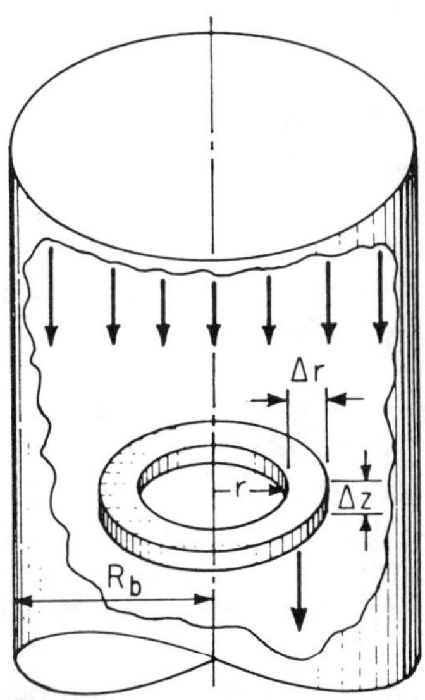

Figure 5.7 Element in packed bed experiencing radial transport and axial flow
(Reproduced by permission: H.F. Rase, "Chemical Reactor Design for Process Plants," copyright 1977, John Wiley & Sons, Inc., New York.)

be equated, respectively, to the rate of consumption or production of the component and to the rate of heat generation or consumption. The following equations can be developed from Eqs. 5.32A and 5.32B by omitting terms I and III from each and assuming that $(d \ln c_p/dZ)/(dT/dZ) = 0$, and recognizing that, for the steady state, $\rho_f u_s = G$ is a constant along the reactor and $C_k = n_{ka} Y_{A_0} \rho_f / M_F$:

$$\frac{\mathbf{G} Y_{A_0}}{M_F} \frac{\partial n_{ka}}{\partial Z} - \frac{\mathscr{D}_r \rho_f Y_{A_0}}{M_F}\left(\frac{\partial^2 n_{ka}}{\partial r_t^2} + \frac{1}{r_t}\frac{\partial n_{ka}}{\partial_{r_t}}\right) = \rho_b \sum_{n=1}^{n_r} \nu_{kn} \hat{r}_n \qquad k = 1, \ldots, k_n \tag{5.53}$$

$$\mathbf{G} c_p \frac{\partial T}{\partial Z} - \lambda_r\left(\frac{\partial^2 T}{\partial r_t^2} + \frac{1}{r_t}\frac{\partial T}{\partial r_t}\right) = \rho_b \sum_{n=1}^{n_r} (-\Delta H_n)\hat{r}_n \tag{5.54}$$

Rearranging Eq. 5.53 gives

$$\frac{\partial n_{ka}}{\partial Z} - \frac{\mathscr{D}_r \rho_f}{\mathbf{G}}\left(\frac{\partial^2 n_{ka}}{\partial r_t^2} + \frac{1}{r_t}\frac{\partial n_{ka}}{\partial r_t}\right) = \frac{\rho_b M_F}{\mathbf{G} Y_{A_0}} \sum_{n=1}^{n_r} \nu_{kn} \hat{r}_n \qquad k = 1, \ldots, k_n \tag{5.55}$$

See variable listing after Eq. 5.56. If each key component selected is unique to one reaction, then it is directly related to the corresponding extent of reaction, and the general equation in terms of extents of reaction is immediately apparent when the following relation is substituted into Eq. 5.55:

$$\frac{\partial n_{ka}}{\nu_{kn}} = \partial X_{na}$$

$$\frac{\partial X_{na}}{\partial Z} - \frac{\mathscr{D}_r \rho_f}{\mathbf{G}}\left(\frac{\partial^2 X_{na}}{\partial r_t^2} + \frac{1}{r_t}\frac{\partial X_{na}}{\partial r_t}\right) = \frac{\rho_b M_F}{\mathbf{G} Y_{A_0}} \hat{r}_n \qquad n = 1, \ldots, n_r \tag{5.55A}$$

Many users prefer this latter form. Rearranging Eq. 5.54, gives

$$\frac{\partial T}{\partial Z} - \frac{\lambda_r}{\mathbf{G} c_p}\left(\frac{\partial^2 T}{\partial r_t^2} + \frac{1}{r_t}\frac{\partial T}{\partial r_t}\right) = \frac{\rho_b}{\mathbf{G} c_p} \sum_{n=1}^{n_r} (-\Delta H_n)\hat{r}_n \tag{5.56}$$

where $\mathscr{D}_r$ and λ_r are the effective radial diffusivity and thermal conductivity, respectively; r_t is the radial distance; n_{ka} is the moles of key component k per mole of limiting reactant A fed; k_n refers to total number of key components, which must be equal to number of reactions n_r; ν_{kn} is the stoichiometric coefficient of key component k in reaction n; ΔH_n is the heat of reaction for reaction n.

The total number of differential equations is $n_r + 2$, the additional ones being the temperature gradient and pressure gradient (Eq. 5.56). For boundary conditions see Table 5.1. Note that the boundary condition at the wall, expressed in terms of a wall heat transfer and the effective radial conductivity, provides for the transfer of heat to the tube wall.

Use of Key Components for Two-Dimensional Model

The use of key components in Eq. 5.55 can be readily comprehended through the following example involving the reaction of air with methanol over an iron-molybdenum catalyst to produce formaldehyde. The reaction is exothermic and is carried out in a multitubular heat-exchange reactor with catalyst in the tubes. Under certain operating conditions, the exothermic temperature rise can be substantial and the two-dimensional model is applicable for such conditions. The reactions are written so that the stoichiometric coefficients of the key components are -1 or $+1$ and so that the limiting reactant is one of the key components:

$$CH_3OH + \frac{1}{2}O_2 \rightarrow CH_2O + H_2O \tag{1}$$

$$CH_2O + \frac{1}{2}O_2 \rightarrow CO + H_2O \tag{2}$$

1. For these two reactions, select two key components. Methanol (M), the limiting reactant, and formaldehyde (F) the desired product make convenient selections, but formaldehyde appears in both reactions 1 and 2, in which case Eq. 5.55 applies.
2. Write the two component DEs and the temperature-gradient DE patterned after Eqs. 5.55 and 5.56. The values of the stoichiometric coefficients for the key components are

$$\nu_{M1} = -1, \quad \nu_{F1} = 1, \quad \nu_{F1} = 0, \quad \text{and } \nu_{F2} = -1$$

$$\frac{\partial n_{Ma}}{\partial Z} = \frac{\mathscr{D}_r \rho_f}{\mathbf{G}} \left(\frac{\partial^2 n_{Ma}}{\partial r_t^2} + \frac{1}{r_t} \frac{dn_{Ma}}{\partial r_t} \right) = \frac{\rho_b M_F}{\mathbf{G} Y_{A_0}} (-\hat{r}_1)$$

$$\frac{\partial n_{Fa}}{\partial Z} = \frac{\mathscr{D}_r \rho_f}{\mathbf{G}} \left(\frac{\partial^2 n_{Fa}}{\partial r_t^2} + \frac{1}{r_t} \frac{\partial n_{Fa}}{\partial r_t} \right) = \frac{\rho_b M_F}{\mathbf{G} Y_{A_0}} (\hat{r}_1 - \hat{r}_2)$$

$$\frac{\partial T}{\partial Z} = \frac{\lambda_r}{\mathbf{G} c_p} \left(\frac{\partial^2 T}{\partial r_t^2} + \frac{1}{r_t} \frac{\partial T}{\partial r_t} \right) = \frac{\rho_b}{\mathbf{G} c_p} [(-\Delta H_1)\hat{r}_1 + (-\Delta H_2)\hat{r}_2]$$

See Table 5.1 for boundary conditions.

3. Write linear algebraic equations relating the other components to the key components.

   Basis: 1 mole methanol in the feed; fresh feed plus recycle values:

$$n_{M_0} = 1.0, \quad n_{F_0} = 0.007, \quad n_{O_0} = 1.10, \quad n_{N_0} = 9.006, \quad n_{CO_0} = 0.011$$

where N refers to nitrogen and O to oxygen

	Moles/Mole Methanol Fed
Methanol	n_{Ma}
Formaldehyde	n_{Fa}
CO	$X_{2A} + 0.011$
H_2O	$X_{1a} + X_{2a}$
O_2	$1.1 - 0.5X_{1a} - 0.5X_{2a}$
N_2	9.006

For convenience, the extents of reactions X_{1a} and X_{2a} are calculated from the values of n_{Ma} and n_{Fa} obtained from the numerical solutions for the DEs:

$$X_{1a} = 1 - n_{Ma}$$

$$X_{2a} = X_{1a} + 0.007 - n_{Fa}$$

Alternatively, Eq. 5.55A can be used:

$$\frac{\partial X_{1a}}{\partial Z} - \frac{\mathscr{D}_r \rho_f}{\mathbf{G}} \left[\frac{\partial X_{1a}^2}{\partial r_t^2} + \frac{1}{r_t} \frac{\partial X_{1a}}{\partial r_t} \right] = \frac{\rho_b M_F}{\mathbf{G} Y_{A_0}} \hat{r}_1$$

$$\frac{\partial X_{za}}{\partial Z} - \frac{\mathscr{D}_r \rho_f}{\mathbf{G}} \left[\frac{\partial X_{2a}}{\partial r_t^2} + \frac{1}{r_t} \frac{\partial X_{2a}}{\partial r_t} \right] = \frac{\rho_b M_F}{\mathbf{G} Y_{A_0}} \hat{r}_2$$

$$\frac{\partial T}{\partial Z} - \frac{\lambda_r}{\mathbf{G} c_p} \left[\frac{\partial^2 T}{\partial r_t^2} + \frac{1}{r_t} \frac{\partial T}{\partial r_t} \right] = \frac{\rho_b}{\mathbf{G} c_p} [(-\Delta H_1)\hat{r}_1 + (-\Delta H_2)\hat{r}_2]$$

The algebraic equations are (basis is same as previous case):

	Moles/Mole Methanol Fed
Methanol	$1 - X_{1a}$
Formaldehyde	$0.007 + X_{1a} - X_{2a}$
CO	$X_{2a} + 0.011$
H_2O	$X_{1a} + X_{2a}$
O_2	$1.1 - 0.5X_{1a} - 0.5X_{2a}$
N_2	9.006

Assessment of Two-Dimensional Model

Unless radial transport parameters are determined by direct experiment with the reaction system of interest, a model based on values estimated from the general correlations is of little value because of large errors in these parameter values. The heat-transfer parameters are the most important because of the effect on temperature. Errors in temperature estimates greatly attenuate the rate error because of the exponential dependence of the rate constant on temperature. The general correlations for effective thermal conductivity and wall heat-transfer coefficient are based on heat transfer in packed beds without reaction. In such apparatus the heat fluxes are usually orders of magnitude less than for reacting systems. Elegant experiments with reacting systems involving precise measurements of radial temperature and concentration profiles at a number of axial positions have been described [1]. Regressed values of λ_r and h_w were found to be 30–50% lower than values calculated from the correlations. Obviously, such error in the correlations is not tolerable, but the correlations are useful for initial estimates required for determining best parameter values from experimental data using optimization code combined with statistical evaluations.

Although the aforementioned and referenced experimental method does produce good data, the equipment is complex and the data required are extensive. The two-dimensional model should be avoided if at all possible by judicious choice of catalyst, mass velocity, and jacket temperature (see Example 7).

SUMMARY OF MODELS

The several models with appropriate initial or boundary conditions along with recommendations for use are summarized in Table 5.1. Since total pressure is an important variable for gaseous reactions, a pressure-drop equation should be included as an additional differential equation, except in the two-dimensional model, for which efficiency in solving at the present suggests calculating pressure drop by a separate routine outside the DE solving procedure. Examples 7, 8, and 9 provide realistic guidance to the usefulness of these models.

DESIGN BY SCALE-UP

Scale-up is in many cases the most reliable procedure for setting reactor size and configuration. In fact, for nonadiabatic reactors, it is the preferred method. For a new process involving a nonadiabatic reactor, a pilot-plant

(*Text continues on page 142.*)

Table 5.1 Models for Fixed-Bed Reactors

Model	*Equation*	*Recommendations*
I. One Dimensional, Ideal Plug Flow, Pseudohomogeneous $T_s = T$ $(n_{ja})_s = n_{ja}$	$\frac{dn_{ja}}{dZ} = \frac{M_F \rho_b}{GY_{A_o}} \sum_{n=1}^{n_r} \nu_{jn} \hat{r}_n \quad j = 1,\ldots,c$ (1) $\frac{dT}{dZ} = \frac{-\sum_{j=1}^{c} \left[\left(H^{\circ}_{f_j\,298} + \int_{298}^{T} c_{p_j}\, dT \right) \sum_{n=1}^{n_r} \nu_{jn} \hat{r}_n \right] \rho_b - U_Z \frac{4}{D} (T - T_h)}{\frac{GY_{A_o}}{M_F} \sum_{j=1}^{c} n_{ja} c_{p_j}}$ (2) $\frac{dP}{dZ} = -\frac{f_k G^2}{C_l D_p \rho_f} \left(\frac{1-\varepsilon}{\varepsilon^3} \right)$ (3) $f_k = \alpha + \beta (1+\varepsilon) / (D_p G/\mu)$ Initial Conditions: At $Z = 0$, $n_{ja} = n^{\circ}_{ja}$, $T = T_o$, $P = P_o$ Suggested DE solver: LSODE	(1) Use for most adiabatic reactors and for nonadiabatic with modest ΔT excursions. (2) Constraints: (a) Negligible interphase gradients (b) Use effective rates based on bulk temperature (c) $L/D_p \geq 50$ (d) Catalyst properly packed (e) Eq. 5.30 and 5.31 are criteria for negligible radial gradients (3) Assumptions: $\frac{d\ln\left(\sum_{j=1}^{c} n_{ja} c_{pj}\right)}{dZ} \Big/ \frac{d\ln T}{dZ} \approx$ 0 for all Z
II. One-Dimensional Heterogeneous Model $T_s \neq T$ $(n_{ja})_s = n_{ja}$	$\frac{dn_{ja}}{dZ} = \frac{M_F \rho_b}{GY_{A_o}} \sum_{n=1}^{n_r} \nu_{jn} (\hat{r}_n)_s \quad j = 1,\ldots,c$ (4) $\frac{dT}{dZ} = \frac{h_s a_m \rho_b (T_s - T) - U_Z \frac{4}{D}(T - T_h)}{\frac{GY_{A_o}}{M_F} \sum_{j=1}^{c} n_{ja} c_{p_j}}$ (5)	(1) Use for adiabatic reactors operating at low G's (usually reactors operating near atmospheric pressure). (2) Constraints: (a) Criteria b, c, d, and e above for Model I apply. (b) Criterion (a) above for Model I is violated.

Table 5.1 *(continued)*

Model	*Equation*	*Recommendations*
	$h_s a_m (T_s - T) = -\sum_{j=1}^{c}\left[\left(\overset{\circ}{H}_{f_j\,298} + \int_{298}^{T} c_{p_j}\, dT\right)\sum_{n=1}^{n_r} \nu_{jn}(\hat{r}_n)_s\right]$ $j = 1,\ldots,c$ (6)	(3) Assumptions:
	$\frac{dP}{dZ}$ = same as (3) above for Model I (7)	$\frac{d\ln\left(\sum_{j=1}^{c} n_{ja} c_{pj}\right)}{dZ} \Big/ \frac{d\ln T}{dZ} \approx$ 0 for all Z
	Initial Conditions: At $Z = 0$, $n_{ja} = n_{ja}^{\circ}$, $T = T_o$, $P = P_o$	
	Suggested solver: LSODI or LSODE with separate subroutine for (6).	
III. Two-Dimensional Pseudohomogeneous $T = T_s$ $n_{ja} = (n_{ja})_s$	$\frac{\partial n_{ka}}{\partial Z} - \varepsilon \frac{\mathcal{D}_r \rho_f}{G}\left[\frac{\partial^2 n_{ka}}{\partial r_t^2} + \frac{1}{r_t}\frac{\partial n_{ka}}{\partial r_t}\right] = \frac{\rho_b M_F}{G Y_{A_o}} \sum_{n=1}^{n_r} \nu_{kn} \hat{r}_n$ (8) $k = 1,\ldots,k_n$ if k is a key component unique to a particular reaction in each case	(1) Use for nonadiabatic reactors with high-activity catalyst and/ or large diameter tubes (> 2 inches).
	$\frac{\partial n_{ka}}{\nu_{kn}} = \partial X_n$ and Eq. 8 can be replaced by:	(2) Constraints: (a) a, b, c, and d above in Model I apply, but criterion (e) is violated.
	$\frac{\partial X_n}{\partial Z} - \varepsilon \frac{\mathcal{D}_r \rho_f}{G}\left(\frac{\partial^2 X_n}{\partial r_t^2} + \frac{1}{r_t}\frac{\partial X_n}{\partial r_t}\right) = \frac{\rho_b M_F}{G Y_{A_o}} \hat{r}_n$ $n = 1, n_r$ (8a)	(3) Assumptions for all Z and r_t: $\left(\frac{\partial \ln c_p}{\partial Z}\right)_{r_t} \Big/ \left(\frac{\partial \ln T}{\partial Z}\right)_{r_t} \approx 0$ $\left(\frac{\partial \ln \mathcal{D}_r}{\partial r_t}\right)_Z \Big/ \left[\frac{\partial \ln(\partial X_n/\partial r_t)}{\partial r_t}\right]_Z \approx 0$
	$\frac{\partial T}{\partial Z} - \frac{\lambda_r}{G c_p}\left(\frac{\partial^2 T}{\partial r_t^2} + \frac{1}{r_t}\frac{\partial T}{\partial r_t}\right) = \frac{\rho_b \sum_{n=1}^{n_r} (-\Delta H_n)\, \hat{r}_n}{G c_p}$ (9)	$\left(\frac{\partial \ln \lambda_r}{\partial r_t}\right)_Z \Big/ \left[\frac{\partial \ln(\partial T/\partial r_t)}{\partial r_t}\right]_Z \approx 0$ $\left(\frac{\partial \rho_f}{\partial r_t}\right)_Z \approx 0$
	$\frac{dP}{dZ}$ = same as (3) Above for Model I	
	Algebraic Equations: Equations relating moles for each component to extent of reaction.	

Boundary Conditions:

At $Z = 0$ and $0 \leq r_t \leq R_t$: $X_n = X_n^\circ$, $T = T_o$, $P = P_o$

At $Z \geq 0$ and $r_t = 0$: $\dfrac{\partial T}{\partial r_t} = 0$

At $Z \geq 0$ and $r_t = 0$: $\dfrac{\partial X_n}{\partial r_t} = 0$

At $Z \geq 0$ and $r_t = R_t$: $\dfrac{\partial X_n}{\partial r_t} = 0$

$$-\lambda_r \left(\frac{\partial T}{\partial r_t}\right)_{r_t = R_t} = h_w \left(T_{R_t} - T_W\right)$$

$$= h_t \left(T_{R_t} - T_h\right)$$

NOTE: λ_r, h_w, and h_t are related to the overall heat-transfer coefficient (U_z) used in the one-dimensional model:

$$h_w\left(T_{R_t} - T_W\right) = h_t\left(T_{R_t} - T_h\right) = U_Z\left(T_m - T_h\right)$$

where T_m is a mean temperature of the bed at a given axial position

Suggested Solution Method: Use collocation method to approximate radial composition and temperature profiles and to convert partial differential equations into a set of ordinary differential equations (ODE's). Then solve ODE's numerically using standard ODE solver. See Refs. 1 and 2 for details.

Table 5.1 *(continued)*

Model	*Equation*	*Recommendations*
IV. One-Dimensional Pseudohomogeneous (Unsteady State) $T = T_S$	Use Eq. 5.32A and B with terms 3 and 4 omitted in each case: $\varepsilon \left(\frac{\partial C_j}{\partial t}\right)_Z + \left(\frac{\partial u_s C_j}{dZ}\right)_t = \alpha\ \rho_b \sum_{n=1}^{n_r} \nu_{jn}\ \hat{r}_n \quad j = 1,...c$ (11) or substitute $\frac{G_o Y_{A_o}}{M_F} \left(\frac{n_{ka} \rho_f}{G}\right)$ for C_j, and G for $\rho_f u_s$. $\varepsilon \frac{G_o Y_{A_o}}{M_F} \left[\frac{\partial (n_{ka} \rho_f / G)}{\partial t}\right]_Z + \frac{G_o Y_{A_o}}{M_F} \left(\frac{\partial n_{ka}}{\partial Z}\right)_t = \alpha\ \rho_b \sum_{n=1}^{n_r} u_{kn}\ \hat{r}_n$ $k = 1,...k_n$ (12) $(\rho c_p)_{fs} \left(\frac{\partial T}{\partial t}\right)_Z + \frac{\partial (G c_p T)}{\partial Z}_t = \alpha\, \rho_b \sum_{n=1}^{n_r} (-\Delta H_n)\ \hat{r}_n$ (13) $(\rho c_p)_{fs} = \varepsilon\ (\rho c_p)_f + (1 - \varepsilon)\ (\rho c_p)_s$ where f refers to fluid and s to solid $\frac{d\alpha}{\partial t} = - r_d$, the deactivation rate in t^{-1}. α = activity factor $\leqq$ 1.0. Boundary conditions: At Z = 0, $n_{ka} = n^{\circ}_{ka}$ (t), $T = T_o^{(t)}$, and $P = P_o^{(t)}$ for all t. At t = 0, $n_{ka} = n^{\circ}_{ka}$ (z) for all $Z \geq 0$. At t = 0, $\alpha = 1.0$ for all $Z \geq 0$. Suggested Solution Method: Use collocation method to appropriate axial composition and temperature profiles and to convert partial differential equations to ordinary differential equations (ODE's). Then solve ODE's numerically, using standard ODE solver.	(1) Use for operating model and for predictive control with deactivating catalyst. (2) Constraints: (a) a, b, c, and d above in Model I apply. (3) Assumptions: $\frac{\partial \ln (\rho c_p)_{fs}}{\partial t} / \frac{\partial \ln T}{\partial t} \approx 0$ for all t and Z

V. One Dimensional Model with Negligible Temperature Gradient in Catalyst and Negligible Interfacial Concentration Gradient; but Significant Concentration Gradients in Catalyst.

$\mathcal{D}_{I_j}$ = cm^3 fluid/ (cm particle) sec

a_m = exterior area of catalyst per unit mass

ρ_b = bed density

ρ_f = fluid density

Fluid:

$$\frac{GY_{Ao}}{M_F}\frac{\partial n_{ja}}{\partial Z} = -\mathcal{D}_{I_j}\, a_m\, \rho_b \frac{Y_{Ao}}{M_F}\left(\frac{\partial n_{ja}\,\rho_f}{\partial r_I}\right)_{r_I = R_p}$$

$$G\, c_p \frac{dT}{dZ} + U_Z \frac{4}{D}(T - T_h) = h_s\, a_m\, \rho_b (T_s - T)$$

Solid:

$$\frac{Y_{Ao}}{M_F}\frac{\mathcal{D}_{I_j}}{r_I^2}\frac{\partial\left(r_I^2 \dfrac{\partial n_{ja}\,\rho_f}{\partial r_I}\right)}{\partial r_I} = \rho_p \sum \nu_{jn}\, \hat{r}_n \qquad j = 1,\ldots,c$$

Boundary Conditions for Solid:

$$\text{At } r_I = 0: \frac{\partial(n_{ja}\,\rho_f)}{\partial r_I} = 0$$

$$\text{At } r_I = R_p: -\mathcal{D}_{I_j}\, a_m\, \rho_b \frac{Y_{Ao}}{M_F}\frac{\partial n_{ja}\,\rho_f}{\partial r_I} = \rho_p \sum \nu_{jn}\left(\hat{r}_n\right)_s$$

$$h_s(T_s - T) = -\sum_{j=1}^{c}\left[\left(H^{o}_{f_j\,298} + \int_{298}^{T} c_{p_j}\, dT\right)\left(-\mathcal{D}_{I_j}\frac{Y_{Ao}}{M_F}\frac{\partial n_{ja}\,\rho_f}{\partial r_I}\right)\right]$$

(1) Use for studying catalysts of high activity and involving multiple reactions. Also use for operating units subject to changes in catalyst size, shape, or activity.

(2) Constraints:
(a) Criteria (c) and (d) apply

(b) Criterion (a) is violated and criterion (e) is satisfied.

(3) Assumptions:

$$\left(\frac{\partial \ln c_p}{\partial Z}\right)_{r_I} \Bigg/ \left(\frac{\partial \ln T}{\partial Z}\right)_{r_I} \approx 0 \text{ for all Z's}$$

$$\left[\frac{\partial \ln \mathcal{D}_{I_j}}{\partial r_I}\right]_Z \Bigg/ \left[\frac{\partial \ln r_I^2\left(\dfrac{\partial n_{ja}\,\rho_f}{\partial r_I}\right)}{\partial r_I}\right]_Z \approx$$

0 for all r_I's.

reactor having one tube can readily be scaled up to a multiple-tube reactor by simpling dividing the required production by the production for the single tube. The same mass velocity would be maintained in both. Modeling often becomes necessary, however, if for some reason, such as space limitations, the pilot reactor is made quite short so that low mass velocities are required. Although scale-up at constant $L/\mathbf{G}$ would seem reasonable, heat transfer would improve at higher mass velocity. In such cases a model with heat-transfer data generated from the pilot-plant studies would be required.

In contrast to the nonadiabatic reactor, the adiabatic reactor for a gaseous system is more reliably designed from bench- or pilot-scale reactor data using a model based on reasonable kinetics. Smaller-scale reactors cannot be operated at the same high mass velocity as that practical for the commercial reactor, and scale-up is not certain. Good kinetic data obtained independent of significant transport gradients is ideal for design. The additional thermodynamic data required are for gases, most accurate and readily available. An operating commercial reactor is most appropriately scaled up at constant $L/\mathbf{G}$ without further effort.

Rules and comments for these valuable techniques of scale-up are summarized in Table 5.2.

Table 5.2 Scale-Up Rules: Gas-Phase Fixed Bed Reactors

Reactor Type	*Scaling Rules (Relation between Small and Large Sizes)*	*Constraints*
Adiabatic	• Dynamic Similarity same G and D_p • Chemical Similarity same $\frac{L}{G}$ and temperature same L and inlet temperature same catalyst size and type • Reactor Diameter: $D = \left(\frac{4W_F}{G\pi}\right)^{0.5}$ where D = inside diameter of reactor W_F = feed flow rate in mass/time	• Proven methods for assuring good distribution in bed at inlet and outlet must be used. $\frac{D}{D_p} < 30.$
Nonadiabatic	• Dynamic Similarity same G, D_p, and D	• Each tube must be packed in identical fashion so that ΔP

Table 5.2 *(continued)*

Reactor Type	*Scaling Rules (Relation between Small and Large Sizes)*	*Constraints*
	• Chemical Similarity same $\frac{L}{G}$ and temperature same L and inlet temperature • Thermal Similarity same heat-transfer media temperature and properties • Number of Tubes: $N_t = W_p/W_t$ where N_t = number of tubes W_p = total desired production in mass/time W_t = production/tube in mass/time (tube)	will be the same across each tube, thereby assuring equal distribution among the tubes. • Temperature of heat-transfer fluid must ideally be essentially constant, a goal reachable with boiling water reactors or high rates of circulation of heat-transfer fluid. • Many high-temperature endothermic reactors require heating by hot gases flowing transverse to tubes, and low heat capacity means large changes in gas temperature. Design multipass exchange for gas to assure close to constant T at any particular axial position [17]. This axial gas temperature profile must be mimicked in the single-tube pilot reactor.

MODEL PARAMETERS AND SCALE-UP FROM BENCH-SCALE, PILOT-PLANT, AND PLANT DATA

Procedures and recommendations for obtaining model parameters or scale-up rules from bench-scale, pilot-plant, or plant data are summarized in Fig. 5.8, which provides a condensed summary of the major issues. The diagram begins with the request to select the desired commercial reactor type. The reader can then follow the appropriate path to the desired conclusion.

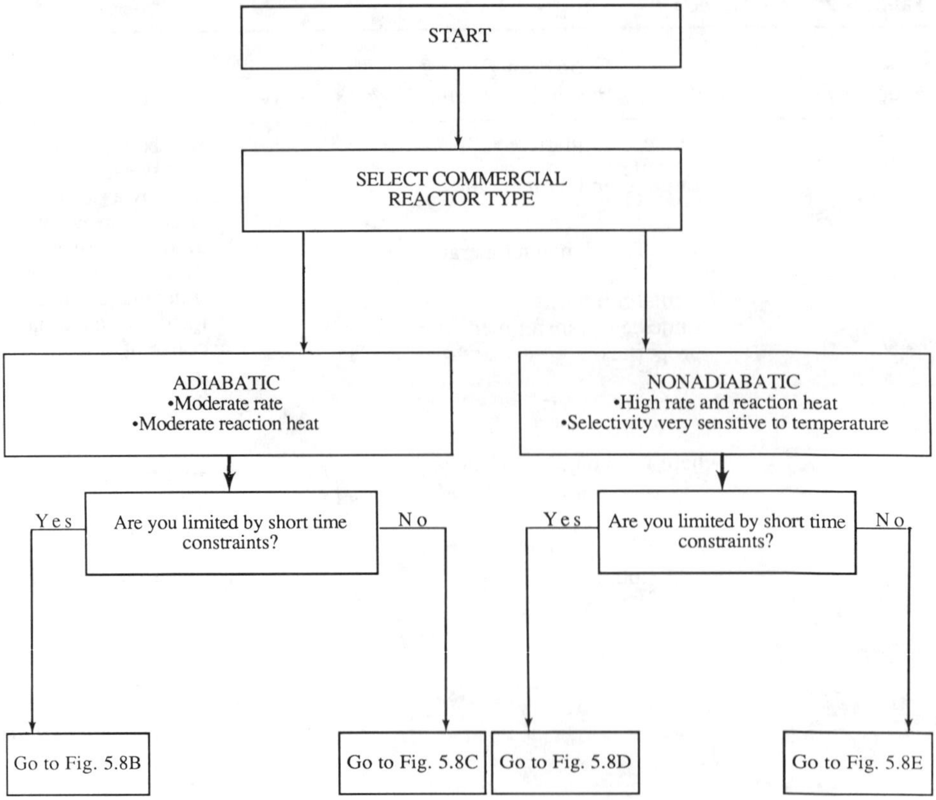

(A)

Figure 5.8 Strategy for determining rate forms and model parameters or scale-up roles using bench-scale, pilot-plant, or plant data
[Follow blocks and arrows that reflect the particular issues of interest.]

Inevitably all calculations involved in designing reactors or analyzing operating reactors require compromises imposed by the realities of time and economics. Ideally, an accurate and well-tested model capable of predicting reactor performance over a wide variety of conditions provides the most potent tool for analyzing all kinds of alternate designs and operating schemes. Unfortunately, such an ideal is not often realized. For new processes with great promise it is often not feasible to budget the necessary experimental laboratory hours for developing reliable kinetic data, particularly if a number of reactions occur simultaneously and consecutively. Although modern software for optimization and statistical evaluation makes data analysis a readily accomplishable task, (see Example 11) the amount of precise experimental data required for complex reaction systems and the time to gather such data are often difficult to justify when compared to the more urgent need to operate pilot-plant tests that are designed to seek conditions for optimum yield and product quality.

From Fig. 5.8A

↓

ADIABATIC PILOT PLANT

- Find G at which $T \approx T_S$. Increase G up to point beyond which outlet composition shows no further change at constant L/G, where L is bed length. If not possible, do not deviate from selected G.
- If required G causes L to be too long for pilot-plant unit, divide into sections. To assure adiabatic operation may need separately controlled zone heaters.
- Determine best operating range from equilibrium and adiabatic temperature plot (see p.). This plot will also indicate need for staging. If required, provide interstage exchangers.
- Run temperature studies at reasonable pressure within thermodynamically favorable region.
- Determine region of highest yield by varying temperature and feed composition.
- If new catalyst, operate at least 30 days to determine life characteristics.
- Scale up (see p.)

↓

MODELING

Even though main emphasis is on rapid scale up, some effort should be placed on developing at least a rudimentary model. For this purpose power-law equations are preferred. Obtain isothermal data on a small scale using a Berty reactor or a differential reactor operating with a very small temperature difference across the bed. See Table 5.4

↓

Use rate forms and parameters thus obtained to model pilot reactor. Adjust if necessary for improved fit. (see Fig. 5.8F)

(B)

Figure 5.8 *(continued)*

The extreme sensitivity of reaction systems to any parameter affected by temperature creates a formidable challenge to even the most seasoned experimentalist with a goal of obtaining precise data. An error of 5% in the value of activation energy, for example, typically can cause an error in the calculated rate of reaction of 75%, an obviously unacceptable deviation from the correct value. Dealing with the possibility of such uncertainties requires a fail-safe approach based on intelligent choice of experimental procedures, as described in Fig. 5.8, that use the appropriate microreactors or pilot reactors (Tables 5.3 and 5.4).

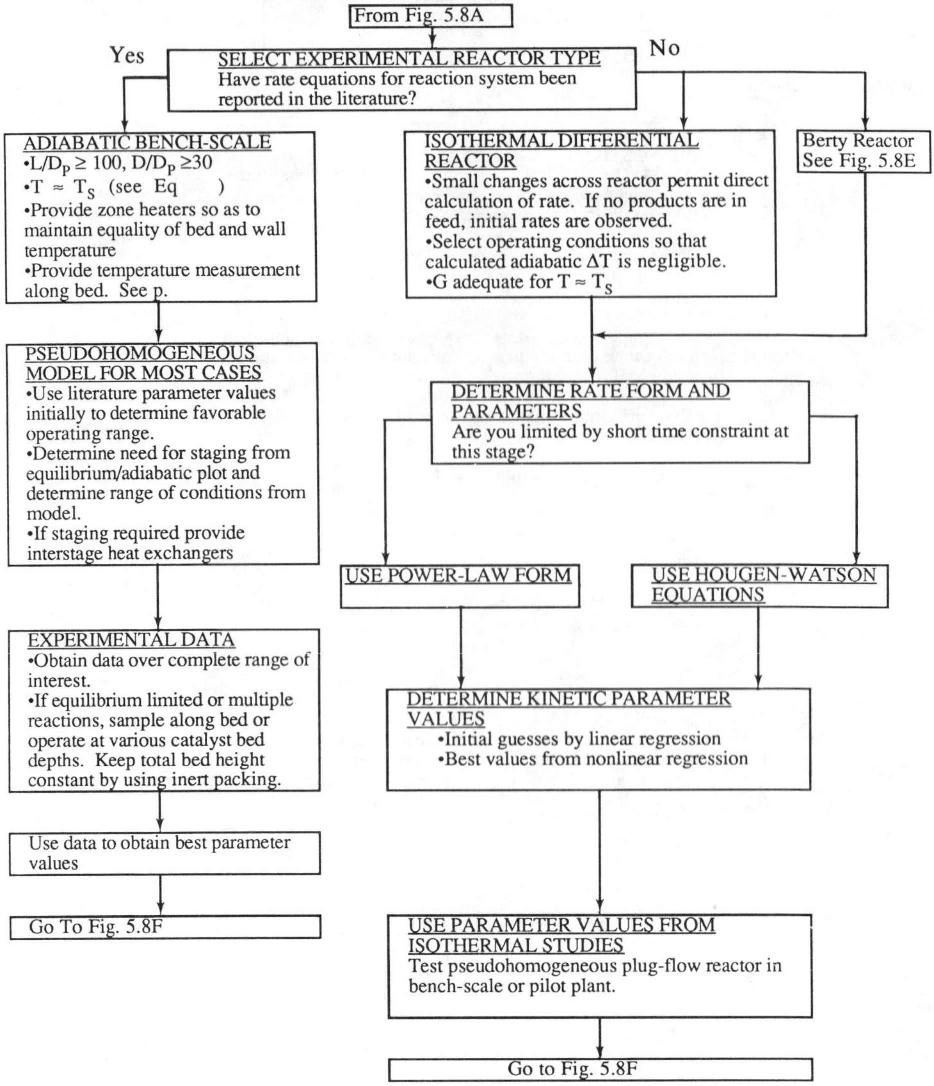

(C)

Figure 5.8 *(continued)*

Experimental data must be taken over a range of partial pressures and temperatures if reliable models are sought. Planning of such experiments is facilitated by factorial design of the data gathering process. Data limited to inlet and outlet measurements for nonadiabatic systems miss the really sensitive and important region of highest rates. This kind of data is commonly produced in pilot plants that can reliably be scaled; but if a request for a model is anticipated at a later date, some effort should be given to obtaining

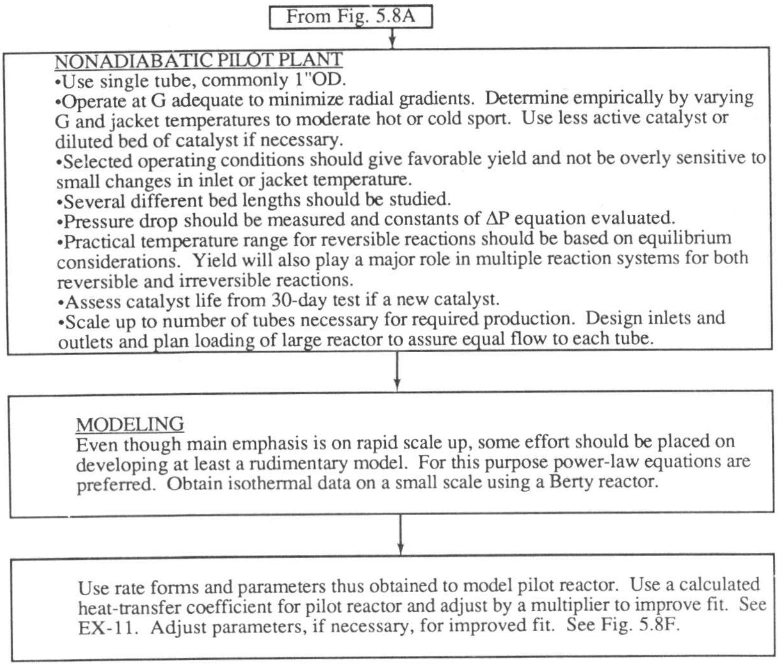

(D)

information in the more active regions. Such data can be obtained by operating with different amounts of catalyst in the reactor while maintaining a mass velocity adequate to avoid significant transport gradients or by providing for intermediate temperature and composition measurements in the pilot reactor.

The strong correlation between A and E for the rate constant that often exists causes further problems in optimal search routines for seeking best-fit values of these parameters from nonisothermal data. The following reparameterization has been suggested [19] as a means for alleviating this problem, and it has been applied successfully in many situations:

Let

$$A'' = A \exp\left(\frac{-E}{RT_m}\right)$$

where T_m is the mean temperature in the reactor or in a portion of the reactor if the reactor is divided into sections. More generally, it can be selected as a convenient mean in the temperature range of interest. Hence,

$$k_p = A'' \exp\left(\frac{E}{RT_m} - \frac{E}{RT}\right)$$

where T is the temperature at any given position in the bed.

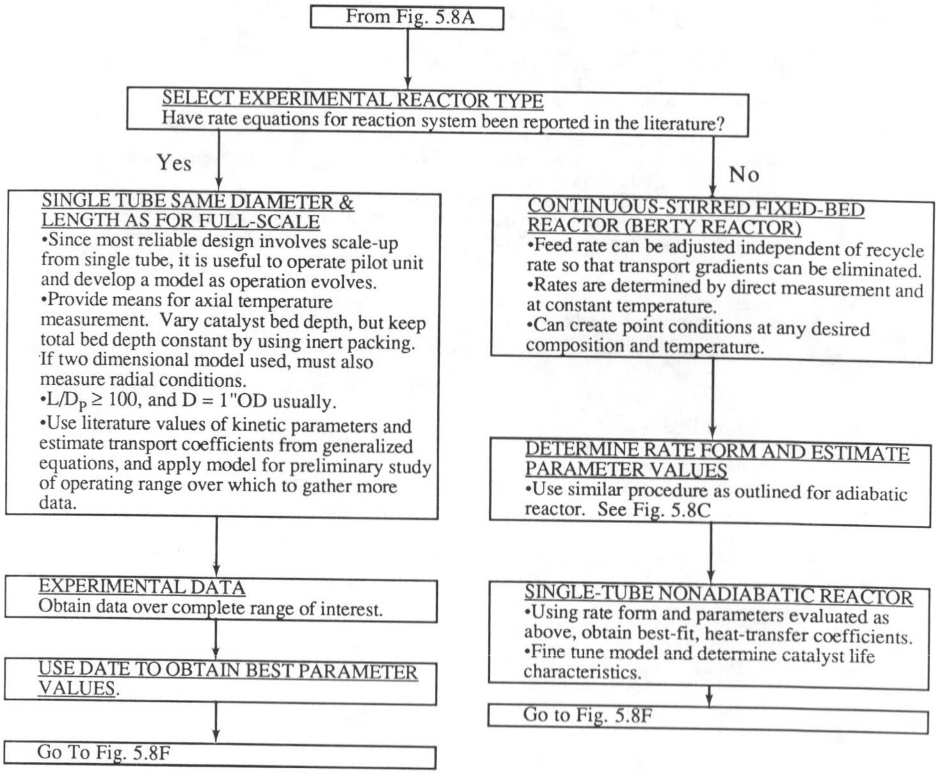

(E)

DETERMINATION OF OPTIMUM PARAMETER VALUES

- Obtain data over full operating range
- Use preliminary values of kinetic parameters from literature or from isothermal reactor studies combined with calculated transport coefficients, if required, as initial guesses for the model. Employ optimization code to find the "best-fit" parameter values that will minimize the error between the appropriate reactor-model output and that of a pilot reactor operated at desired commercial conditions.
- Obtain valid statistical information to enable selection of the most likely parameter values and, if several models are considered, the best model.
- Determine sensitivity of model to each parameter value.
- Plan additional experiments for model improvement.
- As model becomes more accurate by this process use it to optimize the process conditions to maximize conversion for a single reaction and yield for a complex reaction system within necessary constraints of operability, safety, and product quality.
- Test optimum conditions thus calculate in pilot reactor and refine model further if required.
- If new catalyst, do 30-day life test.

Suggested Software

Model Selection and Parameter Estimation:
SIMUSOLV™, the Dow Chemical Company, Midland, Michigan, 1987,1988.
Experimental Design:
ECHIP™, ECHIP, Inc., Hockessin, Delaware

(F)

Figure 5.8 *(continued)*

Table 5.3 Experimental Microreactors for Model Discrimination and Parameter Evaluation

Type	*Description*	*Advantages*	*Disadvantages*	*Rating*
M-1. Differential	Small amount of catalyst in a temperature-controlled environment (radiant heater or sand bath). High flow rates minimize transport resistances and also cause small extents of reaction and ΔT's such that rates may be determined directly as $\Delta X_n/W/F_A$ at the average of inlet and outlet temperatures.	• Easy to construct using standard fittings as reactor [18]. • If reaction extent is low, temperature uncertainty is low, and a single thermocouple in the inlet stream along with one for the reactor wall are adequate. • Good for quick comparisons of various catalysts using initial rates as directly measured.	• For reliable parameter values, need rates over entire concentration range, which makes this procedure tedious, since products must be added to the feed. • Small amount of important components that affect activity may not be known. Recycle of product is required for this case, which is tedious and better handled in Type 3, where it is done more easily.	C
M-2. Isothermal Fixed Bed	Tubular reactor with $L/D_p \geq 100$ and $D/D_p \geq 10$ to assure plug flow and good distribution, to be operated in a temperature controlled environment (radiant heater or sand bath). Isothermal operation possible with moderate reactions with low ΔH_n, especially when catalyst diluted with inert packing.	• Possible to cover full range of composition variables by operating with different flow rates above the minimum required for minimizing transport gradients; or amount of catalyst can be changed while keeping bed height and G constant by means of inert packing.	• Limited to slow to moderate and/or low ΔH_n reactions, or reactions of small amounts of reactants in large quantities of nonreactants.	C

Table 5.3 *(continued)*

Type	*Description*	*Advantages*	*Disadvantages*	*Rating*
M-3. Adiabatic Fixed Bed	Tubular reactor with $L/Dp \geq 10$ to assure plug flow and good distribution, to be fitted with zone-controlled heaters so that each portion of reactor tube can be controlled to the same temperature as the surrounding temperature environment.	• Simple to construct. • If reactor truly adiabatic, temperature is a known function of conversion or composition for a single reaction, and temperature change can be used to obtain rates. This approach avoids analytical problems, but it is not accurate for very rapid or very slow reactions. • For multiple reactions, even though the reactions are not known, the adiabatic reaction temperature at a point calculated from the analysis of the fluid at that point should agree with the measured temperature. • Well-measured temperature profile, along the bed and outlet composition for a check against outlet temperature, provides a wealth of data from one run.	• Difficult to construct suitable reactor and control system.	B

M-4. Recirculating Fixed Bed (Figs. EX-11.1 and EX-11.2)	• Externally Circulated: An ordinary fixed-bed microreactor with compressor for circulating output in a ratio of 25:1 or more to the feed. • Internally Circulated (Berty Reactor): Internal blower circulates reaction mix over fixed-catalyst bed at a high ratio of circulating fluid to feed. • Both behave as CSTR and provide rates as direct measurements. Outlet compositions can be varied by changing feed rate.	• When properly operated, can provide the best measured rate data for any type of gaseous reaction system at known conditions • Excellent for rapid reactions with high heat effects, since transport gradients can be eliminated by high impeller speeds.	• Time to reach steady state for a given run can be long. • Berty reactor limited to pressure of 2 atm or above. At lower pressure, gas density too low for internal impeller to operate effectively.	A

Table 5.4 Pilot Reactors for Scale-Up

Type	*Description*	*Comments*
P-1. Nonadiabatic Pilot Reactor	Use one tube of same diameter and length as for full-scale reactor. Usually 1" OD (2.5 cm) with 0.24 cm wall thickness. D/D_p = 10-20. $L/D_p \geq$ 100. Side entering bare thermocouples installed at various axial positions with junction bent toward direction of flow.	• Required length for pilot reactor estimated from smaller scale studies using same L/G. • Determine sensitive operating regions by varying inlet and jacket temperatures. If rapid exothermic reaction, place reactor within suitable barricade.
P-2. Combination Nonadiabatic Pilot Reactor and Detailed Data Gathering Unit	Reference 1 describes an elegant design that provides an opportunity to measure axial and radial temperatures and compositions so that a model can be developed simultaneously.	• All model parameters including heat transfer coefficients can be determined simultaneously. This procedure, however, may distort kinetic and heat transfer parameters, although the fit may be good in region of interest. • Difficult to construct and often not justified for commercial development work.

P-3. Adiabatic Reactor	Use a well-insulated tube, usually 2-3" (5.0 - 7.6 cm) in diameter, with a length in the range expected for the full-scale unit. This approach assures mass velocities in the same range as in the full-scale unit.	• Because of the large length usually required, the reactor would case a space problem in the pilot-plant area. Often, therefore, a much shorter reactor is selected. • Short reactors, though convenient, seldom mimic full-scale operation because lower mass velocities are necessary to maintain the same L/G. Low velocities increase mass and heat-transfer resistances, which alter the effective rate. • Low velocities and small diameter tubes invariably cause bypassing. • At times a reactor efficiency can be formulated to correct results from small-scale units to full-scale performance.

The model would be solved by using the measured profile and initial conditions. The best values of A'' and E would be determined, from which the best value of A itself could be readily evaluated. Obviously, accurate axial temperature measurements are needed for any evaluation procedure for a nonisothermal reactor. Unfortunately, the convenient axial thermocouple well with movable thermocouple is one of the least accurate methods. The well itself causes fluid bypassing along its walls and temperature measurement errors due to heat loss along the tube. Bare thermocouples inserted radially at various axial positions are preferred for improved accuracy. Bending the thermocouple parallel to the flow minimizes flow-pattern disturbances at the point of measurement.

The various data analyses discussed here and shown in Fig. 5.8 are illustrated in Example 11. A software package called SIMUSOLV™, which was developed especially for producing reliable models from experimental data, is recommended.

OPERATING MODELS

Although most commercial reactors in use today were probably designed by scale-up methods, improvements in operation and troubleshooting have been greatly facilitated through the use of kinetic models, the parameters for which were obtained from the full-scale operating unit. Improvements in optimal control will also depend greatly on such operating models that must take account of the unsteady state, including catalyst activity decline. Such equations are included in Table 5.1 under the heading "Operating Model."

Again, precise measurements of reaction temperatures and compositions of the reacting mixture in the commercial unit are essential. In particular, thermocouples must be properly selected, placed, and maintained. The original plant design should include instrumentation for developing and using an operating model for control and process improvement.

LITERATURE CITED

1. Hoffmann, H., *Ger. Chem. Eng.*, *2*, 258 (1979).
2. Kubicek, M., Hlavacek, V., Holodniok, M. *Chem. Eng. Sci.*, *31*, 727 (1976).
3. Rase, H.F., "Chemical Reactor Design for Process Plants," Wiley, New York, 1977.
4. Lee, H.H., "Heterogeneous Reactor Design," Butterworths, Boston, 1985.
5. Hlavacek, V., and Votrubz, J., *Sci. Papers Prague Inst. Technol.*, *K11*, 143 (1976).

6. Petrovic, L.J., and Thodas, G., *Ind. Eng. Chem. Fundam.*, *7*, 274 (1968).
7. Mears, D.E., *Ind. Eng. Chem. Process Des. Develop.*, *10*, 541 (1971).
8. Aries, R. "Elementary Chemical Reactor Analysis," Prentice-Hall, Englewood Cliffs, N.J., 1969.
9. Bischoff, K.B., and Froment, G.F., "Chemical Reactor Analysis and Design," Wiley, New York, 1979.
10. Weisz, P.B., and Prater, C.D., *Adv. Catal.*, *6*, 143 (1954).
11. Carberry, J.J. *Ind. Eng. Chem. Fundam.*, *14*, 129 (1975).
12. Kelhoe, J.P.G., and Butt, J.B., *AIChE J.*, *18*, 347 (1972).
13. Carberry, J.J., and Wendel, M.M., *AIChE J.*, *9*, 129 (1963).
14. Finlayson, B.A., *Ind. Eng. Chem. Fundam.*, *12*, 412 (1973).
15. Schwartz, C.E., and Smith, J.M., *Ind. Eng. Chem.*, *45*, 129 (1953).
16. Stull, D.R., Westrum, E.F., Jr., and Sinke, G.C., "The Chemical Thermodynamics of Organic Compounds," Wiley, New York, 1969.
17. Eigenberger, G., and Ruppel, W. *Ger. Chem. Eng.* *9*, 74 (1986).
18. Rase, H.F., *Ind. Eng. Chem.*, *57*(1), 18 (1965).
19. Van Damme, P.S., Narayanan, S., and Froment, G.F., *AIChEJ*, *21*, 1065 (1975).

CHAPTER 6

Detailed Recommendations on Reactor Design Strategy

Although, as the time for design of a reactor approaches, experimental and pilot-plant studies may have already indicated rather definite ranges of desirable operating conditions, design strategy should involve an orderly consideration of all important variables and issues. Only in this manner will the resulting design attain the important goals of operability, reliability, and economic viability. It is not uncommon for development programs to be restricted by certain equipment limitations not relevant to full scale but that preclude study of certain ranges of variables that may prove ideal. Systematic analysis will avoid such pitfalls. The following outline provides a guide for such a systematic approach that is useful not only for design using a reaction model but also in planning pilot-plant studies and scale-up.

I. Range of Operating Conditions to Be Studied
   A. Inlet (operating) pressure
      1. Consider elevated pressure for following situations:
         a. Reaction is reversible and produces a decline in moles. Although detailed study may be necessary in some cases, the equilibrium calculations will show the advantage of a higher pressure that can be compared with incremental energy costs and product credits.
         b. At least one gaseous feed component is already available at an elevated pressure, so gas compression costs will be lower if that pressure is used as reactor pressure.
         c. The main reaction is slow, and equipment size can be confined to a practical range by elevated pressure.
         d. Desired reaction is higher order than undesired.
      2. Consider minimum absolute pressure (atmospheric pressure plus pressure drop from inlet of reactor to first pump or compressor) for following situations:
         a. Reaction is reversible and produces an increase in moles.
         b. Large amount of a feed component such as air must be compressed from atmospheric conditions to reactor pressure.
         c. Desired reaction is lower order than undesired.

3. Select a narrow pressure range for more detailed study. Laboratory studies or previous experience may have already set this range.

B. Inlet composition
   1. Use excess of one reactant for following situations:
      a. Desired reaction is higher order than undesired.
      b. A very close approach to equilibrium is required for a reversible reaction. Although the excess reduces the rates in the early part of the reactor, the rates in the high conversion region are favorably increased, so the overall length of the bed is minimized.
   2. Add inerts for following situations:
      a. Suppress side reactions in some cases.
      b. Reduce rapid temperature decline in adiabatic reactors involving endothermic systems.
      c. Control highly exothermic adiabatic reactions when other options are not feasible.

C. Product manipulation
   1. Remove product between stages to increase conversion for a reversible reaction.
   2. Recycle products of undesired reversible side reactions in order to suppress the side reactions.

D. Inlet temperature
   1. Operate above dew point for gaseous reactions.
   2. Select highest temperature for endothermic or irreversible exothermic reactions that will permit operation without exceeding 25°C less than the maximum catalyst use temperature or maximum vessel design temperature (see Tables 6.1 and 6.2 and Fig. 6.1), whichever is lower. Maximum temperature change will occur for adiabatic case, which can be read from the plot described in I E. The 25°C design factor allows for raising the temperature to compensate for activity decline during the life of

Table 6.1 Practical Temperature Limits for Steel Vessels

	Temperature	
	°*F*	°*C*
Carbon steel (C–Si)	850	455
Low alloy (C–0.5Mo)	1000	538
(2.25Cr–1Mo)	1050	566
High-tensile steels for heavy-wall reactors (Mn–0.5Mo)	1000	538
High alloy		
(18Cr–8Ni) with special heat treatment	1200	649
(16Cr–12Ni–2Mo)	1200	649

See Fig. 6.1 for special operating limits in presence of hydrogen.
See Table 6.2 for values of allowable stress for typical reactor steels.

Table 6.2 Selected Plate Steels for Reactor Fabrication

ASTM Designation	Nominal Composition	-20 to 650	700	750	800	850	900	950	1000	1050	1100	1150	1200
		Allowable Stress, Kips/in^2. (ASME) Code - Sect. VIII, Div. 1, 1986) Temperature, °F											
Carbon steels													
A-285 Gr. C.	C-Si	13.8	13.3	12.1	10.4	8.3	6.5	See Note 1 a, b & C.					
							(for intermediate and high temperature service)						
A-515 Gr. 70	C-Si }	17.5	16.6	14.8	12.0	9.3	6.5	4.5	2.5	See Note 1(a).			
							(for temperature service requiring notch toughness)						
A-516 Gr. 70	C-Si }												
Low-alloy steels for resistance to H_2 and H_2S (See Fig. 6.1)													
A-204 Gr. A	C-0.5Mo	16.3	16.3	16.3	15.8	15.3	13.7	8.2	4.8	See Note 2			
Gr. B	C-0.5Mo	17.5	17.5	17.5	17.5	17.1	13.7	8.2	4.8	See Note 2			
Gr. C	C-0.5Mo	18.8	18.8	18.8	18.8	18.3	13.7	8.2	4.8	See Note 2			
A387 Gr. 12, C1.1,	1Cr-0.5Mo	13.8	13.8	13.8	13.8	13.4	12.8	11.0	6.6	4.3	2.6	1.4	1.0
A387 Gr. 22, C1.2,	2.5Cr-1Mo	See Code	17.2	17.2	16.9	16.4	15.8	11.0	7.6	5.8	4.4	2.5	1.3
C1.1,	2.25Cr-1Mo	1.50	15.0	15.0	15.0	14.4	13.1	11.0	7.8	5.8	4.2	3.0	1.6
A387 Gr. 5, C1.1,	5Cr-0.5Mo	See Code	13.7	13.2	12.8	12.1	10.9	8.0	5.8	4.2	3.0	2.0	1.3
High tensile steels for heavy-wall reactors													
A302 Gr. B	Mn-0.5Mo }	20.0	20.0	19.6	18.8	17.9	13.7	8.2	4.8				
Gr. C	Mn-0.5Mo-Ni }												
A533 Gr. A, C1.1,	C-Mn-Mo }	20.0	20.0	20.0	19.1	16.8	13.3	10.0	6.3				
Gr. B, C1.1,	C-Mn-Mo., 0.4-0.7 Ni }									(Quenched and tempered all grades)			
Gr. C, C1.1,	C-Mn-Mo., 07.-1.0 Ni }	20.0	20.0	20.0	19.1								

A542, C1.1 (Same analysis as A387 Gr.22 C1.2 but quenched and tempered. Increases tensile strength from 75 to 105. Allowable stresses listed only under Div. 2 Rules.)

Table 6.2 *(continued)*

ASTM Designation	*Nominal Composition*	*-20 to 650*	*700*	*750*	*800*	*850*	*900*	*950*	*1000*	*1050*	*1100*	*1150*	*1200*
			Allowable Stress, Kips/in^2. (ASME) Code - Sect. VIII, Div. 1, 1986) Temperature, °F										
High-alloy steels for cladding and internal shells for corrosion resistance													
										See Notes 4 and 5			
A240 Gr. 304	18Cr-8 Ni	18.8@ -20°F 11.2 @ 650°F	11.1	10.8	10.6	10.4	10.2	10.0	9.8	9.5	8.9	7.7	6.1
								(4.7 @ 1250F to 1.4 @ 1500F)					
A240 Gr. 304L	18Cr-8 Ni	15.7 @ -20F 9.5 @ 650°F	9.4	9.2	9.1	(Type 304 and 304L for general corrosion resistance)							
				See Note 4									
										See Notes 4 and 5			
A240 Gr 316 (See Note 6)	16Cr-12N-2 Mo	18.8 @ -20°F 11.6 @ 650°F	11.3	11.2	11.0	10.9	10.8	10.7	10.6	10.5	10.3	9.3	7.4
								(5.5 @ 1250°F to 1.2 @ 1500°F)					
A240 Gr 316L (See Note 3)	16Cr-12Ni-2 Mo	15.7 @ -20°F 9.2 @ 650°F	9.0	8.8	8.6	8.4	(Type 316 and 316L for more severe corrosion resistance) See Note 4						
										See Notes 4 and 5			
A240 Gr 321	18Cr-10Ni-Ti	18.8 @ -20°F 11.2 @ 650°F	11.0	10.8	10.8	10.7	10.6	10.6	10.4	9.2	6.9	5.0	3.6
								(2.6 @ 1250°F to 0.3 @ 1500°F)					
		(Stabilized to prevent carbide precipitation in 800 - 1500°F range)											
										See Notes 4 and 5			
A240 Gr 347	18Cr-10Ni-Cb	18.8 @ -20°F 13.2 @ 650°F	12.9	12.8	12.7	12.7	12.6	12.6	12.5	11.9	9.1	6.1	4.4
								(3.3 @ 1250°F to 0.8 @ 1500°F)					

			(Similar to 321)											
A240	Gr 405	12Cr-A1	15 @ -20°F 12.3 @ 650°F	12.1	11.7	11.1	10.4	9.7	8.4	4.0				
				See Note 7 (better weldability and ductility than 410; good for reactor tubing, SA268-TP405)										
A240	Gr 410S	13 Cr	15.0 @ -20°F 12.2 @ 650°F	12.1	11.7	11.1	10.4	9.7	8.4	6.4	4.4	2.9	1.8	1.0
				(used in high-pressure reactor tubing, SA268-410) See Note 8										
A240	Gr 430	17Cr	16.2 @ -20°F 13.2 @ 650°F	13.1	12.7	12.0	11.3	10.5	9.2	6.5	4.5	3.2	2.4	1.8
				(resistant to highly oxidizing atmospheres) See Note 7										

1. (a) Temperature above 800°F for prolonged periods may cause graphitization of carbide phase. (b) Use killed steel above 850°F for A-285 GR.C with not less than 0.1% residual silicon. (c) Do not use in thickness above 2 inches.
2. Temperatures above 875°F for prolonged periods may cause graphitization of carbide phase.
3. L types are specified if welding is done without solution heat treatment.
4. Higher stress values for Types 304 and 304L are listed in the Code when slight deformation is acceptable.
5. These stresses apply above 1000°F if C > 0.04%.
6. Type 317, which is even more corrosion resistant than 316, has same stress values. Type 317L has allowable stress values approximately 25% higher than 316L (see Code).
7. Develops embrittlement after service at moderately elevated temperature.
8. Slightly higher stresses for plate allowed for Type 410.

Choice of steels listed based largely on J. F. Lancaster, *Hydrocarbon Process*, **49** (6) 84 (1970). Refer to this reference for comparison with European Codes. Allowable stress from *ASME Boiler and Pressure Vessel Code*, Section VIII, Div. 1, American Society of Mechanical Engineers, New York, 1986. After H. F. Rase, "Chemical Reactor Design for Process Plants," Wiley-Interscience, New York, 1977. Warning: For preliminary estimates only. See Code for design details.

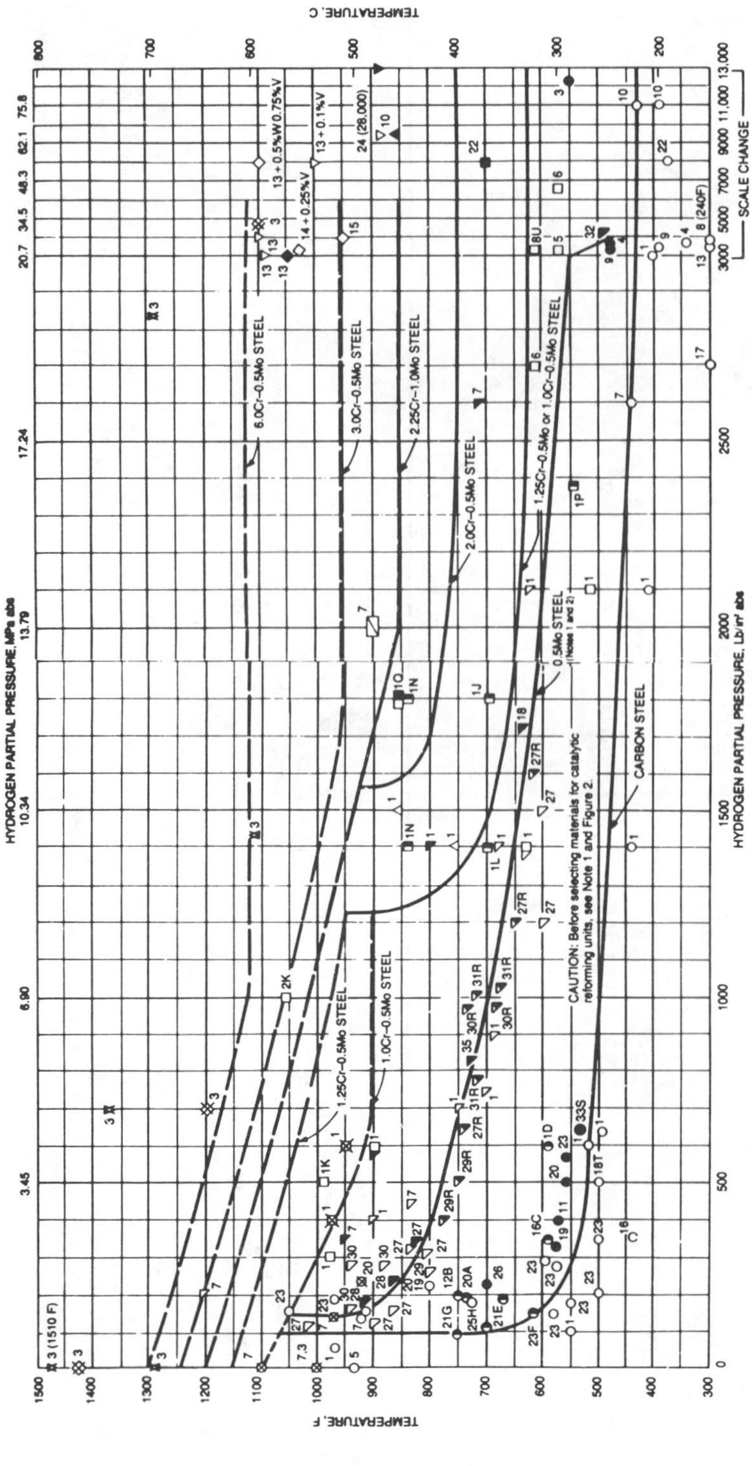
TEMPERATURE, C
TEMPERATURE, F
HYDROGEN PARTIAL PRESSURE, MPa abs
HYDROGEN PARTIAL PRESSURE, Lb/in² abs
SCALE CHANGE
6.0Cr–0.5Mo STEEL
3.0Cr–0.5Mo STEEL
2.25Cr–1.0Mo STEEL
2.0Cr–0.5Mo STEEL
1.25Cr–0.5Mo or 1.0Cr–0.5Mo STEEL
1.25Cr–0.5Mo STEEL
1.0Cr–0.5Mo STEEL
0.5Mo STEEL (Notes 1 and 2)
CARBON STEEL
13 + 0.5%W 0.75%V
13 + 0.1%V
14 + 0.25%V
24 (28,000)
3 (1510 F)
8 (240F)
CAUTION: Before selecting materials for catalytic reforming units, see Note 1 and Figure 2.

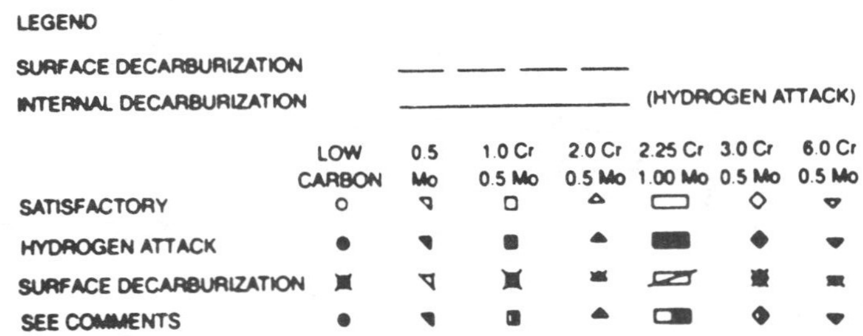

NOTES:

1. After approximately 9–18 years' service, eight occurrences of hydrogen attack have been reported in C-0.5Mo steels operating below the 1977 0.5Mo curve in catalytic reforming units. These steels experienced hydrogen attack under conditions in which extensive decarburization or carbon migration to grain boundaries was not documented or reported prior to 1980. Further use of 0.5Mo steel in catalytic reforming units to resist internal fissuring should be evaluated rigorously until information is available to explain this phenomenon. Refer to 2.3 in the text, Table 1, and Figure 2.
2. The curve for 0.5Mo steel can apply to either C-0.5Mo or Mn-0.5Mo steels.
3. Austenitic stainless steels are generally not decarburized in hydrogen at any temperature or hydrogen pressure.
4. Effect of trace alloying elements (up to 1000 pounds per square inch absolute [6.90 megapascals] hydrogen partial pressure and below 1000 F [538 C]):
   a. Mo has four times the resistance of Cr to hydrogen attack.
   b. Mo is equivalent to V, Ti, or Nb up to 0.1 percent.
   c. Si, Ni, Cu, P, and S do not increase resistance.
5. The limits described by these curves are based on experience with cast steel as well as annealed and normalized steels at stress levels defined by the ASME *Boiler and Pressure Vessel Code*, Section VIII, Division 1. Refer to 2.5 and 2.6 in the text for additional information

Figure 6.1 Operating limits for steels in hydrogen service to avoid decarburization and fissuring.
(Reproduced by permission: API Pub. 941, American Petroleum Institute, 1982; original copyright by G.A. Nelson.)

the catalyst. This is a nominal figure and may be set for the particular catalyst system as follows:

a. Set allowable activity decline: often 50%. Determine temperature rise to compensate

$$\frac{k_{T_2}}{k_{T_1}} = 2 = \frac{Ae^{-E/RT_2}}{Ae^{-E/RT_1}}$$

or

$$\ln 2 = \frac{E}{R}\left(\frac{1}{T_2} - \frac{1}{T_1}\right)$$

b. Special situations may dictate other activity-decline targets.

3. Use a declining temperature profile for exothermic reversible reactions. The reverse reaction becomes more favorable with rising temperature, and a declining profile is optimal. Such a profile can be accomplished by using adiabatic reactors in series with intermediate cooling or by a nonadiabatic reactor with the jacket temperature set at the highest temperature that will provide the desired outlet conversion and yield. The low-rate region toward the outlet of a nonadiabatic reactor requires a major portion of the tube length and, therefore, provides a large amount of heat-transfer area, which ensures a reaction temperature at the outlet very close to the jacket cooling temperature. Again, avoid exceeding design maximum catalyst use temperature recommended by the catalyst supplier or the maximum vessel design temperature (see Tables 6.1 and 6.2 and Fig. 6.1), whichever is lower.
4. Avoid temperatures favorable to undesired side reactions.
5. Use the following approximate criteria as guides. Inlet temperature should be such that the following initial rates are achieved:

Reactor System	*Initial Rate (g moles/g sec)*
Adiabatic, exothermic	0.2×10^{-5} to 0.5×10^{-5}
Adiabatic, endothermic	0.4×10^{-4} to 8×10^{-4}
Nonadiabatic, exothermic	1×10^{-5} to 2×10^{-5}
Nonadiabatic, endothermic	2×10^{-4} to 4×10^{-4}

Often laboratory studies have already suggested inlet temperature ranges, but the laboratory unit may have not been operated at the optimum conditions, and further design studies are needed using a model adjusted to predict correctly the laboratory-reactor results. The initial rate criteria are typical for many practical reactors, although limitations imposed by one or more of the conditions given in the previous parts of this section may be controlling in some cases.

E. Equilibrium compositions and adiabatic reation temperatures
Construct equilibrium and adiabatic reaction-temperature plots. Use the preliminary conclusions from items A through B to set the pressure and temperature ranges. These plots provide an overall view of the behavior of the reaction system. See Example 2.

II. Select Reactor Type

A. Consider equilibrium adiabatic plot as initial basis for selecting reactor type

1. Select adiabatic single bed or multiple bed when necessary in preference to nonadiabatic reactor.
2. Usually three stages is the maximum practical number of adiabatic beds with heating or cooling between stages. Start at inlet temperature suggested by previous experimental work.
3. Step off stages as shown in Example 2. Use the following equation as a conservative approach to equilibrium:

$$\frac{T_{eq} - T_{op}}{T_{eq}T_{op}} = \frac{R}{(-\Delta H_A)} \ln \frac{E + (-\Delta H_A)}{E}$$

where T_{eq} is the equilibrium temperature corresponding to a set composition or conversion, and T_{op} is the optimum adiabatic approach temperature corresponding to that same composition or conversion. This T_{op} is not necessarily the true optimum, since it is possible to approach equilibrium more closely by starting at a higher inlet temperature and effect, thereby, a higher average rate for the process, but this simple calculation provides a quick method for arriving at an approximate value for stepping off stages.

4. Add inerts to feed, if feasible, to moderate temperature change when more than three stages appear necessary to reach desired conversion. In some cases a product may be recycled as a liquid under operating conditions such that it would vaporize along the reaction path.
5. Use heat transfer reactor for reaction systems having heat generation potentials (q_p) exceeding an absolute value of 10.

$$q_p = \frac{(-\Delta H_A)Y_{A_0}}{c_{p_m}} \frac{E}{RT_0^2}$$

(see reference 2, Chapter 5 for derivation)

where Y_{A_0} is the inlet mole fraction of limiting reactant A. If the reaction is equilibrium limited, Y_{A_0} should be multiplied by the expected maximum conversion, which can readily be approximated from the equilibrium adiabatic plot.

B. Consider temperature constraints and effect of temperature change on selectivity and yield
   1. A nonadiabatic reactor may be necessary to provide adequate temperature control if the maximum allowable operating catalyst temperature is close to the inlet temperature for an exothermic reaction.
   2. A goal of high selectivity and yield, if very sensitive to temperature, may be achievable only by means of a nonadiabatic reactor.

III. Determine Minimum Mass Velocity (**G**)

A. Adiabatic reactors—mass velocity
   1. Determine a **G** below which the criterion for minimizing ΔT gradient between catalyst and fluid phase given in Eqs. 5.9 or 5.10 is *not* satisfied.
   2. For multiple beds the first bed is usually controlling.
   3. Inlet values of rate and temperature often control, but not always. In exothermic reactions the maximum rate may occur at the hot spot. For such cases it is necessary to use a model with no interphase gradients in order to determine an approximate value of the maximum rate(s). If another model with gradients already included is available, the most straightforward procedure is to compare the results of the two models and determine the value of **G** at which the difference in calculated rates is approximately 10%.

B. Nonadiabatic reactors—mass velocity
   1. Determine a **G** below which the following criterion for minimizing radial ΔT gradients in the bed is *not* satisfied.

$$\frac{\sum (|\Delta H_n|\hat{r}_n)\rho_p \left(\frac{1-\varepsilon}{1+b_v}\right)\frac{D^2}{4}}{\lambda_r T_w} < 3.2\,\frac{RT_w}{E}$$

   where b_v is the volume of inert solids added per volume of catalyst and ε is the void fraction of the bed. Normally the criterion becomes limiting before the interphase temperature gradient. See Eq. 5.30 for other details.
   2. The controlling region of the bed (maximum rate) usually occurs from 5 to 10 cm before the maximum hot spot. The rate obtained from a one-dimensional model is satisfactory for this initial assessment.
   3. If the criterion is violated only over a small portion of the bed, the one-dimensional model will usually be acceptable.

IV. Evaluate Bed ΔP at Minimum **G** and Select Range for Study

Use the following guidelines in selecting a ΔP range:

A. Bed length

Use a bed length determined from known commercial or pilot-plant space velocities or an $L/\mathbf{G}$ of 500 cm^3 sec/g fluid. A wide variety of practical reactors operate in this range, although variations occur because of operating constraints unique to the process.

B. Normal ΔP range

Commercial units usually operate in a ΔP range of 1 to 10% of the total pressure. The lower range is characteristic for systems involving substantial recycle and/or large flows of feed gas requiring compression to reactor operating condition. Multiple adiabatic beds in series in this category usually involve pressure drops of about 1% per bed. Once-through reactors or those with smaller total flows tend to be designed and operated toward the higher value, 10% of the total pressure. This is especially true of nonadiabatic reactors, where high velocities provide the necessary heat-transfer efficiency for controlling hot spots.

C. Catalyst shape and size

If preliminary ΔP is excessive for the minimum **G**, try larger catalyst, spherical shapes, Raschig-ring forms, or other special shapes designed to lower ΔP. Be certain to check effect of size on rate experimentally. If these changes in catalyst do not reduce the ΔP sufficiently, **G** must be lowered and interphase gradients added to the model for an adiabatic reactor, and radial gradients (two-dimensional model) added for a nonadiabatic reactor. In both cases it must be realized that the models are then much less reliable, especially because of temperature uncertainties created by the introduction of empirical heat-transfer expressions. This uncertainty can be partially overcome by fine-tuning the model using pilot-plant or plant data.

V. Proceed with Detailed Design Study Using Appropriate Model

At this point inlet pressure has usually been set, but it is possible that several inlet pressures require study using differential economics.

A. Previous steps

The previous steps will have set the following:

1. Minimum inlet temperatures for each pressure to be considered
2. One or more feed ratios of interest
3. Maximum design temperature in the reactor
4. Range of mass velocities to consider

B. Pressure

If pressure is to be studied, begin as follows:

1. If equilibrium limited, selected lowest pressure that will provide desired result and lowest excess reactant ratio that is feasible.

2. At constant ratio, study lowest pressure plus several higher pressures to determine economic advantage, if any, of higher-pressure operation.
3. To reduce number of variables, base each case on a design factor of 25°C or other value less than the maximum allowable reactor temperature or a maximum equilibrium approach temperature, whichever is lower.
4. In order to reduce complexity of the economics, design for the same fractional ΔP for each inlet pressure considered. This approach will tend to cause the power consumption for recycle to be the same. The power in any units such as horsepower (HP) is related to the pressures as follows:

$$\mathrm{HP} = f\left[\frac{P_2}{P_1}\exp\left(\frac{n_\mathrm{p}-1}{n_\mathrm{p}}\right)-1\right] = f\left[\frac{1}{1-f_\mathrm{p}}\exp\left(\frac{n_\mathrm{p}-1}{n_\mathrm{p}}\right)-1\right]$$

where P_2 is the reactor inlet operating pressure; P_1 is the reactor outlet pressure; f_p is the fraction of total inlet P_2 lost to friction; n_p is the polytropic exponent.
5. Decision: Based on these pressure studies, select an operating pressure. If pressure is not important, select lowest pressure unless equipment becomes impractically large.

C. Sensitive variable determination

At pressure decided in V B, determine the most sensitive variables. These are usually component ratio in the feed, inlet temperature for an adiabatic reactor, and jacket temperature or heat flux for nonadiabatic reactor.

D. Adiabatic reactors

1. At pressure selected, and starting at minimum temperature and minimum **G** below which gradients will be significant, study operating conditions as follows:

Reversible	*Irreversible*
Based on strategy from equilibrium adiabatic plot, try different ΔT approaches to equilibrium. If beds are staged, approaches can be varied for each stage. Usually, a few cases will show if such changes cause any significant effects. As ΔT approach is changed, inlet T is changed accordingly.	Try several different inlet temperatures.

2. Starting with minimum **G**, vary **G** at the most promising temperatures. $L/\mathbf{G}$ will remain essentially constant at any given inlet T

and conversion. The ΔP will change some with **G** and operating pressure will be affected and could have an effect on selectivity. Avoid **G**'s below which $L/D_p < 100$ so that distribution will be good and back-mixing negligible. The only reason for increasing **G** beyond an operable minimum is to favor reduced equipment size, the advantages of which are opposed by increasing energy costs of feed and recycle compression.

3. Avoid maximum temperatures closer than 25°C (or some other design value) to the maximum allowable operating temperature of vessel, catalyst, or exothermic equilibrium temperature, whichever is less. This temperature difference permits the opportunity of raising temperature to compensate for declining activity. If maximum temperature is governed by vessel design temperature, confine this temperature to that for low-alloy steel if possible (see Tables 6.1 and 6.2 and Fig. 6.1).

E. Nonadiabatic reactor

At pressure selected, develop strategy for minimizing hot spots by studying **G** at minimum value and above.

1. Usually jacket temperature is the controlling variable.
2. If equilibrium is limiting and maximum conversion is required, select a jacket temperature that is close to the desired equilibrium condition based on a feed component ratio previously selected. Vary inlet temperature to observe effect of this variable. Near the range of runaway, inlet temperature is the dominating variable at a constant jacket temperature.
3. For fast reactions consider less active catalyst if necessary in the first portion of the bed followed by a high-activity catalyst in the final half of the bed to increase reaction rate in this normally low-rate region. For some difficult-to-resolve situations it may prove advantageous to use two reactors in series, each with a different jacket temperature.
4. Reject designs with excessive reactor pressure drops. High mass velocity must ensure moderate hot spot without excessive ΔP. Excessive ΔP must be defined in terms of energy costs for a given system.
5. Reject any design condition that proves to be very sensitive to temperature change.

F. Selectivity and yield

In addition to conversion, selectivity and yield are valuable concepts in comparing performance for complex reaction systems.

1. Selectivity of R

$$\mathbf{S}_R = \left(\frac{\text{moles R produced}}{\text{moles A reacted}}\right)_a \div \left(\frac{\text{moles R formed}}{\text{moles A reacted}}\right)_{nsa}$$

where a refers to actual conditions, and nsa to conditions with no side reactions.

or

$$\mathbf{S}_R = \frac{\text{moles A required for R produced if no side reactions}}{\text{moles A actually reacted}}$$

2. Yield of R

$$\mathbf{Y}_R = \left(\frac{\text{moles R produced}}{\text{mole A fed}}\right)_a \div \left(\frac{\text{moles R formed}}{\text{mole A reacted}}\right)_{nsa}$$

or

$$\mathbf{Y}_R = \frac{\text{moles A required for R produced if no side reactions}}{\text{moles A fed}}$$

$$\text{Also, } \mathbf{Y}_R = \mathbf{S}_R X_A$$

where X_A is the conversion of A

3. Example

$$A + 2B \rightarrow 2R + S$$

$$R + B \rightarrow C$$

$$X_{1a} = -(\Delta n_A)_1 = -\frac{(\Delta n_B)_1}{2} = \frac{(\Delta n_R)_1}{2} = (\Delta n_S)_1$$

$$X_{2a} = -(\Delta n_R)_2 = -(\Delta n_B)_2 = (\Delta n_C)_2$$

$$\mathbf{S}_R = \frac{(\Delta n_R)_1 - (\Delta n_R)_2}{(\Delta n_A)_1} \div 2 = \frac{2X_{1a} - X_{2a}}{2X_{1a}}$$

$$\mathbf{Y}_R = [(\Delta n_R)_1 - (\Delta n_R)_2] \div 2 = \frac{2X_{1a} - X_{2a}}{2} = \mathbf{S}_R X_{1a}$$

(Note that in the petroleum-refining industry, where many products are sold by volume, yield is often expressed as volume percent yield:

$$\%\ \text{Yield} = \left(\frac{\text{volume of liquid product @ 60°F}}{\text{volume of liquid feed @ 60°F}}\right) \times 100$$

In addition to yield, a quality standard, such as octane number, is used as a constant operating target.)

G. Economic alternatives

Compare the economics of alternatives with reference to cost of feed, products, reactor, catalyst, and energy in order to select final design.

When substantial compression work is involved, energy costs invariably control. For some products, especially high value-added products, yield often dominates, especially if feed components are costly.

H. Operability

Confirm the operability of the final design by considering various operating possibilities such as

1. How can production be maintained if:
   a. ΔP increases due to partial plugging?
   b. A portion of bed is deactivated or the entire bed is partially deactivated and higher temperature is required?
   c. Heating or cooling heat-transfer rate changes?
2. How might the system be operated at a reduced capacity (turndown)?
3. Sensitivity of reactor to small changes in inlet or jacket temperature. Invariably, if the proposed design conditions cause high sensitivity (large hot spots) when inlet temperature is changed slightly, the design is unsatisfactory and inlet conditions should be changed to overcome this situation.
4. Conduct general "what-if-review," including transient periods and start-up procedures. Give special attention to means for protecting catalyst during start-up and upsets.

I. Safety analysis

1. Safety analysis includes a thorough study of all possible faults that may occur and responsiveness of the system to fail-safe procedures.
2. All process hazards including possible release of toxic materials, flammability limits for process mixtures, and overpressure release systems must be examined.

CHAPTER 7

Shortcut Procedures

As interactive software increasingly becomes the standard tool for process design calculations, the only type of "hand" calculation that remains valuable and crucial is the so-called shortcut calculation. For a shortcut calculation or procedure to be useful, it must be reasonably accurate and meet one or more of the following criteria.

- Can be accomplished by mental arithmetic
- Enables rapid checking of computer output
- Provides a graphical procedure for showing design strategy and identifying practical operational limits

By these criteria the eight procedures presented here in convenient outline form can reasonably be called shortcut procedures and effectively used for that purpose.

I. Adiabatic-Reaction Temperature
   Estimation of the adiabatic-reaction temperature is a potent tool for quick analysis and decisions about reaction systems.
   A. Uses of adiabatic-reaction temperature calculations
      1. Use to check new or existing software after entering data and equations for problem of interest. Even if the reactor is nonadiabatic, a check on the first step is valid, since very little heat can be transferred in a few centimeters. By using the average of the initial rate and that at the first step shown on the printout or the conversion of the first step, one can calculate the adiabatic temperature by hand. It ought to agree closely with the temperature at the first step given on the printout. If not, and the hand calculations are correct, there are errors in the program or the data file. Check both the energy equations and associated data, and the rate expressions. Also use a hand calculation of the initial rate as a comparison with the value on the printout.

2. Use for rapid estimate of maximum temperature for an exothermic reaction or minimum for an endothermic reaction. This estimate should be accurate for the adiabatic case and a limiting condition for the nonadiabatic reactor.
3. Once a model has been written or an adiabatic reaction temperature is calculated, the $\Delta T/\Delta X_A$ can be remembered for a quick estimating tool. Over the range of operating interest for single reaction systems or systems dominated by a single reaction, this slope is constant and can be used for rapid mental calculation concerning temperature changes in an adiabatic reactor.

B. Adiabatic-reaction temperature equation

$$\frac{M_F}{Y_{A_0}} c_p(T_2 - T_1) = \sum_{n=1}^{n_r} (-\Delta H_n)_{T_2} X_{na}$$

where M_F is molecular weight of total feed; Y_{A_0} = moles A initially per mole of total feed; c_p is the heat capacity of the total feed on a unit-mass basis; T_1 and T_2 are the inlet and outlet temperatures, respectively; and X_{na} is the extent of reaction per mole of limiting reactant A.

II. Rate Plots

Plots of rate versus temperature provide a useful overview of a process.

A. Uses of rate plots
1. Valuable for explaining the operating possibilities for a single reaction
2. Also useful for illustrating staging and simultanously illustrating trends in rates

B. Background for rate plots

This plot is accomplished by a simple rate-calculating program and associated graphical software. As shown in Fig. 7.1, rates are plotted at constant conversions and various temperatures. Adiabatic lines can be superimposed, and staging strategy sketched on the graph with the added benefit of an enhanced intuitive understanding of the limits of operating and design alternatives.

III. Optimum Equilibrium Approach

The optimum or practical approach to equilibrium is a valuable tool for gaseous reactions for which excellent thermodynamic data exist.

A. Uses of the equilibrium approach
1. Estimating compositions at reactor outlet when kinetics are not known or reactions are so fast that they closely approach equilibrium (see Example 6)
2. Estimating a range of outlet temperatures for study by reactor model simulation

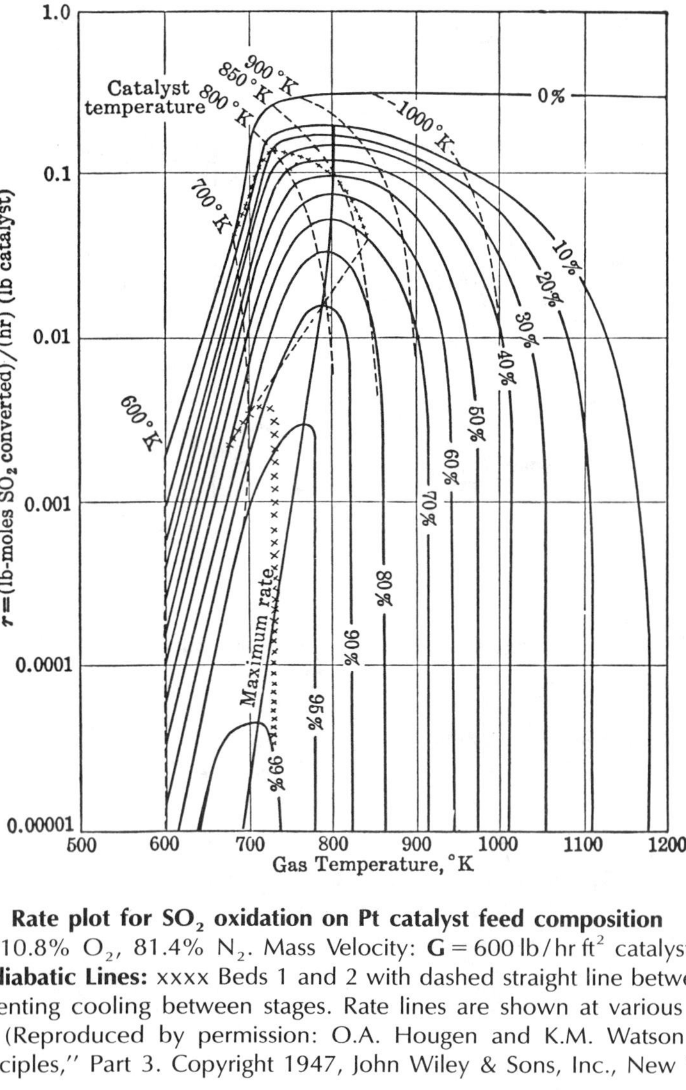

Figure 7.1 Rate plot for SO_2 oxidation on Pt catalyst feed composition 7.8% SO_2, 10.8% O_2, 81.4% N_2. Mass Velocity: $\mathbf{G} = 600\ \text{lb/hr ft}^2$ catalyst. Temperature: ---- **Adiabatic Lines:** xxxx Beds 1 and 2 with dashed straight line between the two parts representing cooling between stages. Rate lines are shown at various percentage conversion. (Reproduced by permission: O.A. Hougen and K.M. Watson "Chemical Process Principles," Part 3. Copyright 1947, John Wiley & Sons, Inc., New York.)

B. Background for equilibrium approach calculations
The equilibrium approach ΔT_a is defined as shown in Fig. 7.2.

$$\Delta T_a = (T_{eq} - T_{op})_{X_A}, \tag{7.1}$$

where T_{eq} is the equilibrium temperature for a given outlet conversion X_A (i.e., the temperature at which the equilibrium constant is equal to the outlet composition quotient) and T_{op} is the operating temperature at the reactor outlet. For many reaction systems the

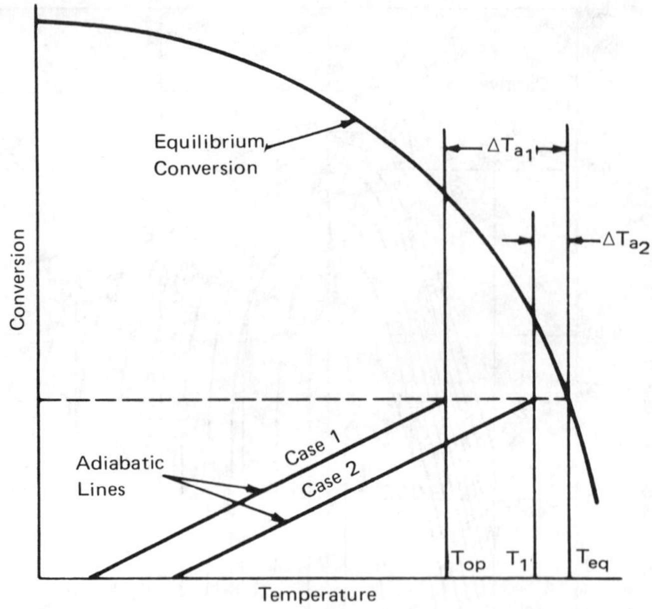

Figure 7.2 Equilibrium conversion and two adiabatic paths for an exothermic reaction ΔT_{a_2} is a close approach; ΔT_{a_1} is a larger approach at T_{op}; both produce the same conversion that can be calculated from equilibrium at $T_{op} + \Delta T_{a_2} = T_1 + \Delta T_{a_1} = T_{eq}$; case 2 has a higher inlet and average temperature than case 1.

equilibrium approach is known based on experience either with plant, pilot-scale, or bench-scale operations. For such cases equilibrium compositions calculated at T_{eq} based on the specified outlet optimum may be used in lieu of a rate model in estimating performance:

$$T_{eq} = T_{op} + (T_{eq} - T_{op})$$

or

$$T_{eq} = T_{op} + \Delta T_a \tag{7.2}$$

Since there is a definable optimum in terms of a maximum rate for an exothermic reversible reaction, it is possible to predict reasonable values of ΔT_a for such cases. As shown in Fig. 7.1, at each conversion level there is a temperature at which the rate is maximized. But the maximization really only becomes important at high conversions when the maximum becomes sharp and rates are already low. It is not unreasonable, therefore, to set the optimum temperature based on a maximum rate for the desired outlet conversion.

For the general power-law form of a hypothetical reaction of $A + B \rightarrow R$,

$$\hat{r}_A = k_p P_A P_B - k_p' P_R$$

$$\hat{r}_A = A e^{-E/RT} P_A P_B - A' e^{-E'/RT} P_R \qquad (7.3)$$

Determine the maximum for the rate with respect to temperature for an equilibrium-limited exothermic reaction [1]:

$$\frac{d\hat{r}_A}{dT} = 0 = \frac{AE}{RT_{op}^2} e^{-E/RT_{op}} P_A P_B - \frac{A'E'}{RT_{op}^2} e^{-E'/RT_{op}} P_R \qquad (7.4)$$

or

$$\frac{A}{A'} \frac{E}{E'} e^{-(E-E')/RT_{op}} = \frac{P_R}{P_A P_B} \qquad (7.5)$$

At equilibrium for this same composition,

$$\frac{A}{A'} e^{-(E-E')/RT_{eq}} = \frac{P_R}{P_A P_B} \qquad (7.6)$$

Since at the same value of conversion, the RHS of Eqs. 7.5 and 7.6 are equal, the LHS may be equated:

$$\frac{E - E'}{R} \left(\frac{1}{T_{op}} - \frac{1}{T_{eq}} \right) = \ln \frac{E}{E'} \qquad (7.7)$$

or

$$\frac{T_{eq} - T_{op}}{T_{eq} T_{op}} = \frac{R}{E - E'} \ln \frac{E}{E'} = \frac{R}{(\Delta H_A)} \ln \frac{E + \Delta H_A}{E} \qquad (7.8)$$

This equation can be solved by using the apparent activation energy and the heat of reaction. The ΔT_a obtained is a useful guide to a reasonable approach temperature, which can be varied around this point in more detailed computer-aided studies. For an adiabatic reactor the T_{op} sets the inlet temperature as well. A lower ΔT_a than the calculated value may actually be preferred in some cases, since that will set a higher inlet temperature (see Fig. 7.2), which could combine with the lower final approach ΔT_a to produce a higher average rate for the entire reaction path.

For endothermic equilibrium-limited reactions the optimum or economically rational approach to equilibrium is not related to a unique maximum rate, since there is none. Instead, rapid reactions may approach equilibrium rather closely in a reasonable tube length. In such cases an empirically determined ΔT_a may be used to predict outlet compositions, conditions, and tube length. The term ΔT_a will be negative for the endothermic case because T_{eq} will be less than T_{op} (see Figs. 7.3 and 7.4 for exothermic and endothermic cases). In each case the conversion at the outlet corresponds to the equilibrium value calculated at $T_{eq} = T_{op} + \Delta T_a$.

IV. Useful Plots

Though not a major technical challenge, plotting a graph has often been avoided because of the tedium involved. The many commerical graphical software packages now available make it possible to use certain graphical procedures that can quickly provide a valuable overview of a reaction system.

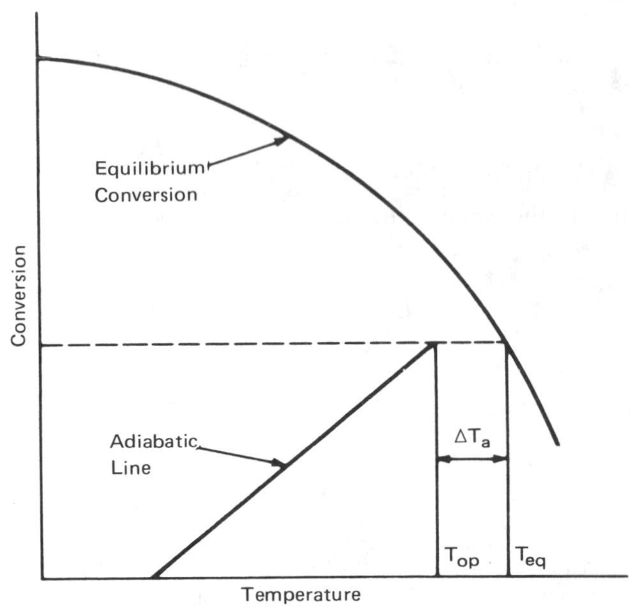

Figure 7.3 Equilibrium conversion and adiabatic path for an exothermic reaction ΔT_a is the equilibrium approach determined as optimum by experience; ΔT_a is a positive number for the exothermic case; T_{op} is the optimum outlet temperature based on experience; T_{eq} is the equilibrium temperature at which equilibrium conversion and composition should be calculated to approximate the observed composition when outlet is at T_{op}.

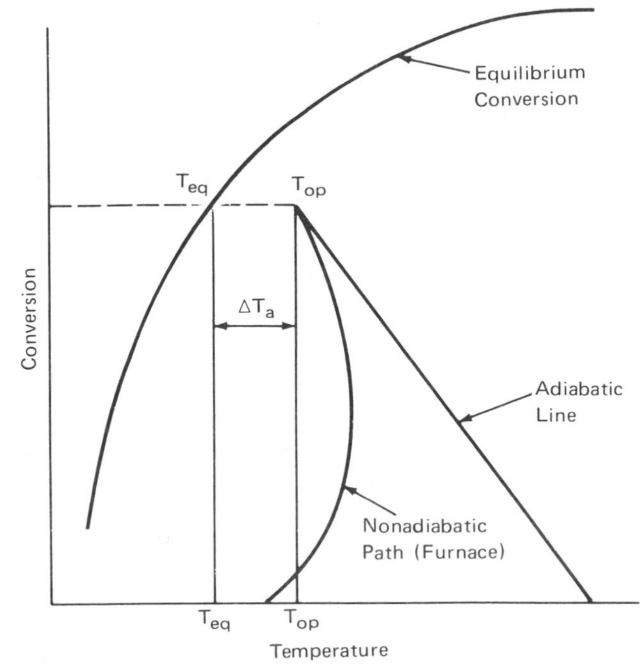

Figure 7.4 Equilibrium conversion with adiabatic and nonadiabatic path for an endothermic reaction

ΔT_a is the equilibrium approach determined as optimum by experience; ΔT_a is a negative number for the endothermic case; T_{op} is the optimum outlet temperature based on experience; T_{eq} is the equilibrium temperature at which equilibrium conversion and composition should be calculated to approximate the observed composition when outlet temperature is T_{op}.

A. Equilibrium and adiabatic-reaction plot
   1. Uses for equilibrium adiabatic plots
      a. Deciding reactor type (adiabatic, staged adiabatic, or nonadiabatic)
      b. Preliminary planning of staged reactors and for explaining such systems (see below).
   2. Background for equilibrium adiabatic reaction plots
      When applicable, nothing is more valuable as a first step in reactor analysis than this simple plot of equilibrium conversion versus temperature and conversion versus adiabatic temperature. It is most useful for a single reaction or a main reaction that dominates the action in a system using a highly selective catalyst. See Example 3.

a. Procedure for computer graphing
Write code using a data source such as the works of Stull et al. [2] for equilibrium and Reid et al. [3] for heat capacity equations. Alternatively, use existing software after first checking the validity of the data used and assumptions made (See Table 3.3 for summary of equilibrium equations and Example 3 for an illustrative problem). Use at least three values of K from Stull et al. in the range of interest and fit to the equation

$$\ln K = -\frac{\mathrm{A}}{T} + \mathrm{B}$$

Use ideal-gas heat capacity equations for adiabatic lines (see Example 2). For very high pressure, enthalpy corrections are required. Use generalized correlations or use simulation packages with corrections based on equations of state.

b. Procedure for hand calculation
Calculate one adiabatic line by using three values based on average heat capacity. Draw best straight line, since variation of heat capacity with temperature is linear over temperature increments of usual interest. Draw other needed lines parallel to the calculated line. Calculate equilibrium lines using logarithms of equilibrium constants of formation (K_f) from Stull et al. at temperatures given. If interpolating between 100°C increments, do so on the basis of a linear relation between $\log K_f$ and the reciprocal absolute temperature and a linear relation between c_p and T.

B. Adiabatic-reactor staging graphs.
1. Uses of staging graphs
   a. Illustrating and planning reactor staging
   b. Planning case studies and operating changes
2. Indirect intercooling
For this operation, cooling or heating does not change composition, and a proposed staging sequence would begin at a selected inlet temperature and proceed to some reasonable approach to equilibrium, which may be known based on previous experience or estimated for exothermic reactions by using Eq. 7.8. The cooling or heating stage then follows a constant-conversion (composition) line to some predetermined inlet temperature for the next stage, often the same temperature as the first-stage inlet. Then the next stage follows a parallel adiabatic line, again approaching the equilibrium line. See Fig. 7.5 and Example 3.

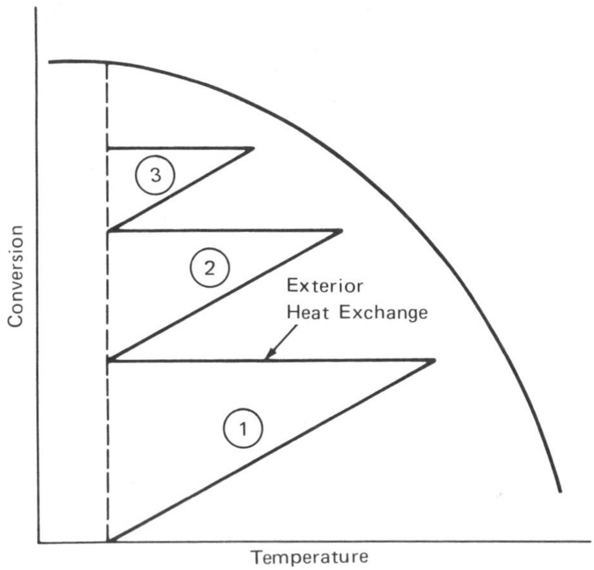

Figure 7.5 Adiabatic reactor staging with interstage cooling in exterior heat exchanger Stages are numbered along reaction lines; cooling occurs at constant conversion in exchanger exterior to the reactor.

3. Direct contact quench
   This procedure is used for exothermic reactions in which a portion of the feed stream (combined fresh feed and recycle) is conducted to a mixing region between stages (beds), effecting thereby rapid cooling while diluting the composition of the reaction stream and lowering the conversion level. This series of events may also be rapidly illustrated on a graph by using the equation developed below [4].
   a. Heat balance at point of mixing flow from bed 1 and quench

$$(\mathscr{F}_1 M_1 c_{p_1} + \mathscr{F}_Q M_F c_{p_F})(T_M - T_Q) = \mathscr{F}_1 M_1 c_{p_1}(T_1 - T_Q)$$

   where $\mathscr{F}_1$ is the total moles leaving stage 1; $\mathscr{F}_Q$ is the total moles of feed quench added at exit of stage 1; M is the indicated molecular weight; subscripts 1, F, M, Q refer, respectively, to bed 1 exit, feed, mixture of quench, and bed 1 outlet and quench.
   b. For equal heat capacities and molecular weights

$$\frac{\mathscr{F}_1}{\mathscr{F}_1 + \mathscr{F}_Q} = \frac{T_M - T_Q}{T_1 - T_Q} \tag{7.9}$$

where M denotes the condition for mixture of quench plus discharge from stage 1, Q refers to the quench, and 1 refers to the exit from bed 1.

c. Mole balance for product R

$$(\mathscr{F}_1 + \mathscr{F}_Q)Y_{R_m} = \mathscr{F}_1 Y_{R_1} + \mathscr{F}_Q Y_{R_Q}$$

where Y_R denotes the mole fraction of product R in the indicated stream. Rearranging gives

$$\frac{\mathscr{F}_1 + \mathscr{F}_Q}{\mathscr{F}_1} = \frac{Y_{R_1} - Y_{R_Q}}{Y_{R_m} - Y_{R_Q}} \tag{7.10}$$

Combine Eqs. 7.9 and 7.10 to get

$$\frac{Y_{R_m} - Y_{R_Q}}{T_M - T_Q} = \frac{Y_{R_1} - Y_{R_Q}}{T_1 - T_Q} \tag{7.11}$$

or, in terms of conversion,

$$\frac{X_{A_m} - X_{A_Q}}{T_M - T_Q} = \frac{X_{A_1} - X_{A_Q}}{T_1 - T_Q} \tag{7.12}$$

d. Procedure (see Fig. 7.6)
Draw adiabatic line to desired approach. Then draw quench line from this point (Y_{R_1} or X_{A_1} and T_1) to Y_{R_Q} or X_{A_Q} and T_Q. The composition or conversion level of the inlet stream to the next bed is found at any mixture temperature desired. The procedure is repeated for the next stage.

V. Equivalent Isothermal Temperature
The equivalent isothermal temperature is an approximate temperature at which the same result would be accomplished as for the actual non-isothermal reactor.

A. Uses of equivalent isothermal temperature
   1. Quick estimate of the performance of an existing reactor at a new condition
   2. Estimate reactor size for new conditions

B. Background for equivalent isothermal temperature
Many reactions approximate first-order disappearance of the main reactant over a reasonable range of interest. From Eq. 5.37, there follows

$$\frac{dX_A}{dZ} = \frac{\alpha \hat{r}_A \rho_b M_F}{\mathbf{G} Y_{A_0}}$$

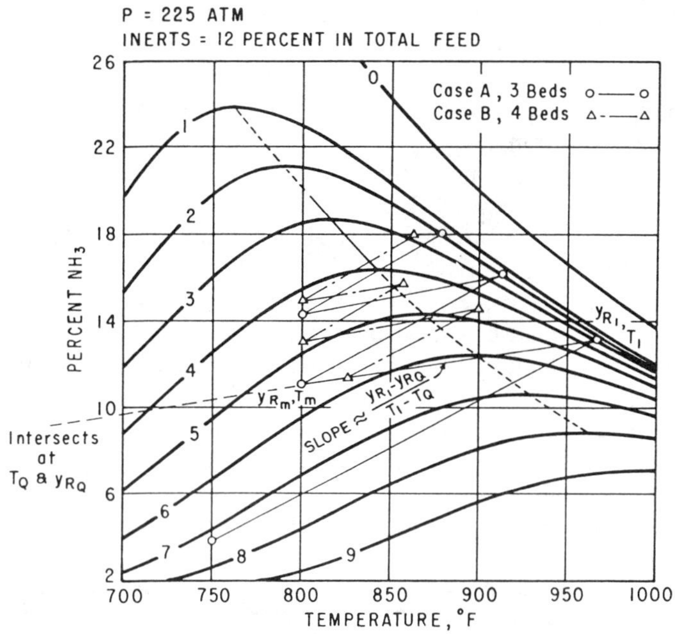

Figure 7.6 Staging graph for adiabatic beds with direct-contact cooling between beds (cold-shot cooling) using colder recycle or feed
Computer-generated rate plot for ammonia synthesis with graphical construction superimposed by hand or by means of computer program with light pen or "mouse" used to designate beginning and ending points. (Reproduced by permission: H.F. Rase, "Chemical Reactor Design for Process Plants," copyright 1977, John Wiley & Sons, Inc., New York.)

or

$$\int_{Z=0}^{Z=L} \frac{\rho_b M_F}{\mathbf{G} Y_{A_0}} \, dZ = \int_0^{X_A} \frac{dX_A}{\alpha \hat{r}_A} = \int_0^{X_A} \frac{dX_A}{\alpha k_o((1 - X_A)/\bar{n}_{T_a})P} \tag{7.13}$$

Integrating at an equivalent isothermal temperature gives

$$\frac{\alpha \mathbf{m} k_0 \rho_b M_F \bar{P} L}{\mathbf{G} Y_{A_0} \bar{n}_{T_a}} = \ln \frac{1}{1 - X_A} \quad \text{or} \quad \ln \frac{1}{1 - X_A / X_A^*} \tag{7.14}$$

where X_A^* is the equilibrium conversion; k_o is the apparent first-order rate constant evaluated at reactor inlet or outlet temperature; **m** is the factor for the particular system, either applicable to inlet or to outlet temperature, by which to multiply k_o to obtain a value based on an equivalent isothermal temperature; $\bar{P}$ is the average operating pressure; and α is the activity factor, $\leqq 1$.

This equation provides a simple severity factor in terms of conversion on the RHS and various operating parameters on the LHS. Once **m** is observed for a particular system, it can be used for brief speculation on the effect of changing operating or design parameters for a given conversion.

1. Increasing temperature and thus k_o requires decreased P or L; or else **G** may be increased.
2. Increasing conversion requires the opposite.
3. Declining activity (α) requires lowering **G** and/or increasing temperature, P, or L. Temperature is the most sensitive variable because of its exponential effect on rate.

In using calculations based on the primary reaction only, one must make decisions resulting therefrom cautiously, since pressure and temperature changes can affect the selectivity by accelerating secondary more than primary reactions. This unfavorable event will occur as follows.

1. Increase in pressure for higher-order irreversible secondary reactions
2. Increase in pressure for a reversible undesired reaction exhibiting a decline in moles
3. Increase in pressure for a reversible main reaction with an increase in moles
4. Increase in temperature where E (desired reaction) $<$ E (undesired)

When multiple reactions occur, a ratio between equations of the form of Eq. 7.14 for any reaction and the main reaction provides a very simple and useful result for performance in the same reactor:

$$\frac{k_{B_o}}{k_{A_o}} = \frac{\ln(1 - X_A)}{\ln(1 - X_B)} \tag{7.15}$$

For feeds containing members of a homologous series, the activation energies are often the same for each entity, and the ratio of k_o for each is a constant value independent of temperature.

VI. Extent of Temperature Increase Required for Deactivating Catalyst

As activity (α) declines, the production can be preserved by increasing the inlet temperature when such temperature increase does not damage selectivity excessively. For the same activity throughout the bed,

$$\alpha_2 k_2 = \alpha_1 k_1$$

$$\frac{k_2}{k_1} = \frac{\alpha_1}{\alpha_2} \tag{7.16}$$

where k_1 and k_2 represent the apparent rate constants at conditions 1 and 2, respectively. But

$$k = Ae^{-E/RT}$$

and

$$\frac{k_2}{k_1} = \frac{e^{-E/RT_2}}{e^{-E/RT_1}} = \exp \frac{E}{R}\left[\frac{1}{T_1} - \frac{1}{T_2}\right]$$

$$\ln \frac{\alpha_1}{\alpha_2} = \frac{E}{R}\left[\frac{1}{T_1} - \frac{1}{T_2}\right] \tag{7.17}$$

or

$$T_2 = \left[\frac{1}{T_1} - \frac{R}{E}\ln \frac{\alpha_1}{\alpha_2}\right]^{-1}$$

As an approximation

$$\frac{\alpha_1}{\alpha_2} = \frac{X_{A_1}}{X_{A_2}} \tag{7.18}$$

where X_{A_2} is the conversion observed after some deactivation has occurred, T_1 is the original inlet temperature prior to deactivation, and T_2 is the new inlet temperature required to preserve the same rate of the main reaction after deactivation as that existing before. The strategy of maintaining production despite activity decline by raising temperature is commonly practiced, and control of the rate of increase is often built into the control system. Unless the activity decline is catastrophic, the change in temperature required per unit time is so small that the increase is usually governed simply by trial and error, using a well-designed feedback control system. The most common type of activity decline involves depletion of activity beginning at the inlet and progressing along the bed with time. In this instance, the effective length of the bed is reduced progressively with time. Equation 7.17 also applies to this situation:

$$T_2 = \left[\frac{1}{T_1} - \frac{R}{E}\ln \frac{\alpha_1}{\alpha_2}\right]^{-1} \tag{7.19}$$

$$\frac{\alpha_1}{\alpha_2} \propto \frac{X_{A_1}}{X_{A_2}} \propto \frac{L_1}{L_2} \tag{7.20}$$

where L_1 and L_2 are active bed lengths. *Caution:* In following these lines of reasoning and implementing operating changes thereby indicated, one must avoid the following undesired changes that can result from increasing reaction temperature:

1. The rate of sintering may increase if the heightened temperature enters range of rapid agglomeration of dispersed active components or sintering of the support.
2. Selectivity to the desired product will decline if activation energies for undesired reactions are greater than that for the desired reaction.
3. Design temperature limits of equipment may be exceeded.

VII. The Art of Ratioing

After computer calculations have been made and conclusions and recommendations are being presented, questions often arise concerning operating conditions different from those studied. It is always nice to be able to respond with reasonable quantitative answers without having to be excused to allow time for additional computer studies. Ratioing a basic equation for a known condition with that for a new condition can be most revealing and simplifying.

A. Background

The following equations may be used for ratioing:

1. General plug-flow continuity equation for limiting reactant

$$\frac{\mathbf{G}Y_{\mathrm{A_o}}\,dX_{\mathrm{A}}}{\rho_{\mathrm{b}}M_{\mathrm{F}}\,dZ} = \alpha\eta\hat{r}_{\mathrm{A}} = \alpha\eta f(T, X_{\mathrm{A}}, P) \tag{7.21}$$

$$\frac{\rho_{\mathrm{b}}M_{\mathrm{F}}L}{\mathbf{G}Y_{\mathrm{A_o}}} = \int_0^{X_{\mathrm{A}}} \frac{dX_{\mathrm{A}}}{\alpha\eta f(T, X_{\mathrm{A}}, P)} \tag{7.22}$$

2. Pressure drop

$$-\Delta P = \frac{Lf_k\mathbf{G}^2}{D_{\mathrm{p}}\rho_{\mathrm{f}}}\left(\frac{1-\varepsilon}{\varepsilon^3}\right) = \frac{Lf_k\mathbf{G}^2P}{D_{\mathrm{p}}MRT}\left(\frac{1-\varepsilon}{\varepsilon^3}\right) \tag{7.23}$$

B. Quick estimates by ratioing

1. Change **G** while holding conversion, temperature, and pressure essentially the same. From Eq. 7.22,

$$\frac{L_2}{\mathbf{G}_2} = \frac{L_1}{\mathbf{G}_1} \quad \text{or} \quad \frac{L_2}{L_1} = \frac{\mathbf{G}_2}{\mathbf{G}_1} \tag{7.24}$$

Must change L to maintain constant $L/\mathbf{G}$.

2. Effect on pressure drop

$$\frac{\Delta P_2}{\Delta P_1} = \frac{\mathbf{G}_2^2 L_2 D_{\mathrm{p_1}}}{\mathbf{G}_1^2 L_1 D_{\mathrm{p_2}}} \tag{7.25}$$

An increase in **G** by a factor f_{m} requires an increase in L by the same factor to maintain approximately the same conversion. The pressure drop will increase by the factor f_{m}^3. This situation can be partially ameliorated by increasing particle size if the effectiveness factor remains unchanged. When the particle size is doubled,

$$\frac{\Delta P_2}{\Delta P_1} = \frac{f_m^3}{2} \tag{7.26}$$

If the effectiveness factor is initially low, doubling the size will half the value ($\eta \propto 1/D_p$ for low values), which would require another doubling of L, bringing the situation back to the original case

$$\frac{\Delta P_2}{\Delta P_1} = f_m^3 \tag{7.27}$$

These arguments are most valid for adiabatic reactors, where the larger cross section causes bed void fraction to be unaffected by catalyst size. See Example 7.

C. Equivalent isothermal temperature
   Equation 7.13 is also a valuable ratioing tool for nonadiabatic or adiabatic reactors when the pseudo-first-order assumption is applicable. At different conditions m will cancel, as will those variables that remain reasonably constant.

VIII. Inflection Point of Axial Temperature Profile for Nonadiabatic Reaction
Since at the maximum or minimum temperature of a nonadiabatic reactor, $dT/dZ = 0$, the general heat balance becomes simplified.

A. Uses of inflection-point estimate
   1. Quick estimate of heat-transfer coefficient from operating data. Assume conversion level is 30% of total or other experiment-based value at inflection point.
   2. Estimate conversion level at measured maximum or minimum temperature point, using a calculated or known value of U_Z.
   3. Confirm the temperature measured at inflection point. Assume $X_A = 0.30$ or other value based on experience.

B. Background for inflection-point estimate
   With $dT/dZ = 0$, the general energy balance equation becomes

$$\frac{4U_Z}{D}(T - T_h) = \hat{r}_A \rho_b(-\Delta H_A) = \frac{X_A \mathbf{G} Y_{A_0}}{M_F Z}(-\Delta H_A) \tag{7.28}$$

CONCLUSION

The eight procedures for quick estimates and checking supplemented by experience with both actual plant operation and simulation can be invaluable in quickly assessing proposal design decisions and operating strategies.

LITERATURE CITED

1. Denbigh, K.G., *Trans Faraday Soc.*, *40*, 352 (1944).
2. Stull, D.R., Westrum, E.F., Jr., and Sinke, G.C., "The Chemical Thermodynamics of Organic Compounds," Wiley, New York, 1969.
3. Reid, R.C., Prausnitz, J.M., and Polina, B.E., "The Properties of Gases and Liquids," 3rd Edition, McGraw-Hill, New York, 1987.
4. Brötz, W., "Fundamentals of Chemical Reaction Engineering," Addison-Wesley, Reading, Mass., 1965.

CHAPTER

8

Diagnosing Reactor Operating Problems

Diagnosis, troubleshooting, fault detection, malfunction detection, failure analysis, fault-free analysis, and cause-and-effect relation analysis are part of the growing list of words and phrases used to describe the age-old practice that probably had its beginning as a conscious act with the early practitioners of the healing arts. Until recently members of the medical profession and all other professions developed expertise in diagnosis through observing experienced practitioners and by working under the critical observation of such practitioners.

A SYSTEM OF LOGIC FOR DIAGNOSTICS

Little formal instruction based on some system of logic existed, and much of the learning involved remembering symptoms and corresponding procedures related to appropriate diagnostics. The use of computers as replacements for memorization as well as for installing the expertise of gifted diagnosticians as expert systems has sharply focused interest on the logical constructs that are used by experts in establishing causal connections. Surprising to some perhaps is the fact that John Stuart Mill published in his "A System of Logic," in 1843, methods of experimental inquiry based on discovering causal connections. Although establishing causal connections, causes for observed events, no longer would be considered the sole procedure for research, it remains a valuable part of observation in conducting experiments and diagnosing problems both in experimental equipment and operating units, as well as all other professional diagnoses. The causal connection or causal hypothesis, when confirmed, allows for some remedial action that will solve or ameliorate the problem. It does not necessarily lead directly to more creative hypotheses that can generate new ideas and inventions that will circumvent the problem originally confronted and prevent its recurrence. Certainly, however, a rational approach to diagnostics encourages and nourishes creative thinking that can lead to innovative products, devices, and procedures.

It is worthwhile, therefore, to consider briefly the system of logic proposed by John Stuart Mill [1] as a guide to the establishment of causal connections and successful diagnosis. He proposed four rules, which are quoted here with the caveat "may be" added as an effort to face reality in a complex technological world. They are adapted from an earlier discussion that employed a general example related to catalysts [2].

A major value of these rules is that they verbalize seemingly recognizable but largely undefined procedures used by many professionals, and they strongly fortify the intuitive logic that urges gathering facts and comparing a troubled situation with trouble-free cases as the best means for rapid diagnosis.

Mill's Rules of Logic

1. *Method of agreement*: If two or more instances of an event under examination have only one circumstance in common, that circumstance (may be) the cause of the event.
2. *Method of difference*: If two instances have all circumstances in common save one, and that one circumstance occurs only in the instance in which the phenomenon of interest is observed, then the circumstance in which alone the two instances differ may be the cause of the phenomenon.
3. *Combined method of agreement and differences*: If two or more instances in which the phenomenon occurs have only one circumstance in common, while two or more instances in which it does occur have nothing in common, save the absence of that circumstance, the circumstance in which alone the two sets of instances differ (may be) the cause of the phenomenon.
4. *Method of concomitant variations*: Whatever phenomenon varies in any manner whenever another phenomenon varies in some particular manner, is either a cause or an effect of that phenomenon, or is connected with it through some fact of causation.

Example of the Application of Mill's Rules

On three operating days the catalyst in a hydrogenation reactor declined in activity suddenly. The following circumstances were deemed relevant on days for which information was verifiable without any uncertainties.

Date	*Circumstances*	*Observations*
Oct. 13	b c j	No deactivation
Oct. 15	e f h	No deactivation
Oct. 16	k j g	No deactivation
Oct. 30	g c d	No deactivation
Nov. 4	a b c d	Deactivation
Nov. 10	a e f g	Deactivation
Dec. 4	a h j k	Deactivation

Circumstances Defined

a. Temperature rose above 1200°F during several half-hour periods.
b. Chief operator was sick.
c. Feed stock was from new purification unit.
d. New catalyst had been charged to unit three days previous.
e. Feed flow recorder being repaired.
f. Insulation on reactor damaged by painters.
g. Heavy rains caused temporary backup in drainage system.
h. Possibility of steam leak into feed stream through faulty steam-out valve.
j. Temporary upset in feed preparation plant increased light components in feed.
k. Brief low-voltage power failure affected instrument chart drives and potentiometric instruments.

Diagnostics as Suggested by Mill's Rules

a. The *method of agreement* rule would suggest that a is related to the cause, since November 4 and 10 and December 4, the days when catalyst activity declined suddenly, each have a in common.
b. The *method of difference* suggests a to be related to the cause, using the observations on October 30 and November 4.
c. The application of the *combined method of agreement and differences* rule is the most compelling argument, since it involves all the data gathered, with the October data having an absence of a and the November-December data clearly involving a.
d. Although not illustrated by the example involving either deactivation or no deactivation, the *method of concomitant variations* rule is valuable when, for example, deactivation is continuously observed but variations in the degree can be linked to an observed circumstance.

GENERAL DIAGNOSTICS FOR FIXED-BED REACTORS

As illustrated by Example 10, computer-based expert systems apply these rules in diagnostics by constructing rules based on the presence or absence of a combination of various symptoms. There remains the need to provide some accumulated wisdom of experience of many professionals with fixed-bed catalytic reactors. For convenience, general symptoms and related diagnoses are summarized in Table 8.1, which is divided into three categories, related to catalysts, catalyst-bed conditions, and recycle systems. One should enter the table based on symptoms that seem to agree with those being observed. By comparing various observed conditions with operating days when the symptoms did not appear, one should be able to put forward the most logical diagnoses, which will in turn suggest the cause. Since often the conclusions

Table 8.1 Diagnostics for Fixed-Bed Reactors

Symptoms	*Diagnosis*		*Confirmation*	*Action*
	Possible Causes	*Distinguishing Features*		
		Activity-Decline Diagnosis		
1. Gradual decline in production or gradual increase in temperature required to offset decline in activity. Rate of change increasing or decreasing depending on particular system.	(a) Coking	Deactivation rate highest during early portion of run, then tends to change slowly with time.	Repeat conditions on laboratory reactor. Regenerate and test performance. Most of activity should be restored. In cases where bed may be sampled measure activity before and after regeneration of sample.	Regenerate as needed using great care so as to realize even burn and avoid excessive temperatures.
	(b) Trace poisons in feed or generated by side reactions	(b.1) Axial temperature constant in a region of an adiabatic bed, the length of which region increases with time. (b.2) Hot spot in non-adiabatic exothermic reactor moves with time toward outlet. (b.3) Axial gradients in both types of reactors similar to that before poisoning occurred, but reduced in magnitude caused by even reduction of activity throughout bed. (b.4) Bed ultimately is no longer capable of accomplishing desired production rate even at higher temperatures.	Conditions b.1–b.2 are easy to spot and occur when poison lays down progressively from inlet to outlet. An essentially totally deactivated region thus increases with time until it covers entire bed. Condition b.3 can be reproduced by model using a constant activity factor approximately equal to new production/original production. For all cases analyze feedstock and compare with previous feed that did not cause problem. Identification of trace compounds may be possible. Separate tests on bench-scale equipment will confirm poisoning action and allow observations appropriate to the catalyst being used.	• Look for upsets in operation of units providing feedstock. • If a temporary poison, purging with an inert or other gas such as hydrogen at operating temperature will remove it. Also clean feedstock will cause temporary poison to desorb after operation for an hour or two.

	(b.4) Loss of selectivity occurs at time poison first detected in the system. Occurs on multifunction catalysts when poison is strongly adsorbed on one function that normally leads to desired reaction sequence.		• If permanent poison, will not be possible to remove it. Dump catalyst and recharge with fresh catalyst. Must have found source of poison and eliminated it.
(c) Gradual sintering and/or agglomeration of active component	(c.1) Activity decline or temperature increase required to maintain activity accelerates as operating temperature is increased into region where rate of sintering becomes high.	Only certain procedure involves determining total surface area by BET and/or active area by SEM or TEM procedures on representative catalyst sample.	Usually catalyst must be replaced except in some cases where active components or promoters agglomerate and a means has been found for redispersion.
	(c.2) Selectivity declines as temperature is increased in a high-temperature region only—due to agglomeration of an active component or promoter that inhibited side reactions.	Compare operating results with normal results at same operating conditions from previous operating history or from calculations based on a reliable model. In this way normal effects of temperature on selectivity may be compared with total effects caused by both temperature and agglomeration.	
(d) Volatilization of an active compound and transport downstream	Portion of bed evidences no axial temperature change due to loss of activity by utilization of active species in the region of constant *T*.	Samples from various portions of bed will show different concentration of active component.	Ultimately catalyst must be replaced unless a method for redispersion is known.

Table 8.1 *(continued)*

Symptoms	*Diagnosis*		*Confirmation*	*Action*
	Possible Causes	*Distinguishing Features*		
		Activity-Decline Diagnosis		
2. Rapid decline in production and yield above a certain temperature range.	(a) Activity decline requires temperature increases that ultimately approach unfavorable equilibrium conditions for exothermic reaction.	Temperature increase to improve production rapidly reaches point of no improvement due to equilibrium limitations. Lag in observed effect equivalent to holding time of reactor system.	Compare conversion with calculated equilibrium conversion.	Find new optimum approach temperature. If production rate still unacceptable, shut down and replace catalyst.
	(b) Sintering and agglomeration	Temperature has been increased into a region where sintering and agglomeration occur rapidly, causing marked decline in surface area of support and/or active components.	Show that extent of change is not explainable by equilibrium arguments. Then show by separate tests or examination of catalyst from reactor that surface area has declined. Also look for decline in active area (see this column under 1(c.1).	Replace catalyst.
3. Lowered activity observed after regeneration	(a) Sintering and/or agglomeration	Regeneration not properly executed. Oxygen concentration too high or excessive reactants and products were left adsorbed on catalyst. Excessive combustion rate produced high temperatures in bed that sintered catalyst support or agglomerated active components.	See item 1(c).	If activity decline excessive, replace catalyst. Otherwise use lower oxygen concentration during regeneration and better control of burn.

	(b) Volatilization of an active component	Activity is markedly reduced in part of bed due to loss of active component on catalyst. Little temperature change occurs through this portion of bed.	Recreate phenomenon by heating a bench-scale unit in presence of nitrogen flow to same temperature as hot spot. Test activity and determine active component concentration in different portions of bed.	Same as above.
	(c) Inadequate regeneration	(c.1) Oxygen content of regeneration gas so low that burn is incomplete both in the bed and in particles.	Samples from bed will reveal portions of bed still black with coke and paritally burned pellets exhibiting uniform gray or black color throughout pellet.	Find optimum regeneration conditions by pilot-plant or bench-scale studies or from catalyst manufacturer or licensor.
		(c.2) High oxygen content causes too rapid a burn and leaves interior of pellets unoxidized.	Pellets will be free of coke toward exterior but black toward center. High chemical oxidation rate causes burn to be controlled by oxygen diffusion.	
4. Sudden decline in production after switching feed tanks. No change in operating conditions.	(a) Permanent poison in new feed (b) Temporary poison in new feed	See these columns in item 1(b).		
5. Temporary runaway in nonadiabatic reactor evidenced by pressure surge, which is overcome by lowering pressure on boiling jacket fluid, lowering feed temperature, and use when feasible of a quench gas. Activity is reduced by this event.	(a) Sintering of support in region where hot spot occurred (b) Agglomeration of active component or promoter	See these columns in 1(c), (d).		

Table 8.1 *(continued)*

Symptoms	*Diagnosis*		*Confirmation*	*Action*
	Possible Causes	*Distinguishing Features*		
		Activity-Decline Diagnosis		
6. Poor selectivity following turnaround and replacement of catalyst.	(a) New catalyst may not be same as previous catalyst or preconditioning may have been faulty	Although batches can vary, commercial catalysts are usually quite uniform. The problem is often associated with improper procedures in pretreating catalyst such as inadequate drying or presence of liquid water, incomplete reduction or excessively high-temperature reduction, excessive reaction severity during early highly active phase of the catalyst, and similar events unique for each particular catalyst.	Check manufacturer's start-up recommendation. If no guidance available, must study conditions on bench-scale reactor, comparing earlier successful start-up with new start-up.	Must obtain or develop rules for start-up that will be easy to implement and ensure success.
	(b) If multitubular reactor, walls could have been activated for unwanted side reaction.	Wall passivation treatment, normally practiced was not redone prior to reloading, or a new procedure was tried for the first time and the walls are actively catalyzing unwanted side reactions because new method is not effective or was not done correctly.	Repeat steps on a one-tube pilot unit and also study various passivating techniques used or being considered.	
	(c) New operating conditions may favor side reaction.	Pressure is higher or temperature is markedly different, causing poor selectivity to desired product.	If model available, outcomes for successful conditions can be compared with new conditions.	Find from model best conditions and use them.

		Catalyst-Bed Diagnosis		
1. Pressure drop increasing with time on stream at constant flow rate	(a) Rapid coking, some of it noncatalytic, causing reduction in bed void fraction	(a.1) New feedstock has higher quantity of coke precursors, as shown by standard test or analysis.	Regenerate bed and see if ΔP is reduced.	Stop using new feedstock or try a guard bed ahead of main reactor.
	(b) Dust or fine scale from upstream processes or from oxidizing airstream	(b.1) Corrosion fines from furnace or other equipment being carried into feed-line.	Install in-line filter at outlet of suspected equipment. Inspect after period of use. Also inspect catalyst bed after shutdown. Bulk of fines should be at top of bed.	Remove top portion of bed and replace. Install in-line filters permanently at main feedline to reactor.
		(b.2) New atmospheric conditions causing dust to reach air intake.	Compare air samples for existing situation with earlier samples. Examine bed after shutdown.	Increase height of air intake or install filter or improved filter.
		(b.3) Air filter defective	Continue operating and work with air filter to repair or improve its operation.	Permanent solution for air filter operating problem.
2. Pressure drop and production lower than for previous run at same conditions	(a) Catalyst not properly loaded and channeling is occurring	(a.1) Axial temperature profile varies markedly in portions of cross section indicating channeling	Can be tested by residence time distribution studies. Outlet signal will show two peaks, the bypassed material and the remainder.	Reload large reactor according to procedures given in Table 5.2.
	(b) Larger-size catalyst used for the new batch, causing a lower effectiveness factor and ΔP.	(b.1) Check previous laboratory studies on effect of catalyst size. If previous catalyst had low effectiveness factor, then ratio of production new/old will be $(D_p)_{old}/(D_p)_{new}$.	Compare two catalysts in bench-scale tests.	Try same size catalyst in Rasching-ring form.

Table 8.1 *(continued)*

Symptoms	*Diagnosis*		*Confirmation*	*Action*
	Possible Causes	*Distinguishing Features*		
		Catalyst-Bed Diagnosis		
3. Same as above, but ΔP increasing with time.	(a) Lower flow rate in portion of bed causing excessive side reaction and polymer production or other deposit and further plugging of this region.	(a.1) Polymer and coke precursors found in effluent. Plugging produced by polymerization and coking. (a.2) Axial temperature gradients not flat. Some regions of bed at higher temperature than others. Additional plugging due to side reactions further restricts flow and diverts flow to remaining cross section.	Study low flow rates on bench-scale tests. Observe deposits in bed after tests complete.	Reload reactor with care.
4. Pressure drop higher than for previous catalyst batch. All other conditions and catalysts the same.	(a) Catalyst fines not removed during loading process. (b) Rough handling during loading process.	(a.1) Unused catalyst runs show high fines content, which caused plugging in the reactor. (b.1) Reports indicate poor loading procedures followed, such as dumping catalyst from drums at long distance from resting place, walking by personnel on catalyst while loading, and excessive raking and handling.	Examine contents of bed after it is dumped.	Replace catalyst or screen catalyst and reload.

5. Lower production in multitubular reactor with increased side-reaction products observed.	(a) catalyst not loaded properly and ΔP not same in all tubes. Lower flow rate in more densely packed tubes causes longer residence time, more side reactions, and lower production.	(a.1) ΔP not checked for each tube when catalyst loaded, and some tubes are packed more densely, causing longer residence time in these tubes and encouraging side-reaction products. (a.2) Higher total flow should reduce side-reaction products, but ΔP will be higher than that predicted for identically loaded tubes.	Observe effect of low flow on side reactions using single-tube pilot reactor.	Reload tubes after first screening catalyst for fines removal.
6. Lower production in new adiabatic reactor per unit mass of catalyst than for older smaller reactor at same G and operating conditions. Also more side-reaction products.	Distributor at inlet and collector at outlet improperly designed so that significant portion of bed is bypassed. Some flow occurs in bypassed region, causing excessive contact time and production of side-reaction products.	Axial temperature profile not flat and ΔP higher because portion of bed is not used. Also drawings of reactor internals indicate inadequate design.	Study existing distributor design on water table to demonstrate the character of the distribution system.	Develop new design through water-table studies.

might require major changes in the operation, including costly shutdowns, an independent means for confirming the cause is essential. These suggestions are also given. In the last column a course of action is provided, assuming successful confirmation of the causal connection that has now been established.

Table 8.1 is a useful guide to steps necessary in the logical resolution of a reactor problem. Once a causal connection is established, such as circumstance a in the previous example, an increase in temperature to 1200°F for 30 minutes, the diagnostic procedure is completed by a formal diagnosis consistent with the cause selected. The formal diagnosis includes a statement related to known characteristics of the catalyst or the reactor system and, like all hypotheses, capable of confirmation or rejection. According to Table 8.1, symptom 2 fits the observed operating anomaly. Diagnosis 2(b) provides the necessary hypothesis based on this symptom, the change occurring in the catalyst (sintering and/or agglomeration), and measurable and/or observable changes caused by the change (total surface area and/or active surface area decline). Then the next column suggests means for confirming the diagnosis (the hypothesis), and the final column suggests a course of action for returning to acceptable operation.

Since no general operational theory of catalysis exists, the success of the more important task of preventing recurrence of previous troubles rests on gaining more understanding about the particular catalyst being used, in the following ways:

- If a merchant catalyst, ask the technical representative of the manufacturer for information. The representative has strong incentive to provide every possible help to ensure successful use of a catalyst.
- Study the published literature on the catalyst.
- Do studies on a small scale in the laboratory that will enhance understanding of the behavior of the catalyst. Include studies of surface morphology, active component dispersion, component migration, nature of catalyst deposits, and susceptibility to poisons, all as a function of operating, pretreating, and regenerating conditions. Out of such studies new ideas will grow, including those suggesting much improved catalysts.

EXPERT SYSTEMS

The use of the programming language LISP has accelerated the development of general-purpose software useful in developing expert systems. Many of the packages do not require any knowledge of LISP. An overlay of interactive input provides the necessary structure for easy development of a specific system for a process.

When it is recognized that over a period of several years problems can recur in a reaction system, the value of an organized codification of symptoms and successful diagnostics related thereto becomes eminently clear. During a period of several years, personnel changes and fading memories can contrive to make an old problem appear totally new and require many hours before a successful solution is rationalized. The human brain with its miraculous abilities for parallel reasoning, innovation, invention, and judgment has one weakness—imperfect recall. Unfortunately, written records are seldom complete, and it makes perfect sense to structure previous problems and their solutions into expert system by means of artificial intelligence software in the form of symptoms, rules, causes, and courses of action so that a problem can be identified by means of the same chain of reasoning used in its first successful solution.

Building an effective expert system requires careful planning and execution of the following major steps:

1. Assembling a narrative account of all known problems that have occurred together with the symptoms, causes, actions, and explanations of the phenomena related to all the observed facts. This task involves the combination of
   a. First-hand experiences and information from the technical literature relevant to the observations.
   b. Accounts by experienced operators, experts on the particular process, and R&D personnel familiar with the chemistry and catalyst behavior.
   c. Background knowledge from catalyst manufacturer concerning characteristics of catalyst being used.
   d. Opinions of instrument technicians and analytical laboratory personnel on the reliability of reported measurements.
2. Prepare for each problem, using the narrative material, a tabular summary of the symptoms, causes, and recommended courses of action.
3. Enter into the expert system all symptoms or abnormal conditions that have been observed, all known causes for any type of problems that have been previously determined, and a series of rules for each problem structured as *if* statements for abnormal conditions and *then* statements for causes, and recommended courses of action. Added to all this should be a clearly and concisely written explanation of the reasons for the cause(s) and courses of action recommended. This portion should make it possible for an intelligent user not only to have some understanding of the problem but also to understand the consequences of the action to be taken.
4. Test the finished system involving each expert who originally provided input for the background material. It should also be tested by all operators involved with the process unit, and they should be urged to ask questions and give their reaction and opinions. If a question indicates a failure to understand a diagnostic for a problem, that material should be redone so that it is clear.

Example 10 illustrates the preliminary development of an expert system for a steam reformer. Step 4 has not been undertaken, and the system will be complete only after thorough testing and review as described above.

LITERATURE CITED

1. Mill, John Stuart, "A System of Logic," 1843, republished by Longmans, Green, New York, 1930.
2. Rase, Howard F., "Philosophy and Logic of Chemical Engineering," Gulf, Houston, 1961.

PART III

EXAMPLES

The examples in this part were selected to illustrate modern techniques in solving realistic problems related to fixed-bed reactors. Various types of professional software were used and are described in the examples. The objective is stated at the beginning of each example, and general conclusions relative to the problem type are summarized at the end of each example.

Because of the vast amount of readily available, reliable thermodynamic data for gases, a series of examples is included to show the utility of thermodynamic calculations in the design-decision process. These data require no experimental effort or expense by the user, yet they yield highly useful results.

Caution: Kinetic data used in the examples are taken from the open literature, in which the exact details on catalyst characteristics are often not provided. Where necessary for illustrative purposes, constants have been altered to provide a closer representation of actual performance. When using kinetic models for design, operation, or control, one should obtain data from tests on the actual catalyst to be used in the commercial unit. For such purposes, the form of the equations that may be found in the literature for a reaction system will usually be applicable. Only the kinetic parameters must be evaluated independently.

EXAMPLE

1

Comparison of Equilibrium Constant Calculation Methods

OBJECTIVE

Accurate equilibrium constants are important for reliable models involving reversible reactions. The method of estimating equilibrium constants from tabulated data can, in regions removed from nearly complete conversion, have a significant effect on the equilibrium conversion. This example compares several methods in common use.

PROBLEM STATEMENT

Determine equilibrium constants at 400, 500, 600°K by the three methods described here and compare with tabulated values for hydrogenation of benzene to cyclohexane.

$$C_6H_6 + 3\,H_2 \leftrightarrow C_6H_{12}$$

BASE CASE

Values determined from logarithms of the equilibrium constants of formation ($\ln K_f$) from the elements at 400, 500 and 600°K as tabulated by Stull et al. [1] are used as criteria for comparison.

$$aA + bB \leftrightarrow rR$$

$$\ln K = r(\ln K_f)_R - [a(\ln K_f)_A + b(\ln K_f)_B]$$

For the example case

$$\ln K = (\ln K_f)_{C_6H_{12}} - (\ln K_f)_{C_6H_6}$$

Table EX-1.1 Comparison of Calculated Equilibrium Constants

Temp. °K	*"Correct" Values by Direct Calculation from Tabulated Data (1)*		*METHOD 1 Calculated by Interpolation between 400° and 600°, and 500° and 600° Values, Respectively*		*METHOD 2 $\ln K = A + B/T$ Calculation from Fitted Equation Based on Values in Range*		*METHOD 3 Calculated from Single Value of K at 298 and $\Delta H°$ @ 298, and Then Corrected by Eq. EX-1.15*	
	K	X	K	X	K	X	K	X
400	69.5025E6	1.0			71.1347E6	1.0	66.4696	1.0000
500	181.551	0.99996	166.2	0.99995	171.282	0.99995	172.6069	0.999954
600	0.0297117	0.8579	0.0287078	0.8547	0.0307704	0.8611	0.02845	0.8538
					A = 25,873.5 B = -46.6038			

Note: Values of conversion based on 10 atm 4:1 $H_2:C_6H_6$, and zero inerts

Hydrogen does not appear because it is an element. The Stull values for each compound are given as $\log_{10} K_f$. They were converted to natural logarithms for convenience for later use. Values of K thus determined are given in the first column of Table EX–1.1.

METHOD 1: VALUES OBTAINED BY INTERPOLATION

For illustration, the $\log_{10} K$ values from the tabulation at 400 and 600°K and at 500 and 700°K are used to obtain interpolated values at 500 and 600°K, which can then be compared with the tabulated values at 500 and 600°K.

		$Log_{10} K_f$		
Temp (°K)	*I/T*	*Benzene*	*Cyclohexane*	log_{10} *K*
400	2.5×10^{-3}	−19.126	−11.284	7.842
500	2×10^{-3}	−17.152	−14.893	2.259
600	1.666667×10^{-3}	−15.901	−17.428	−1.527
700	1.4285714×10^{-3}	−15.051	−19.308	−4.257

Since $\log K$ is a linear function of T^{-1}, $\log_{10} K$ at 500°K is found by the following interpolation

$$\log_{10} K = [7.842 - (-1.527)] \frac{500^{-1} - 600^{-1}}{400^{-1} - 600^{-1}} - 1.527 = 2.2206$$

or $K = 166.2$

Similarly, K at 600°K is found by interpolation to be $K = 0.0287078$.

This procedure is rapid and always useful for quick estimates.

METHOD 2: EQUATION FITTED USING TABULATED VALUES IN TEMPERATURE RANGE OF INTEREST

1. When an equation for K is needed for a computer program, the most logical procedure is to use heat capacity equations as shown in Method 3. But if such equations are not conveniently available, values of K from the tables of Stull in the range of temperature interest can be used to develop a linear approximation.

2. Fit values from Stull et al. [1] to the form $\ln K = A + B/T$ using least-squares approximation. There are many convenient programs for this

purpose, ranging from those for pocket calculators to standard FORTRAN routines such as the IMSL routine IFLSQ, as used for this example.

The form $\ln K = A + B/T$ is based on the defining equation

$$\ln K = \frac{-\Delta G^\circ}{RT} = -\left(\frac{\Delta H^\circ - T\,\Delta S^\circ}{RT}\right) \tag{EX-1.1}$$

$$\ln K = \frac{\Delta S^\circ}{R} - \frac{\Delta H^\circ}{RT} = A + \frac{B}{T} \tag{EX-1.2}$$

When constant values of A and B are used over a temperature range of interest, the assumption is that an average value of ΔH° and ΔS° applies sufficiently to permit a straight-line correlation with minimal error. That such is the case has been confirmed by a host of reaction systems over short temperature ranges.

Since the minimum of the square of the deviation in $\ln K$ is obtained by the least-squares procedure, the deviation for K values will be a minimum percentage square deviation, which is useful for K values, since they usually range over many orders of magnitude within a temperature range of interest.

3. Use the resulting equation to calculate values of K at any temperature in the range.

The range of interest is selected at 400 to 600°K and values of $\log K$ of formation at 400, 500 and 600 are used to obtain values of A and B. Results are shown in Table EX–1.1. For comparison, "correct" values calculated from the tabulated values are given in the second column. Obviously, the deviations produced by Method 2 are due to the faulty assumption of constant ΔH° over a 200° temperature range.

Method 2 provides a rapid means for estimating reasonable values of K at any temperature over a wide range of temperatures.

METHOD 3: SINGLE VALUE OF *K* AT 298°K CORRECTED TO OTHER VALUES, USING EMPIRICAL HEAT-CAPACITY EQUATIONS

This procedure is widely used, since in some cases only a single value of K at 298°K is known. Equations 3.29 and 3.30 developed in Chapter 3 provide the necessary relationships.

$$\frac{d\ln K}{dT} = \frac{\Delta H^\circ}{RT^2} \tag{EX-1.3}$$

$$\Delta H^\circ = I_H + \sum_{j=1}^{c} \nu_j\left(A_jT + \frac{B_jT^2}{2} + \frac{C_jT^3}{3} + \frac{D_jT^4}{4}\right) \tag{EX-1.4}$$

$$R \ln K = \frac{I_H}{T} + I_K + \sum_{j=1}^{c} \nu_j \left(A_j \ln T + \frac{B_j T}{2} + \frac{C_j T^2}{6} + \frac{D_j T^3}{12} \right) \tag{EX-1.5}$$

Let $\Sigma\ \nu_j A_j = \Delta A_j$, and similarly for other constants.

$$R \ln K = \frac{I_H}{T} + I_K + \Delta A_j \ln T A_j \ln T + \frac{\Delta B_j T}{2} + \frac{\Delta C_j T^2}{6} + \frac{\Delta D_j T^3}{12} \tag{EX-1.6}$$

$$\Delta H^\circ = I_H + \Delta A_j T + \frac{\Delta B_j T^2}{2} + \frac{\Delta C_j T^3}{3} + \frac{\Delta D_j T^4}{4} \tag{EX-1.7}$$

The integration constant I_K for a given reaction can be calculated by substituting the value of K at 298°K into Eq. EX-1.4 and solving for I_K after first solving for I_H in Eq. EX-1.5, using the value of ΔH° at 298°K.

The following results for the cyclohexane reaction were obtained by using constants from Reid et al. [2].

$$C_6H_6 + 3\,H_2 \leftrightarrow C_6H_{12}$$

Heat-Capacity Constants for c_p (cal/mole K)

	A_j	B_j	C_j	D_j	$H^\circ_{f_j}$	$\ln K_f$
C_6H_{12}	−13.03	0.1460	−6.027E − 5	3.156E − 9	−29.43	−12.8024
C_6H_6	− 8.101	0.1133	−7.206E − 5	1.703E − 8	19.82	−52.3009
H_2	6.483	0.00215	−3.298E − 6	1.826E − 9	0	0

$$\Delta H^\circ_{298} = -29.43 - (19.82) = -49.250\ \frac{\text{kcal}}{\text{g mole}} \text{ or } -49,250\ \frac{\text{cal}}{\text{g mole}}$$

$$\Delta A_j = -13.03 - [-8.101 + 3(6.483)] = -24.378$$

Similarly,

$$\Delta B_j = 0.02623$$

$$\Delta C_j = 2.1684\text{E} - 5$$

$$\Delta D_j = -1.9352\text{E} - 8$$

$$\ln K\ @\ 298 = -12.8024 - (-52.3009) = 39.4985$$

From Eq. EX-1.7 @ 298, $I_H = -43304$.
From Eq. EX-1.6 @ 298, $I_K = 67.86235$.
From Eq. EX-1.6 @ 400, $K = 66.4696$.
Values of K at 500 and 600 were obtained using the same I_H and I_K values in Eq. EX-1.6.

Results by this method are given in column 4 of Table EX–1.1.

CONCLUSIONS

Equilibrium conversions were calculated to aid in the comparisons at each of the temperatures for the several methods, using a nonlinear equation solver (IMSL NEQNF).* The value of K determined using tabulated values is best compared at 600°K, where it can affect the calculated conversion. All the methods give reasonable values. Method 1 is ideal for hand calculation, but Method 3 is most convenient for general design programs for which only enthalpies and entropies of formation at 298°K need be stored along with heat-capacity equations. The log K_f at 298°K can readily be calculated from $H^\circ_{f_j}$, and $S^\circ_{f_j}$ and corrected as shown by Method 3 to values at any temperatures of interest.

REFERENCES

1. Stull, D.R., Westrum, E.F., Jr., and Sinke, C.C., "The Chemical Thermodynamics of Organic Compounds," Wiley, New York, 1969.
2. Reid, R.C., Prausnitz, J.M., and Poling, B.E., "The Properties of Gases and Liquids," 4th Edition, McGraw-Hill, New York, 1987.

* Copyright by International Mathematical and Statistical Library, Houston, 1987.

EXAMPLE 2

Generation of Equilibrium Conversion and Adiabatic-Reaction Temperature Plots

OBJECTIVE

The value of equilibrium conversion and adiabatic temperature plots in selecting reactor type and operating conditions is demonstrated.

PROBLEM STATEMENT

Hydrogenation of benzene to cyclohexane is typical of many exothermic, reversible reactions where the desire is to approach complete conversion (99.95% in this case) so that a difficult separation step can be avoided. Favorable equilibrium is in the lower temperature range where rates are slower. Produce plots of equilibrium conversion versus temperature range of 400–600°K, and at various pressures and H_2/benzene ratios. Suggest favorable operating conditions for achieving 99.95% conversion, and also recommend type of fixed-bed reactor. Consider the effect of inerts for the case where lower-purity hydrogen must be used.

CONVENIENT SOFTWARE

The engineer should be able to quickly generate these plots using interactive software, so this important overview should never be ignored because of the press of time. Such a goal is readily realized by creating a simple driver program that will invoke a nonlinear algebraic equation solver for the equilibrium expression(s), solve the simple algebraic equation for the adiabatic heat balance, and plot the results. The IMSL* routine NEQNF is an efficient

* Copyright by International Mathematical and Statistical Library, Houston, 1987.

equation solver and was used for this example. (See Appendixes A.1 and A.2.) Results were dumped to a plotter.

THE EQUILIBRIUM RELATION

The equilibrium relation for entering into the program template is developed as follows:

$$C_6H_6 + 3\,H_2 \rightarrow C_6H_{12}$$

$$K = \frac{\hat{f}_C}{\hat{f}_B \hat{f}_H^3} = \left[\frac{P_C}{P_B P_H^3}\right]\left[\frac{\phi_C}{\phi_B \phi_H^3}\right] = \frac{P_C}{P_B P_H^3} K_\phi \qquad \text{(EX-2.1)}$$

where C, B, and H designate cyclohexane, benzene, and hydrogen, respectively; ϕ_C, ϕ_B, ϕ_H are the fugacity coefficients and K_ϕ is the fugacity quotient. First check the fugacity coefficient terms.

	$T_c(K)$	P_c *(atm)*	T_r *@ 500°K*	P_r *@ 500°K*	ϕ_i *from Fig. 3.2*
H_2	33.3	12.8	15	0.8	1.0
C_6H_6	562	48.6	0.84	0.21	0.92
C_6H_{12}	553	40.25	0.94	0.25	0.9

$$K_\phi = \frac{\phi_B}{\phi_C \phi_H^3} = 1.02$$

A 10% deviation of this ratio from unity generally produces less than 1% error in the conversion. Hence,

$$K = \frac{P_C}{P_B p_H^3} = \left[\frac{n_B}{n_C n_H^3}\right]\left[\frac{n_T^3}{P^3}\right] \qquad \text{(EX-2.2)}$$

where n_B, n_C, n_H are moles of indicated component per mole of benzene fed, and n_T is total moles on the same basis.

Basis: 1 mole benzene fed

	Moles
n_B	$1 - X_B$
n_H	$n_{H_0} - 3X_B$
n_C	X_B
Inerts	n_{I_0}
Total	$n_T = 1 + n_{I_0} + n_{H_0} - 3X_B$

$$K = \frac{X_B(1 + n_{I_0} + n_{H_0} - 3X_B)^3}{(1 - X_B)(n_{H_0} - 3X_B)^3 P^3} \qquad \text{(EX-2.3)}$$

$$F(1) = \text{RHS} - \text{K}$$

or in Fortran,

$$\text{F(1)} = \text{X(1)} * (1 + \text{PAR(1)} + \text{PAR(2)} - 3 * \text{X(1)}) ** 3/((1 - \text{X(1)}) * (\text{PAR(2)} - 3 * \text{X(1)} ** 3) * \text{PAR(3)} ** 3) - \text{PAR(4)} \qquad \text{(EX-2.4)}$$

Assign n_{I_0}, n_{H_0}, P, and K adjustable parameter names PAR(1) through PAR(4) so that various values can be entered interactively as desired.

For cases involving high pressure, such as ammonia synthesis, the K_ϕ term will become significant and must be included with the equilibrium constant as K/K_ϕ, as shown in Table 3.3. Values of the coefficients can most readily be determined from equations of state available in most professional process simulator programs.

THE HEAT BALANCE

For plotting purposes, Eq. 5.50 can be used for the heat balance.

$$[(\bar{c}_{p_B} + n_{H_0}\bar{c}_{p_H})(T_2 - T_1)]_{\text{In}} = (-\Delta H_B)_{T_2} X_B \qquad \text{(EX-2.5)}$$

The conversion X_B can be determined for various values of T_2 by using average heat capacities calculated from heat-capacity equations for T_1 to T_2. Heats of reaction at 25°C combined with heat-capacity equations yield the required values of ΔH at various T_2 values. Adiabatic lines tend to be linear and parallel over the practical range of interest. For quick hand calculation, therefore, one need calculate only one X_{B_2}, T_2 set of values, since the inlet temperature and zero conversion are given.

RESULTS

The program was used to calculate and plot curves for a range of pressures, temperatures, and feed compositions. The resulting plots are shown in Figs. EX–2.1 and 2.2. Tabular values of the conversions are given in Table EX–2.1 for a more exact comparison of the results at high conversions.

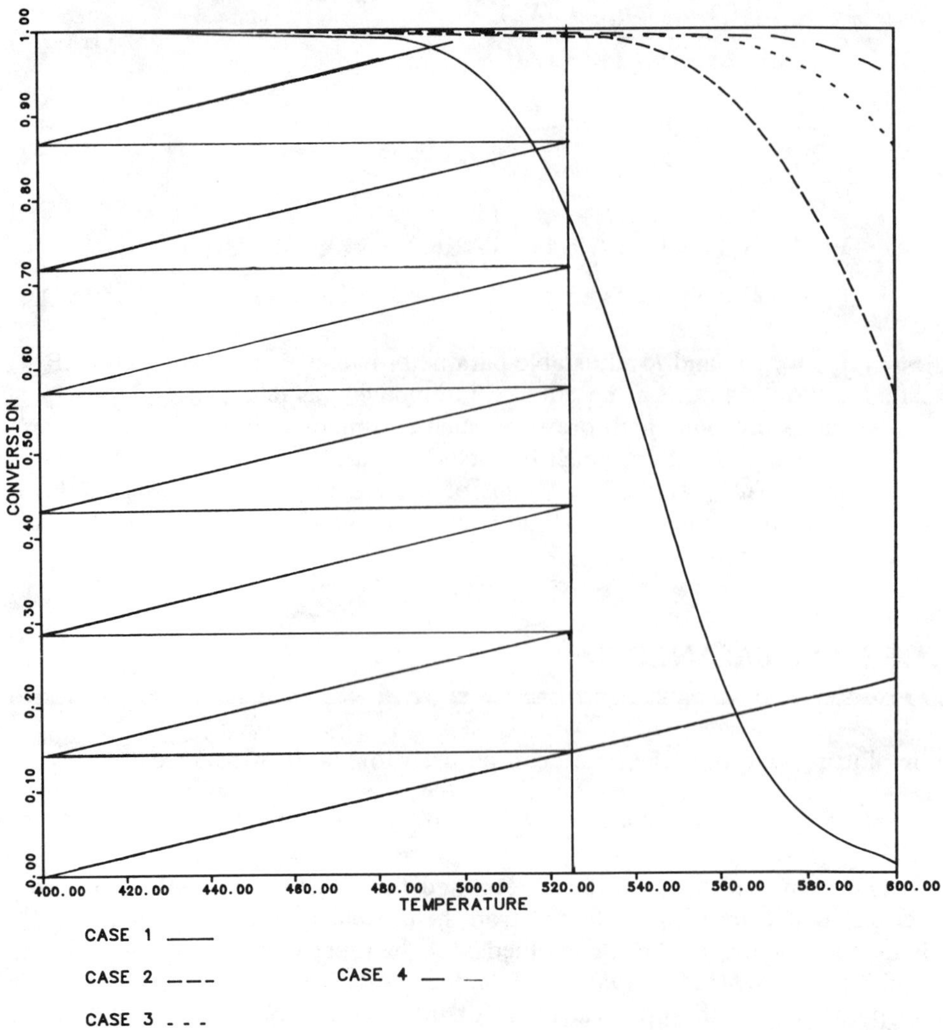

Figure EX-2.1 Cyclohexane equilibrium and adiabatic reaction temperature at H_2/ benzene ratio of 4 and various pressures

Multiple beds with maximum outlet of 525°K and inlet of 400°K are stepped off by hand-drawing adiabatic lines parallel to the one produced by the plotting routine and a horizontal line back to the 400°K position. At least seven stages are required to reach close to 100% conversion; pressures in atm: case 1 = 1, case 2 = 5, case 3 = 10, case 4 = 15; example stages at 10 atm.

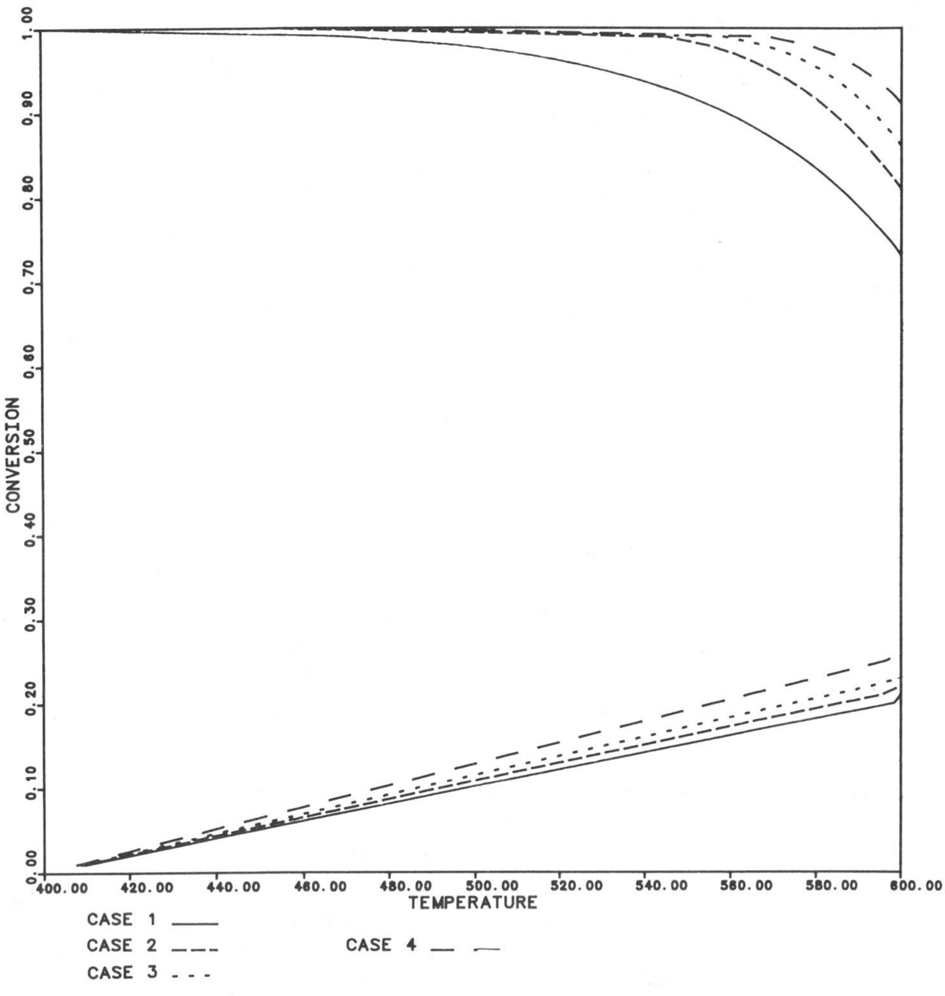

Figure EX-2.2 Cyclohexane equilibrium and adiabatic reaction temperature at 10 atm and various H_2/benzene ratios
H_2/benzene ratios: case 1 = 5, case 2 = 4, case 3 = 3.5, case 4 = 3.

CONCLUSIONS

Study of Figs. EX–2.1 and EX–2.2 and Table EX–2.1 reveals the following useful insights based on the need to attain 99.95% conversion of benzene.

- More that the stoichiometric amount of hydrogen is needed to attain the desired conversion. An excess will certainly increase the rate toward the outlet.

Table EX-2.1 Equilibrium Conversions for Benzene Hydrogenation (10 atm and Various H_2 and Inert Flows)

	EQUILIBRIUM CONVERSION (NO INERTS) H_2/C_6H_6 in Feed:				
	3	3.25	3.5	3.75	4.0
500°K	0.9775		0.9984		1.0000
600°K	0.7318	0.7748	0.8097	0.8384	0.8611

	EQUILIBRIUM CONVERSION (H_2:C_6H_6 in feed = 4) Inerts / C_6H_6 in Feed:			
	0	0.5(9.1%)	1.0(16.7%)	1.5(23%)
500°K	1.0000	0.9999	0.9998	0.9998
600°K	0.8611	0.8062	0.7551	0.7080

- Temperatures below ~460°K at the outlet appear necessary for the target conversion for H_2: benzene of 3.5 and above. Table EX–2.1 suggests a ratio of 4 for complete equilibrium conversion at 500°K.
- Inerts up to 20% do not affect equilibrium conversion significantly in the favorable temperature range of 500°K and below.
- If a completely gas-phase reaction is desired, the dew point at both inlet and outlet must be checked.
- A pressure of 10 atm appears to be reasonable, particularly if hydrogen is available in that pressure range.
- Because of the high exothermicity of the reaction, a heat-exchanger reactor should be considered for the gas-phase design.The goal must be to operate as much as possible below 500°K and not approach the range of hydrocracking. The possibility of reaching this goal can only be evaluated by pilot-plant or model studies (see Example 7).
- Adiabatic reactors in series with intermediate cooling require too many reactors for a practical system, as shown in Fig. EX–2.1, unless the slope of the adiabatic line is increased. Such a change can be accomplished by adding inerts or recycling product cyclohexane (see Example 7).
- High-pressure operation that would allow mixed-phase operation with vaporization of recycle cyclohexane would provide very effective direct-contact cooling in each adiabatic reactor.

- Clearly, a diagram of the type illustrated is most valuable for setting operating and design strategies for further investigation for a single-reaction system or a reaction system with a single dominating reaction.

CHECKING PROGRAM RESULTS

The user of a program should always do some shortcut calculations to check for reasonableness of the results of a program. In this case checking is quite simple, since the hand calculations of Eq. EX-2.3 and EX-2.5 are straightforward. For $n_H = 4$, $P = 10$, and no inerts, X_B was calculated by the program at 600°K to be 0.8611. Substituting X_B into Eq. EX-2.3 gives the RHS as 0.03077, which is the value of K. The graph also reads a conversion of 0.86 at these conditions.

The graphed adiabatic temperature line can be checked quickly by using Eq. EX-2.5 and heat capacities at the intermediate temperature of 500°K.

Basis: 1 mole benzene

$$\left[\sum n_j c_{p_j}(T_2 - T_1)\right]_{\text{In}} = (-\Delta H)_{T_2} X_B$$

$$[(1)(32.6) + (4)(6.993)]\,(600 - 400) = -(-35{,}570 - 16{,}710)\,0.23$$

$$12{,}114 = 12{,}024 \quad \text{which is close enough}$$

EXAMPLE 3

Equilibrium Composition for Multiple Reactions Using Simultaneous Solution of Equilibrium Expressions

OBJECTIVE

This example demonstrates the setting up of multiple equilibrium expressions for complex reactions and the simultaneous solution of such equations by nonlinear algebraic solvers.

PROBLEM STATEMENT

Calculate equilibrium conversions for synthesis gas ($CO + H_2$) from methane by steam reforming at 1000°K, 12 atm, and 2 : 1 steam/methane ratio in the feed.

General Statement of Reaction System

CH_4 (g) and H_2O (g)$\rightarrow$$CO$ (g), H_2 (g), and CO_2 (g)

ESTABLISHING STOICHIOMETRIC EQUATIONS

There is much lore and research on the synthesis-gas reaction system, but for purposes of illustrating how one should approach a new system, let us assume that all we know are the reactants and products as given above. Following the procedure on p. 72, write stoichiometric equations that yield the listed products. For example, using reasonable knowledge of chemistry and the minimum number of equations, yields

$$CH_4 + H_2O \rightleftarrows 3\,H_2 + CO \tag{1}$$

$$CH_4 + 2\,H_2O \rightleftarrows 4\,H_2 + CO_2 \tag{2}$$

$$K_1 = (K_f)_{CO} - (K_f)_{CH_4} - (K_f)_{H_2O} = 26.1722$$

$$K_2 = (K_f)_{CO_2} - (K_f)_{CH_4} - (K_f)_{H_2O} = 35.727$$

Values of K_f were obtained from Stull [1]. These equations do not imply any mechanistic conclusions. They represent overall reactions. In fact, other overall reactions can be written to yield the same products from the same reactants,

Table EX-3.1 Solutions to Equilibrium for Methane-Steam Reforming

	Case A	*Case B*
	$CH_4 + H_2O \rightleftarrows 3H_2 + CO$ (1)	$CH_4 + H_2O \rightleftarrows 3H_2 + CO$ (1)
	$CH_4 + 2H_2O \rightleftarrows 4H_2 + CO_2$ (2)	$CO + H_2O \rightleftarrows H_2 + CO_2$ (2)
Extent of Reaction		
X_1	0.2872	0.8581
X_2	0.5709	0.5709
Mole Fractions at Equilibrium		
CH_4	0.0163	
H_2O	0.5244	
H_2	0.3609	
CO	0.0329	
CO_2	0.0655	
Computation Time*	0.25-0.28 seconds	0.25-0.28 seconds
Typical Successful Initial Guesses: X_1, X_2	0.3, 0.6; 0.2, 0.5; 0.1, 0.2; 0.4, 0.5, 0.2, 0.7	0.8, 0.5; 0.5, 0.1; 0.9, 0.1; 0.6, 0.2; 6.7, 0.3; 0.4, 0.2

*Cyber 170/750

and the following reaction is often used in place of reaction (2) because it corresponds more closely to an accepted empirical kinetic scheme for the oxidation of CO by steam:

$$CO + H_2O \rightleftarrows H_2 + CO_2 \tag{3}$$

$$K_3 = (K_f)_{CO_2} - (K_f)_{CO} - (K_f)_{H_2O} = 1.368$$

Based on the rule from p. 72, reaction (3) is not required along with reactions (1) and (2), since it, like reaction (2), describes the production of CO_2. The redundancy of both reactions (2) and (3) together can also be shown by dividing K_2 by K_1, which yields K_3.

$$\frac{K_2}{K_1} = 1.368 = K_3$$

In industry, reactions (1) and (3) are normally used in calculations and discussions about this system, since they are the basis of most empirical kinetic schemes. We will use both combinations (case A: 1 and 2; case B: 1 and 3) to illustrate that either pair will provide the same result for equilibrium conditions as shown in Table EX–3.1.

EQUILIBRIUM EXPRESSIONS

After writing the moles of each component in terms of extent of reaction, one may formulate the equilibrium expressions in terms of the extents of reaction on a per mole of CH_4 fed.

Basis: 1 mole CH_4, n_{w_0} moles steam

Component	Case A: Reactions 1 and 2 *moles/mole CH_4 fed*	Case B: Reactions 1 and 3 *moles/mole CH_4 fed*
$CH_4 = M$	$1 - X_{1a} - X_{2a}$	$1 - X_{1a}$
$H_2O = W$	$n_{W_0} - X_{1a} - 2X_{2a}$	$n_{W_0} - X_{1a} - X_{2a}$
$H_2 = H$	$3X_{1a} + 4X_{2a}$	$3X_{1a} + X_{2a}$
$CO = CO$	X_{1a}	$X_{1a} - X_{2a}$
$CO_2 = CO_2$	X_{2a}	X_{2a}
Total $= T$	$1 + n_{W_0} + 2X_{1a} + 2X_{2a}$	$1 + n_{W_0} + 2X_{1a}$

$$K_1 = \frac{n_H^3}{n_M} \frac{n_{CO}}{n_W} \frac{P^2}{n_T^2} \qquad \text{(EX-3.1)}$$

$$K_2 = \frac{n_H^4}{n_M} \frac{n_{CO_2}}{n_W^2} \frac{P^2}{n_T^2} \qquad \text{(EX-3.2)}$$

$$K_3 = \frac{n_H n_{CO_2}}{n_{CO} n_W} \qquad \text{(EX-3.3)}$$

The second reaction for case B is reaction (3), but the extent of reaction is called X_{2a} since it is for the second equation for case B. At conditions of interest, the fugacity coefficient term is essentially unity in each case.

METHOD OF CALCULATION

An interactive driver program (see Appendix A.1) was employed to solve the simultaneous equations by using the IMSL routine NEQNF [2]. The following equations were entered. The respective mole subscripts are used as symbols for the moles of a component per mole CH_4 fed.

Case A

Fortran statements for Eqs. 3.1 and 3.2:

```
F(1) = H ** 3 * CO * PAR(1) ** 2/(M * W * T ** 2) - PAR (3)
F(2) = H ** 4 * CO2 * PAR(1) ** 2/(M * W ** 2 * T ** 2) - PAR (4)
```

where PAR (1) = P, PAR (2) = n_{W_0}, PAR (3) = K_1, and PAR (4) = K_2. H, CO, M, W, T refer to moles per mole of methane fed.

Case B

Fortran statements for Eqs. 3.1 and 3.3:

```
F(1) = H ** 3 * CO * PAR (1) ** 2/(M * W * T ** 2) - PAR (3)
F(2) = CO2 * H/(CO * W) - PAR (4)
```

where PAR(4) now represents K_3. The values of the parameters could be changed interactively in order to calculate results rapidly at other conditions.

The mole quantities per mole of methane fed are given above in terms of extents of reaction for the two reactions being considered for each case. These quantities were placed in a subroutine for easy transfer to the functional relations shown for cases A and B.

INITIAL GUESSES

Nonlinear equation solvers require initial guesses. These do not have to be accurate for efficient solvers, but they must be constrained for physical realities. In this case inspection of the mole balances and the magnitude of the equilibrium constants reveal the following constraints on extents of reaction:

Mole Balances	Magnitude of Equilibrium Extents of Reaction
$X_{1a} + X_{2a}$ must be less than 1	$X_{1a} < X_{2a}$ for case A
$X_{1a} \neq X_{2a}$	$X_{1a} > X_{2a}$ for case B

Within the confines of these constraints a wide range of initial guesses gave essentially instant solution, as shown in Table EX–3.1.

RESULTS

The two methods of formulating stoichiometric statements for production of the products gave identical compositions for the equilibrium mixtures, as shown in Table EX–3.1.

CONCLUSIONS

When possible, overall reactions can be reasonably stated based on general knowledge of the chemistry. Solution of the resulting algebraic equations is simple and straightforward with modern nonlinear algebraic equation solving software. In setting up the equations, follow the rules on p. 72.

REFERENCES

1. Stull, D.R., Westrum, E.F., Jr., and Sinke, G.C., "The Chemical Thermodynamics of Organic Compounds," Wiley, New York, 1969.
2. IMSL MATH/LIBRARY, "NEQNF, Nonlinear Equation Solver," IMSL, Houston, 1987.

EXAMPLE

4

Use of Equilibrium Calculations to Estimate Possible Contaminants

OBJECTIVE

This example demonstrates how equilibrium calculations can be used to determine if an operation may be conducive to the formation of an unwanted side product in small but unacceptable amounts.

PROBLEM STATEMENT

Isopropyl alcohol is to be heated in copper tubes to 400°C at 1 atmosphere. It is known that copper is a catalyst for production of acetone and propionaldehyde, along with perhaps some isomerization of isopropyl to normal propyl alcohol. These three products can be shown with the following independent reactions in the gas phase.

$$\underset{\text{IP}}{iC_3H_7OH} \rightleftarrows \underset{\text{NP}}{nC_3H_7OH} \qquad K_1 = 0.064$$

$$\underset{\text{IP}}{iC_3H_7OH} \rightleftarrows \underset{\text{AC}}{(CH_3)_2CO} + \underset{\text{H}}{H_2} \qquad K_2 = 0.076$$

$$\underset{\text{IP}}{iC_3H_7OH} \rightleftarrows \underset{\text{PR}}{C_3H_5CHO} + \underset{\text{H}}{H_2} \qquad K_3 = 0.00012$$

Subsequent processing steps require that the alcohol, although it may contain acetone, not contain more than 0.05 mol% of propionaldehyde. Is there any possibility that this specification will be violated at the reaction conditions described?

EQUILIBRIUM EXPRESSIONS

As in Example 3, the compositions can readily be written in terms of extents of reaction.

Basis: 1 mole of isopropyl alcohol fed

$$\mathrm{IP} = 1 - X_{1a} - X_{2a} - X_{3a}$$
$$\mathrm{NP} = X_{1a}$$
$$\mathrm{AC} = X_{2a}$$
$$\mathrm{PR} = X_{3a}$$
$$\mathrm{H} = X_{1a} + X_{2a}$$
$$\text{Total} = T = 1 + X_{2a} + X_{3a}$$

EQUILIBRIUM FUNCTIONS

Fortran Statements:

```
F(1) = NP/IP - PAR(1)
F(2) = AC * H/(NP * T) - PAR(2)
F(3) = PR * H/(NP * T) - PAR(3)
```

where PAR(1) = K_1, PAR(2) = K_2, PAR(3) = K_3. The values of the components in the above equations are read from a subroutine, and the solver is executed to yield the results in 0.25 sec. Almost any initial guesses worked as long as $(X_{1a} + X_{2a} + X_{3a}) < 1.0$. If $(X_{1a} + X_{2a} + X_{3a}) \geq 1.0$, the denominator would be zero or negative.

RESULTS

$$X_{1a} = 0.0446$$
$$X_{2a} = 0.2579$$
$$X_{3a} = 0.0004$$
$$\%\text{Propionaldehyde} = \frac{(X_{3a})(100)}{1 + X_{1a} + X_{2a} + X_{3a}} = \frac{0.0004}{1.2583}(100) = 0.03\%$$

The specification is not violated even at equilibrium, and probably even less will be produced in the actual heat exchanger. This statement is based on conditions at equilibrium. If the mechanism of the reaction system were such that propionaldehyde formed as a short-lived intermediate, it is conceivable that larger amounts might be present for a short contact-time heater. This scenario does not seem reasonable for this system, and the direct formation of propionaldehyde is a reasonable path that would be deterred by the greater tendency for the reverse reaction to occur, which would be apparent if kinetic data were available.

CONCLUSIONS

Equilibrium calculations can be useful in estimating the possibility of the appearance of certain small amounts of undesired side reaction products in the total product. Care must be exercised in drawing conclusions, based on mechanistic arguments and other knowledge, on the chemistry and kinetics of the system, since other compositions will apply away from equilibrium. At the very least, however, the equilibrium calculations provide a useful and simple means for speculating about and comparing different operating conditions.

EXAMPLE 5

Equilibrium Composition for Multiple Reactions Using Free-Energy Minimization

OBJECTIVE

This example illustrates the technique of free-energy minimization using professional software. The ease of use and the advantages and disadvantages of the method are demonstrated.

PROBLEM STATEMENT

The example selected here is the highly successful catalytic reforming process (not to be confused with steam reforming) that is used to make base stock for unleaded gasoline by converting with Pt-Re/alumina catalyst a low-octane paraffinic feedstock to a more aromatic, high-octane product in the same approximate boiling range. Although the feedstock is made up of C_6–C_9 hydrocarbons, only the C_7 family will be used to demonstrate a typical equilibrium calculation that can be useful in planning operating conditions. The following compounds may be present in the reaction system. The numbers shown are the percentages from a gas chromatographic analysis of a light naphtha. They do not add up to 100, because only the C_7's are considered for this example.

2,2-Dimethylpentane	0.3	2-Methylhexane	9.1
2,3-Dimethylpentane	3.6	3-Methylhexane	11.1
2,4-Dimethlypentane	1.0	2,2,3-Trimethylbutane	0
3,3-Dimethylpentane	0.2	1,cis-2-Dimethylcyclopentane	0.3
3-Ethylpentane	1.0	1,cis-3-Dimethylcyclopentane	2.3
Heptane	20.3	1,1-Dimethylcyclopentane	1.5

Ethylcyclopentane	0.6	1,trans-3-Dimethylcyclopentane	2.1
Methylcyclohexane	7.7	Toluene	2.9
1,trans-2-Dimethylcyclopentane	3.8		

The feed components are H_2 plus the liquid feed. It is important to include all expected products, even though they are not in the feed. Toluene, which is the major aromatic product, and all the paraffinic isomers are present in the feed and are included in the feed list. Light paraffinic hydrocarbons, C_1–C_5 will also be present in the product, but these are obviously produced by cracking reactions, which are irreversible under the operating conditions and contact time for commercial units. Thus, the C_1–C_5 paraffinic products should not be included, since the calculations will force all reactions to equilibrium or completion and produce a trivial result in terms of cracking [1].

Determine the effect of temperature in the range (700–1000°F) at the following conditions.

200 psia and 3 : 1 H_2/oil in feed
400 psia and 7 : 1 H_2/oil in feed

The equilibrium compositions can be readily calculated by commercial simulation software such as CHEMSHARE DESIGN II™ [2]. A user-friendly PC program is used to enter all components (both reactants and products) designated in accordance with the specific system ID names and the amounts of each in the feed. Again, all products of interest must be designated whether present in the feed or not. The desired temperature and pressure are set, and the program is transferred to a VAX/VMS computer for execution. The program on the VAX determines the composition of the reaction mixture at equilibrium by determining the composition that minimizes the total free energy as described with Eq. 3.38.

RESULTS

The results with the CHEMSHARE™ program, General Equilibrium Reactor, are summarized in Fig. EX–5.1 and EX–5.2 for the desired operating conditions. Table EX–5.1 presents a typical printout of the product mole fraction. The spaces between curves give the mole fraction of the indicated component in the hydrocarbon product mixture. The lower-pressure operation with less hydrogen is favorable as is clearly shown, for example, by the reaction stoichiometry in the single reaction of heptane, which involves an increase in

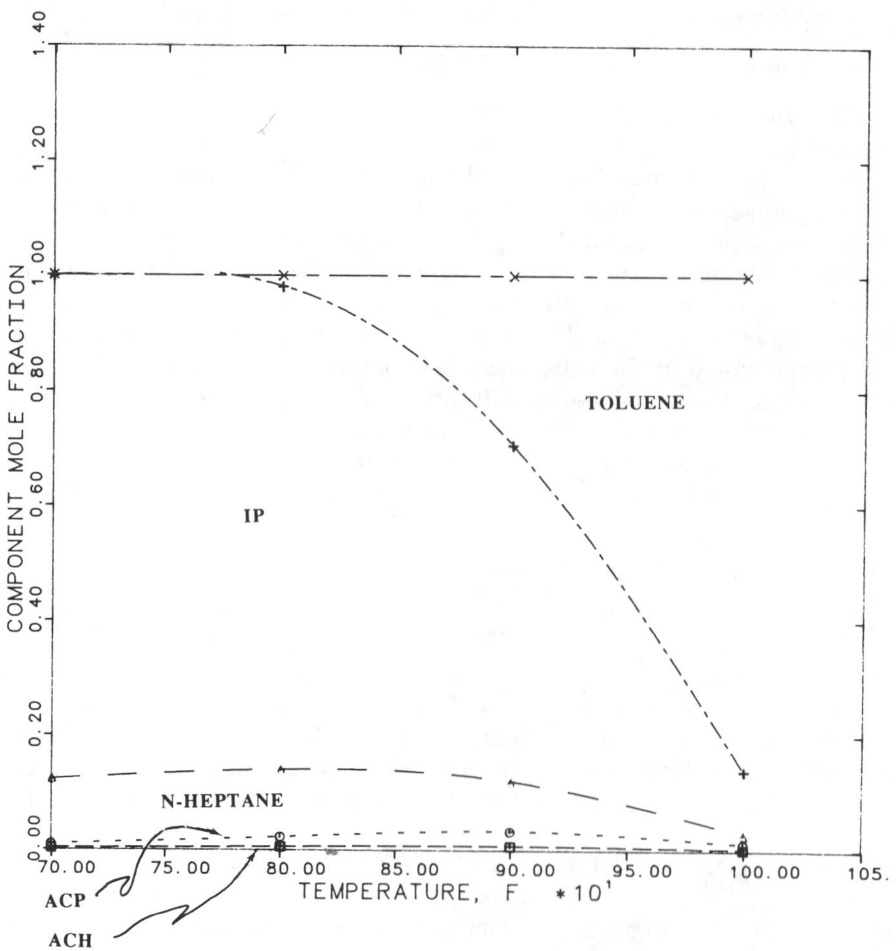

Figure EX-5.1 C_7 reformer equilibrium
P = 400 psia (27.2 atm); H_2/oil = 7; region between curves represents the mole fraction of the indicated component; ACH = alkylcyclohexanes, ACP = alkylcyclopentanes, IP = isoparaffins.

both total moles and hydrogen as a product.

$$C_7H_{16} \rightleftarrows CH_3C_6H_5 + 4\,H_2$$

The equilibrium plot shows clearly the magnitude of the improvement in equilibrium conversion at the lower pressure, compared with operation at higher pressure and higher hydrogen ratio. The savings in energy costs are

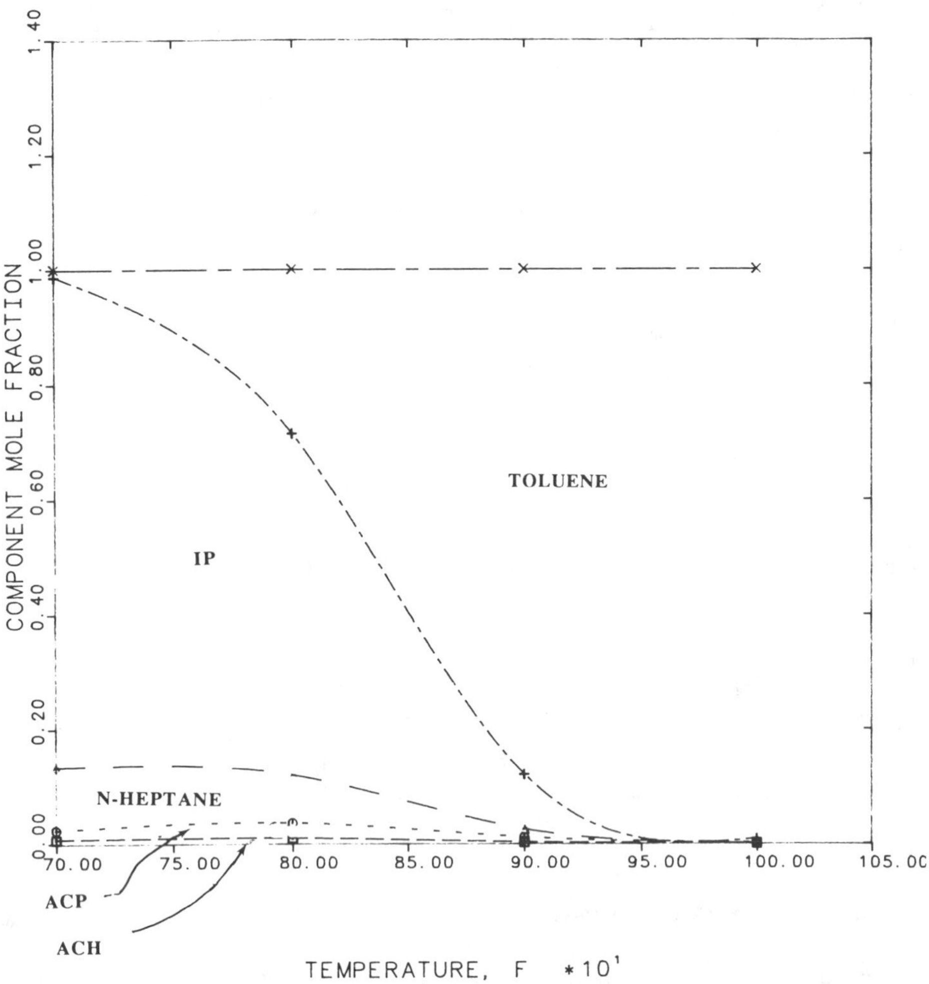

Figure EX-5.2 C_7 reformer equilibrium
P = 200 psia (13.6 atm); H_2/oil = 3; see note on Fig. EX-5.1.

considerable. This possibility encouraged development of the newer catalyst with both Pt and Re on alumina that made lower hydrogen-pressure operation possible. The Pt/Re-alumina catalyst not only forms coke more slowly but also tolerates more coke before experiencing significant deactivation than does Pt-alumina catalyst. Thus it is possible to use less hydrogen for coke inhibition with Pt/Re and gain the more favorable equilibrium conditions for aromatics formation.

Table EX-5.1 C_7 Reforming Equilibrium Mole Fraction

	700°F	*800°F*	*900°F*	*1000°F*
Pressure: 400 psia	H_2/Oil = 7			
Alkylcyclopentanes	0.0070	0.0169	0.0255	0.0093
Methylcyclohexane	0.0034	0.0064	0.0076	0.0023
Heptane	0.1132	0.1176	0.0866	0.0170
i-C_7's	0.8759	0.8403	0.5870	0.1102
Toluene	0.0005	0.0169	0.2933	0.8604
Pressure: 200 psia	H_2/Oil = 3			
Aklylcyclopentanes	0.0163	0.0265	0.0086	0.0011
Methylcyclohexane	0.0080	0.0099	0.0026	0.0003
Heptane	0.1105	0.0834	0.0142	0.0009
i-C_7's	0.8533	0.5963	0.0959	0.0065
Toluene	0.0119	0.2840	0.8788	0.9910

CHECKING THE SOFTWARE

It is always desirable to compare the results of computer calculations with a calculation by some independent means. In this case extent of reaction for a single reaction of heptane to toluene can be compared by using the free-energy minimization software and the equilibrium constant method.

$$C_7H_{16} \leftrightarrows CH_3C_6H_5 + 4\,H_2$$

Conditions: 200 psia; H_2/oil = 3

C_7	$1 - X$
Toluene	X
H_2	$\underline{3 + 4X}$
Total	$4 + 4X$

$$K = \frac{X(3+4X)^4}{(1-X)(4+4X)^4} P^4$$

Using the nonlinear solver described in Examples 1 and 3, one compares the resulting conversions with those obtained from the free-energy minimization software for the same single reaction in Table EX–5.2. The agreement is good, especially when the exponential effect of errors in ΔG_f° on K values is considered. At 700°K the difference between the apparent K (46,105), obtained from the mole fractions determined by the free-energy minimization, and that calculated from the data of Stull et al. [3] (38,636) is 19%. The corresponding ΔG_f° differ by only 1.7% (14,936 and 14,690), which demonstrates the exponential effect of error on the equilibrium constant. The resulting error in conversions, of course, declines rapidly at increasing values of conversion (see Table Ex–5.2).

Table EX-5.2 Comparison of Conversions by Two Calculational Procedures

$$C_7H_{16} \rightarrow CH_3C_6H_5 + 4H_2$$

Temperatures, °K	*Conversions, % from ΔG Minimization*	*Conversion, % from K Values*	*Δ%*
700	71.64	68.21	5
800	99.82	99.78	0.04

REFERENCES

1. Kugelman, A.M. *Hydrocarbon Process*, *35* (1), 95 (1976).
2. CHEMSHARE DESIGN II™, "General Equilibrium Reactor," ChemShare Corp., Houston, 1986.
3. Stull, D.R., Westrum, Jr., E.F., and Sinke, G.C., "The Chemical Thermodynamics of Organic Compounds," Wiley, New York, 1969.

EXAMPLE 6

Design of Reactors Based on Equilibrium Approach

OBJECTIVE

In this example a procedure is presented for designing a fixed-bed reactor in which the several reactions approach equilibrium closely by a known ΔT based on plant experience.

PROBLEM STATEMENT

The endothermic steam-reforming reaction is used to illustrate the technique

$$CH_4 + H_2O \leftrightarrow 3\,H_2 + CO \tag{1}$$

$$CO + H_2O \leftrightarrow H_2 + CO_2 \tag{2}$$

BACKGROUND

Steam reforming of light hydrocarbons, primarily CH_4, is a major source of hydrogen for use in producing products as varied as edible vegetable fats, nylon precursors, and many specialty products. It is also used to make synthesis gas (CO/H_2) for methanol and OXO synthesis, and N_2/H_2 mixtures in conjunction with a secondary reformer to oxidize excess CH_4 to CO and CO_2 with air, and thereby provide the required amount of N_2 from the air to yield a 3:1 H_2/N_2 mixture for ammonia synthesis.

A direct-fired furnace reactor is used (see Fig. 1.5) with multiple tubes containing Ni deposited on pellets having a Raschig-ring shape, produced from a silica-free, ceramic support.

PROBLEM DETAILS

(a) Do the process design for a reactor for this process involving the determination of the number of 5″ ID (5.75″ OD) × 40′-0″ long tubes required to produce 50 million SCFD (60°F, 1 atm) of H_2 with 95% purity. Estimate product composition using the empirically observed equilibrium approach, ΔT, for the particular catalyst used as 25°C (45°F) for reaction 1 and 0°C for reaction 2. Design conditions at the outlet are 1460°F (1067°K) and 12.2 atmospheres. Compare results with data from an operating plant (Table EX–6.1), as has been previously shown [1].

(b) Consider the effect of changing to a more active catalyst, and recommend operating conditions that will minimize steam usage.

Table EX-6.1 Steam Reformer Operation-Predicted and Observed Performance

Gas Inlet Temperature	687°F (637°K)
Gas Outlet Temperature	1460°F (1067°K)
Outlet Pressure	12.2 atm
Approximate Mean Heat Flux (based on outside tube area)	17,000 BTU/1hr ft^2

H_2/CH_4 Molar Ratio in Feed: 5.7

	Calculated	Operating Plant (2)
% CH_4 Reacted	91.65%	91.7%
No. of Tubes	227	260
Mass Velocity		
Inlet Pressure	14.33 atm	14.3 atm
Outlet Pressure	Use 12.2 atm	12.2 atm

Discharge Composition (Calculated)

$X_{1a} = 0.9165$

$X_{2a} = 0.5305$

	Moles/Mole CH_4	Mole Fraction
CH_4	0.0835	0.0098
W	4.2530	0.4984
H_2	3.2799	0.3844
CO	0.3860	0.0452
CO_2	0.5305	0.0622
Total	8.5329	1.0000

CH_4 (dry gas basis): 1.95%

DESIGN DETAILS

As shown in Section IIIB of Chapter 8 the equilibrium approach for an endothermic reaction is related to the operating temperature T_{op} as follows:

$$\Delta T_a = T_{eq} - T_{op}$$

where T_{eq} is the temperature at which the equilibrium is equal to the reactor outlet composition quotient, and, therefore, the temperature at which the equilibrium conversion is the same as the outlet conversion.

OUTLET COMPOSITION AND CONVERSION

In order to compare the calculated results with the actual results, inlet and outlet temperature and outlet pressure of the operating unit will be used as input parameter values, as shown in Table EX–6.1.

Solve the equilibrium expressions for reactions 1 and 2 as shown in Example 3 with $P = 12.2$ atm, $T_{op} = 1460$°F (1067°C) at outlet and K values as follows:

$K_1 = 78.36$ @ 1415°F
$\Delta T_a = -45$°F
T_{eq} for reaction $1 = T_{op} + \Delta T_a = 1460 - 45 = 1415$°F
$K_2 = 1.06$ @ 1460°F, since $\Delta T_a = 0$ for this reaction

Results are summarized in Table EX–6.1. The operating data included only a CH_4 conversion, which is essentially the same as that obtained from the calculation.

REQUIRED NUMBER OF TUBES

The number of 5″ ID (5.75″ OD) × 40-ft tubes with 37 ft of heated length required can be calculated based on a heat balance [1].

Basis: 1 mole CH_4 fed, 5.7 moles steam, and inlet temperature T_1

$$q_m = \left[\sum n_j \bar{c}_{p_j}(T_2 - T_1)\right]_{\text{Feed}} + \sum(\Delta H_n)_{T_2} X_{na}$$

where q_m = cal/g mole CH_4 fed added by the furnace burners.

$$q_{m} = (\bar{c}_{p_M} + 5.7\bar{c}_{p_W})(1067 - 637) + (\Delta H_1)_{T_2} X_{1a} + (\Delta H_2) X_{2a}$$

where subscripts M and W refer to CH_4 and H_2O, respectively.

Data from Stull (1)	H_{f_j} @ 1067°K	$\bar{c}_{p_j}$ @ (1067 + 637)/2
CO	−26857	
H_2O	−59347	9.42
CO_2	−94345	
CH_4	−21568	15.49

$$\Delta H_1 = 54058 \qquad \Delta H_2 = -8150$$

$$q_m = [15.49 + 5.7(9.42)](1067 - 637) + (54058)(0.9165) + (-8150)(0.5305)$$

$$q_m = 74969.7 \text{ cal/g mole } CH_4 \text{ fed}$$

Required CH_4 flow (basis: 50×10^6 SCFD, 95% H_2):

The product from the reformer is purified by using a two-stage shift converter to react CO to CO_2, and subsequently most of the CO_2 is removed by monoethanolamine absorption.

From Table EX–6.1 the moles of H_2 in product per mole methane fed is 3.2799.

$$CH_4 \text{ flow} = \frac{(50 \times 10^6)(0.95)}{(24)(379.48)(3.2799)} = 1590 \text{ lb mole/hr}$$

$$\text{No. of tubes} = \frac{\text{heat load}}{(\text{heat flux})(\text{heat transfer area/tube})}$$

$$= \frac{(1.8)(74969.7)(1590)}{(17{,}000)(\pi)(5.75/12)(37)} = 227$$

This result compares favorably (5% error) with the actual number of 260. The actual furnace may have been slightly overdesigned. The smaller number of tubes, if used in another plant having the same capacity, will produce a higher ΔP.

PREDICTING PERFORMANCE OF MORE ACTIVE CATALYST

Assuming equilibrium conversion for both reactions when using the more active catalyst, plots of equilibrium methane percentage in the product were readily generated for various pressure, temperature, and steam-to-methane ratios by using the program described in Example 3. Results are shown in Fig. EX–6.1.

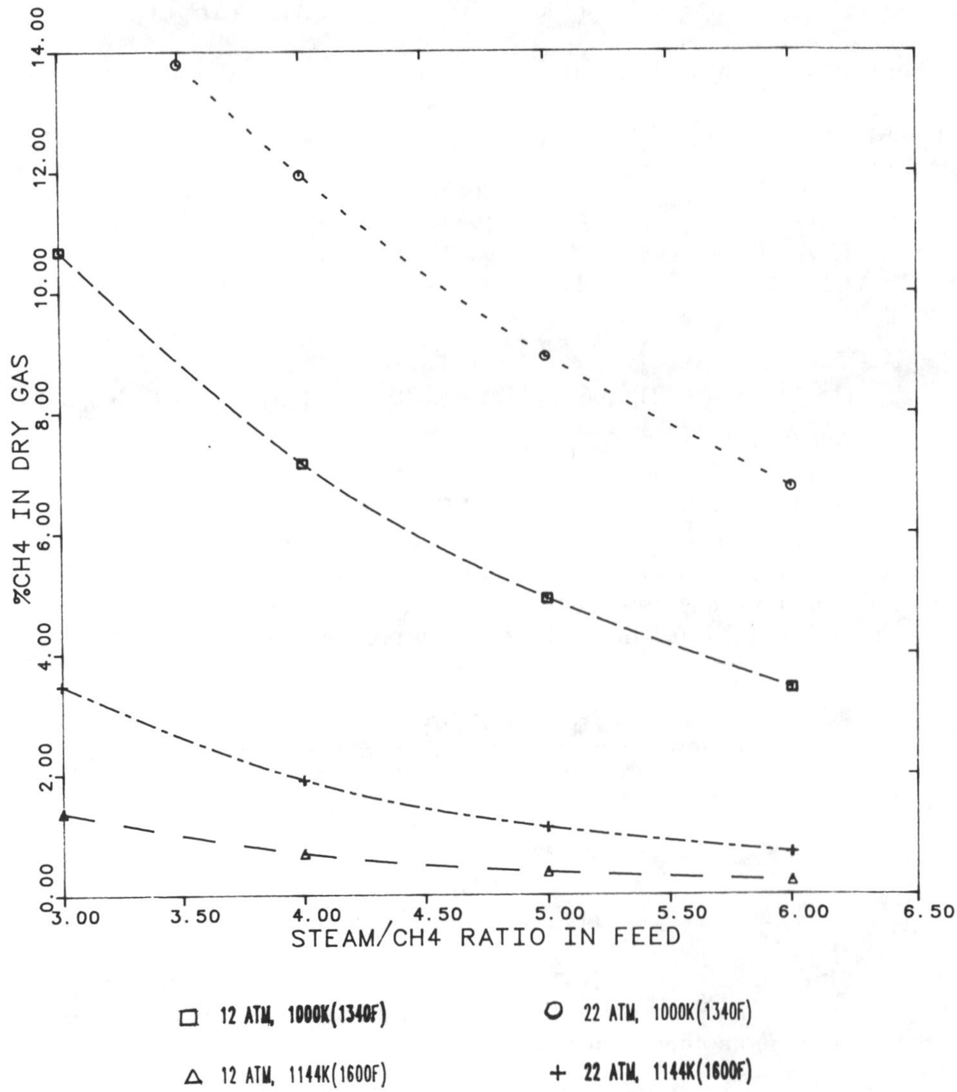

Figure EX-6.1 Equilibrium CH_4 in product of steam reforming: dry gas basis
Characteristic of a high-activity catalyst for which equilibrium for major reactions is a reasonable assumption.

With reference to Fig. EX–6.1, higher pressure can be seen to inhibit the reaction and thus cause more residual CH_4 in the product at the same temperature. Higher pressure (22–25 atm) is attractive in the many cases where natural gas is delivered at these pressures, since less energy will be needed to compress product hydrogen to pressures at or above this level.

Clearly, the disadvantage of increased pressure on conversion can be overcome by using higher temperature, which will also allow lower steam requirements as well. The design case for the less-active catalyst, summarized in Table EX–6.1 but not shown in Fig. EX–6.1, used a steam-to-CH_4 ratio of 5.7:1 at 12 atm and 1460°F, and produced a methane residual of 1.95% in the dry product gas. This same residual can be attained at 1600°F, 4:1 steam-to-CH_4 ratio, and 22 atm by using the high-activity catalyst for which equilibrium for both reactions is approximated (see Fig. EX–6.1). A savings in steam and compression energy costs would thus be realized by changing to the high-activity catalyst.

There is great incentive, because of the high cost of steam, to lower the steam ratio even further, but that must be done with the willingness to accept higher CH_4 residuals in the reformer outlet stream. If high-purity hydrogen is required, pressure-swing adsorption with molecular sieves will be necessary to lower the CH_4 content to an acceptable level. A steam-to-CH_4 ratio as low as 2.5 is economically viable when pressure-swing adsorption is used.

AVOIDING CARBON FORMATION

As the steam-to-CH_4 ratio is lowered, there is danger of entering a regime where carbon formation becomes thermodynamically favorable, and such a situation must be avoided. This dilemma affords another opportunity to apply equilibrium data. It is possible to set operating conditions under which undesired side reactions can be avoided when such reactions are restricted to a specific region beginning at a condition where a reaction just begins to proceed in the undesired direction, as determined from equilibrium considerations.

In steam reforming, an undesired product is a carbon deposit that can form as whiskers on the catalyst and cause fracturing of the catalyst, which in turn produces increased pressure drop. Equilibrium values of the carbon-forming reactions prove to be useful guides to operating conditions that will not be conducive to carbon production.

Since equilibrium of the two reactions forming CO and CO_2 is nearly reached in practical reactors and carbon does not usually form, it is reasonable to assume equilibrium of these two reactions near the catalyst surface and then determine the tendency to form carbon by comparing the equilibrium gaseous component quotient (Q) with the equilibrium constant for any carbon-forming reactions.

$$2\,\mathrm{CO} = \mathrm{C} + \mathrm{CO_2} \qquad Q_4 = \frac{P^*_{\mathrm{CO_2}}}{(P^*_{\mathrm{CO}})^2} \tag{4}$$

$$\mathrm{CH_4} = \mathrm{C} + 2\,\mathrm{H_2} \qquad Q_5 = \frac{(P^*_{\mathrm{H_2}})^2}{P^*_{\mathrm{CH_4}}} \tag{5}$$

where the asterisk indicates the equilibrium value of the indicated partial-pressure terms based on the main reactions (1 and 2). Since the activity of graphite is taken as unity, the quotients should be equal to the respective equilibrium constants when these reactions reach equilibrium. When the respective quotients are greater than the appropriate equilibrium ratios, there is no driving force to form carbon.

Thus for conditions under which carbon will not form,

$$\frac{P^*_{CO_2}}{(P^*_{CO})^2} > K_4$$

$$\frac{(P^*_{H_2})^2}{P^*_{CH_4}} > K_5$$

Either criterion may be used. Consider the calculations summarized in Table EX–6.2 at 1600°F (1144°K), 22 atm, and various H_2O/CH_4 ratios. As the H_2O/CH_4 ratio approaches unity and below, the tendency to produce carbon becomes a possibility. There are several reasons for not approaching such regions too closely [2].

Table EX-6.2 Effect of H_2O/CH_4 Ratio in Feed on Equilibrium Quotient, $(P^*_{H_2})^2/P^*_{CH_4}$ at 22 atm and Indicated Outlet Temperature

Feed H_2O/CH_4	*Quotient,* $(P^*_{H_2})^2/P^*_{CH_4}$ 1144°K (1600°F)	1089°K (1500°F)
6	947	
5	635	
4	404	151.35
3	238	98.54
2	125	
1	49.5	
Carbon Forming Reaction:	$CH_4 \rightleftarrows C + 2H_2$	$K_5 = 41.0$ @ 1600°F,
Goal:	$\frac{(P^*_{H_2})^2}{P^*_{CH_4}} > K_5$	25.35 @ 1500°F

1. Equilibrium for the reaction in question may not occur, and the rate process may dominate.
2. The criterion should be calculated along the entire temperature gradient of the reactor.
3. Radial gradients or other gradients may cause the catalyst to be at a higher temperature at some point. (In this example the temperature is higher at the tube wall, causing K_5 to increase.)
4. The form or characteristic of a solid product may be different than the one upon which thermodynamic data are based. In this case the carbon is graphitic, but it is in the form of very thin whiskers, which changes its surface energy characteristics.
5. The undesired product may form by other means. In this case thermal pyrolysis produces a different kind of coke that covers the active nickel and reduces its activity. The coke-forming rate of pyrolysis is enhanced as pressure is increased. It is also inhibited by the presence of steam.

The plant operator must therefore not rely entirely on thermodynamic considerations, but they do show that, in the case of steam-reforming, catalytic carbon formation will not be a problem at steam ratios as low as 2:1. Of course, much experience has been accumulated by gradually lowering the ratio and observing the result. In well-designed furnaces with good control of heat flux, operation at a 2:1 ratio has been successful.

MAXIMUM OPERATING TEMPERATURE

The advantage of high-temperature operation must be considered together with disadvantages of higher tubewall temperatures. Tube life is an important factor in furnace economics, and tube material and thickness, along with furnace design, play an important role in setting the hours of service and reliability essential for minimizing production costs over a long period.

There are a number of high-temperature centrifugal-cast tube types, most of which are high Ni-Cr steels with various other elements in smaller amounts, such as Co and Nb. Operating factors that affect tube life include temperature, pressure, and frequency of shutdown. Increases in any one of these reduces tube life, but temperature plays a dominant role with even a 10° increase over a prolonged period producing as much as a 30% drop in tube life [2].

A typical design goal is a tube life of 100,000 hours. Since the maximum tube temperature will be as high as 100°F above the maximum outlet gas temperature, a typical maximum outlet value is in the range of 1500–1600°F, depending on the tube material and furnace design. Care should be exercised in selecting a catalyst with adequate life (3–5 years) under the planned operating conditions.

CONCLUSIONS

The techniques for using equilibrium calculations for reactor planning and design are applicable to other reaction systems that approach or reach equilibrium. They are also useful in diagnosing plant difficulties. Charts such as Fig. EX–6.1 are useful in getting an idea of the extent of deviation from equilibrium with time of operation. As this deviation increases, the degree of catalyst deactivation increases proportionately.

Unlike calculations based on rate equations that will provide estimated temperature and concentration profiles along the reactor length, equilibrium calculations can only provide information on the outlet composition. Simulation of hot spots and other reactor anomalies must be done by using models based on kinetics. If such studies are not required, the equilibrium calculations are readily accomplished and, in the case of gases, require no experimental data.

Well-known professional simulation software such as SIMSCI PROCESS™ and CHEMSHARE™ include equilibrium reactor models based on the same approach used here. They provide an opportunity to apply approach temperatures based on experience with a given reaction system.

REFERENCES

1. Rase, H.F., "Chemical Reactor Design for Process Plants," Wiley, New York, 1977.
2. Rostrup-Nielsen, J.R. "Catalytic Steam Reforming" in *Catalysis-Science and Technology* Vol. 5, Ed. J.R. Anderson and M. Boudart, Springer-Verlag, New York, 1983.

EXAMPLE

7

Highly Exothermic Reaction: Comparison of Reactor Types

OBJECTIVE

The purpose of this example is to illustrate the use of models in comparing reactor types for the classical case of an initially rapid and highly exothermic reversible reaction that must be driven to almost complete conversion. It demonstrates not only the use of reactor design software but also the implementation of convenient and time-saving professional simulator software for determining recycle quantity and composition, along with requirements for auxiliary equipment.

PROBLEM STATEMENT

Compare both a multitubular nonadiabatic reactor and a series of adiabatic beds for the hydrogenation of benzene to produce cyclohexane as a raw material for producing adipic acid and caprolactam needed in the manufacture of several types of nylon. Recommend a final design in accordance with the following specifications:

Cyclohexane production: 50×10^6 gal/hr (6.522 lb/gal @ 0.60°F)

Minimum cyclohexane purity: 99.8 wt%

Maximum benzene content: 0.05 wt%

Feed: Benzene (99.96 wt%, 0.04 wt% cyclohexane)

Make-up Hydrogen (10 atm, 297°K)	*Mole%*
H_2	85.0
CH_4	12.0
C_2H_6	2.0
C_3H_8	1.0
	100.0

Catalyst

Ni-on-kieselguhr 1/8″ × 1/8″ tablets
Four levels of activity (types A, B, BL, and C). See following section under Rate Equation. (Ni is in form of NiO, which is reduced to optimum value in situ.)

Bed void fraction 0.36 in small tube and 0.33 in large-diameter bed

Poisons Compounds of S and P are permanent poisons.
Chlorine forms $NiCl_2$, which leaves the catalyst surface. Nitrogen is a temporary poison.

Fouling Carbonaceous deposit (usually slow in forming).

Life 1700 gal C_6H_6/lb of catalyst

Sintering Reduces active Ni area and thus catalyst activity
Caused by prolonged operation at high temperatures

Maximum Operating Temperature 525°K
Above this temperature hydrocracking of cyclohexane occurs at significant rates, causing rapid and possibly uncontrolled increases in reactor temperature. Since this reaction is not included in the model to be described, it is essential that this temperature limit not be violated.

Tube size for nonadiabatic reactor 2.0574 cm ID, 2.54 cm OD, and thickness 0.2413 cm.
This tube diameter is the practical minimum for a packed tube. It maximizes heat-transfer area per unit volume while avoiding excessive wall effects.

RATE EQUATION

A rate equation for the hydrogenation of benzene to cyclohexane in the gas phase has been reported for a supported nickel catalyst in the form of a Hougen-Watson equation [1]:

$$r = \frac{k_P K_H^3 K_B (P_H^3 P_B - (P_C/K)}{(1 + K_H P_H + K_B P_B + K_C P_C)^4} \qquad \text{(EX-7.1)}$$

$$\ln k_p = \frac{-13300}{RT} + \frac{33.5}{R}$$

$$\ln K_H = \frac{15500}{RT} - \frac{31.9}{R}$$

$$\ln K_B = \frac{11200}{RT} - \frac{23.1}{R}$$

$$\ln K_C = \frac{8900}{RT} - \frac{19.4}{R}$$

$$K = \exp\,(25980/T - 46.87)$$

where K_B, K_H, and K_C are apparent adsorption equilibrium constants for benzene, hydrogen, and cyclohexane, respectively. K is the reaction equilibrium constant.

Insufficient details about the supported nickel catalyst were presented to allow dependence on these data for reactor design. In an actual design the form of the equation can be retained and parameter values obtained from data gathered on the specific catalyst to be used in the plant. For purposes of illustrating the design methods and decision making, however, the original parameter values given above will be used to represent an active catalyst with 60% nickel. A total of four catalysts of high, medium, and low activity are assumed available as follows:

Designation	*Type*	*Rate Extension*
Catalyst A	High Activity	Use Eq. EX-7.1
Catalyst B	Medium Activity	(0.25)(Eq. EX-7.1)
Catalyst BL	Low-Medium Activity	(0.20)(Eq. EX-7.1)
Catalyst C	Low Activity	(0.025)(Eq. EX-1)

Since the kinetics are based on gas-phase reaction, conditions at the inlet and throughout the reactor must be above the dew point.

SOLVING RECYCLE LOOP PRIOR TO REACTOR DESIGN

Recycle of a valuable reactant used in excess, such as unreacted hydrogen in this case as shown in Fig. EX–7.1, is common practice. The simulation of such plants is greatly simplified, and computer time is saved if the recycle loop is solved separately from considerations of reaction path in the reactor. This technique is possible because it is normal practice to set the extent of reaction (conversion in this case) and yield at constant target values. The recycle loop will consist of a compressor, feed streams, purge, heat exchangers, a reactor "black box," flash drum, and a stabilizer column. The reactor "black box" simply sets the benzene conversion value so as to yield a liquid product with 0.05 wt% maximum of benzene. Solution of this loop will provide the desired total reactor feed amount and composition. Professional simulator routines

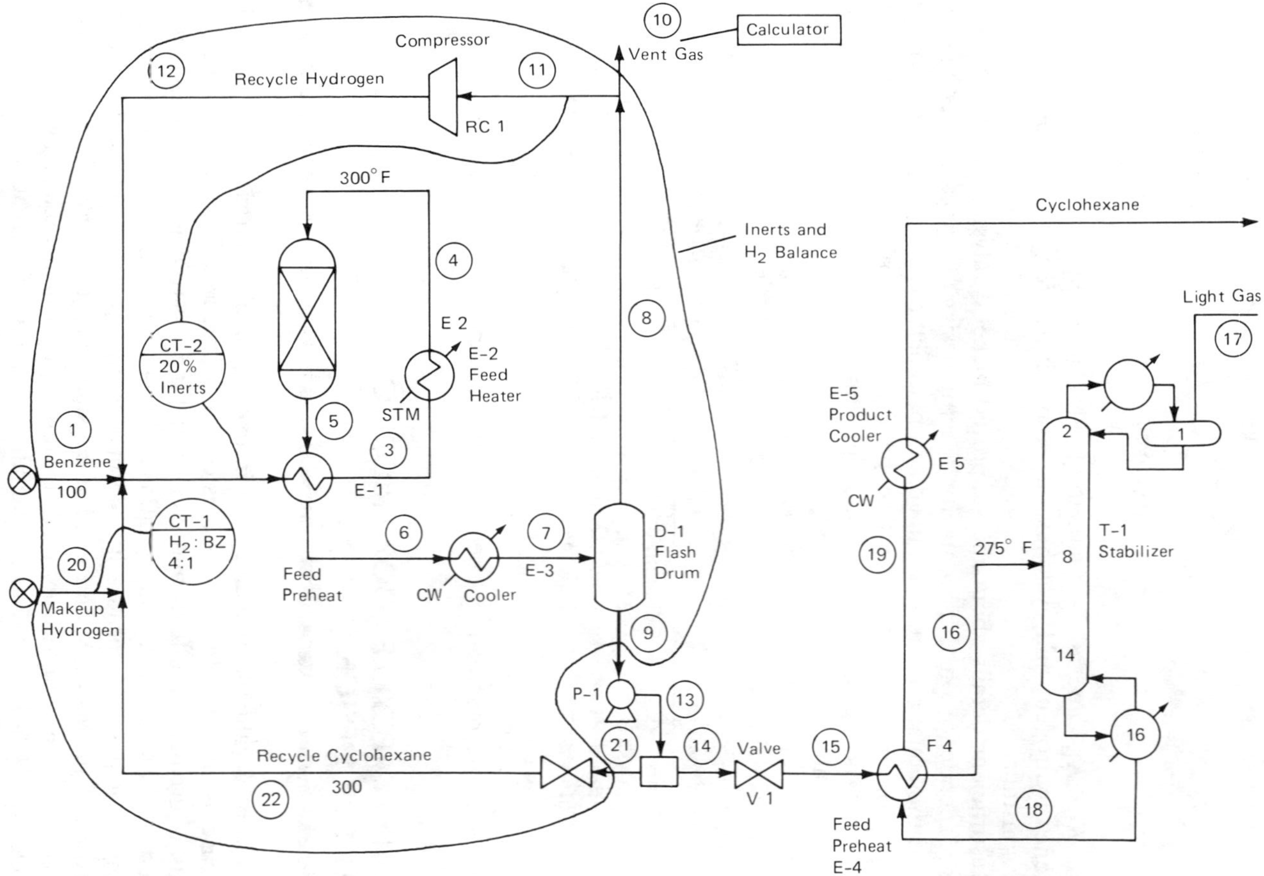

Figure EX-7.1 Process flow diagram for cyclohexane unit
(Adapted by permission: PROCESS™ Input Manual, p 81.1, Simulation Science, Fullerton, California, 1985.)

such as SIMSCI PROCESS™ [2] make such calculations quite effortless and reliable. An interactive graphics input program is used to create a driver program with desired components, inerts composition constraints, feed H_2/benzene ratio, outlet reactor conversion, and operating conditions. The subroutine thus assembled is sent to the mainframe (VAX-VMS in this case), and the results are rapidly obtained. The following conditions and specifications are needed.

Conditions and Specifications

Set the various conditions and specifications for the recycle loop.

Target Conversion

Basis: 100 kg moles/hr benzene feed (0.04 wt% cyclohexane), 0.05 wt% benzene in product cyclohexane

Let n = grams benzene in product

$$\frac{n}{n + (99.86)(84.156/78.108) + 0.04} = 0.0005$$

$$n = 0.05387 \text{ g benzene remaining}$$

Target Conversion of Benzene

$$\frac{99.96 - 0.05387}{99.96} = 0.99946 \quad \text{or} \quad 99.95\%$$

By setting target conversion to meet product specification on benzene residual, one avoids a costly separation of benzene from cyclohexane. All that is needed is a stabilizer to remove light ends introduced with the hydrogen.

Cases to Be Considered and Set Conditions

This example will be restricted to two cases of flash-bottoms recycle:

1. Nonadiabatic reactor: no flash-bottoms recycle
2. Adiabatic-reactor series: 3:1 ratio of flash bottoms to benzene feed. This recycle ratio is about the highest possible for vapor-phase operation in the range of conditions that are to be considered.

Other reasonable set conditions are

- H_2-benzene recycle ratio: 4
- Inerts (C_1, C_2, C_3) in reactor feed: 20%
- ΔP for nonadiabatic reactor: approximately 1 atm
- ΔP for adiabatic reaction system: 20 psi (1.36 atm) overall; 2 psi (0.136 atm) between beds
- Maximum allowable temperature: 525°K

- Inlet pressure to reactor (initial bed in the case of reactors in series): 10 atm
- Conversion: 99.95%

Benzene Feed Rate

The simulation of a typical loop as shown in Fig. EX-7.1 for purposes of establishing feed composition for reactor simulation is most efficiently and conveniently handled by selecting a feed basis of 100 moles (for simulator use it is 100 kg moles/hr). The heat-exchange units in the loop use the heat-balance equation $[UA\,\Delta T_m = m_f\,c_p(T_2 - T_1)]$ to supply the duty, and the heat flux or area if a heat-transfer coefficient is provided. The compressor unit calculates an outlet temperature and the work required if a compressor efficiency is provided. Other units include the flash drum and product stabilizer for light ends removal. A stabilizer of this type would have around 20 trays or for a typical efficiency, 15 theoretical trays, including partial condenser and reboiler. The results for all units and streams can be scaled up by using the stream rates corresponding to the desired product rate.

The design rate $(F_n)_D$, for any stream is determined simply by multiplying the ratio of stream to cyclohexane product, $(F_n/F_p)_S$, obtained by the simulator by the actual cyclohexane product rate taken for the design basis:

$$(F_n)_D = (F_p)_D\left(\frac{F_n}{F_p}\right)_S \tag{Ex–7.2}$$

where F_p is the product rate; D refers to the design value: S refers to the simulator values; and n refers to the stream number.

Estimate of Reactor Outlet Rate and Composition

All recycle calculations require an initial estimate of some stream in the loop, and the simplest one for that purpose is the reactor-exit stream (see Fig. Ex–7.1). The reactor feed and product can be easily estimated using as a basis 100 moles of benzene fed, a 4:1 H_2-benzene ratio, a desired 20% inert content (C_1, C_2, C_3), and 0.9995 fractional conversion of benzene.

Basis: 100 moles benzene fed (no cyclohexane recycle)

	Benzene Feed	*Reactor Feed*	*Reactor Outlet* Moles	*Reactor Outlet* Mole Percent
H_2		399.85152	100.11282	30.80
C_1		99.97030[a]	99.97030	30.75
C_2		16.66172[a]	16.66172	5.125
C_3		8.33086[a]	8.33086	2.56
Benzene	99.96288	99.96288	0.04998	0.015
Cyclohexane	0.03712	0.03712	99.95002	30.75
	100.00000	624.81440	325.07570	100.00

[a]Apportioned as in fresh makeup hydrogen to total 20% inerts. This, of course, is only an estimate since the recycle stream will contain slightly different relative amounts of inerts because of the flash at D-1.

A similar calculation for recycle of cyclohexane at 3:1 flash liquid-to-benzene ratio follows:

	Benzene Feed	*Hydrogen*	*Recycle Flash Liquid*	*Reactor Feed*	*Reactor Moles*	*Product Mole %*
H_2		399.85152		400.00	100.00	14.53
C_1		56.44962	5.00[b]	148.12[a]	148.12	21.54
C_2		9.40827	3.00[b]	31.60[a]	31.60	4.60
C_3		4.70414	2.00[b]	17.78[a]	17.78	2.60
C_6H_6	99.96288			99.96		
C_6H_{12}	0.03712		290.00	290.04	390.00	56.7
	100.000	470.41355	300.00	987.50	687.50	100.00

[a]Apportioned as in fresh makeup hydrogen.
[b]Guesses.

Since the reactor effluent goes to the entrance of D-1 unaltered in composition and amount, it is best to select stream 7 as the entry or tear stream for the calculations. If stream 5 were selected, a problem would arise since the cross-exchanger flow for stream 2 is not known until the loop is completed. The estimated composition and flow rate can be entered for stream 7 along with the desired temperature and the pressure set by the allowable reactor and exchanger pressure drops. Refer to Fig. EX–7.2, which is a block diagram provided by the computer outlet, and to Appendix B.1, which is a copy of the driver program developed from interactive graphic displays on an IBM PC-AT or PS-2, and then transferred to a VAX/VMS for solution.

Calculation Procedure

The program must be executed in a proper sequence and with the desired constraints applied by means of controllers and material balances.

Sequence

The calculation sequence must be selected so that it begins and ends with a unit and returns thereby to the starting point, stream 7 as shown in Figs. EX–7.1 and EX–7.2. In this manner recycle convergence can be tested by stream 7, the tear stream. According to Fig. EX–7.2, the sequence of units is given such that E-2 precedes the reactor, and then E-1 and E-3 follow the reactor, thereby completing the loop. An examination of the units and the stream properties given will make it clear that all streams are defined by a unit, a preceding unit, or a calculator.

Calculation of Purge from H and Inert Balances

The PROCESS™ [2] software permits the user to introduce calculations into the loops so that simple analytical solutions may be used where possible to avoid unnecessary use of computer time. This feature is valuable for establish-

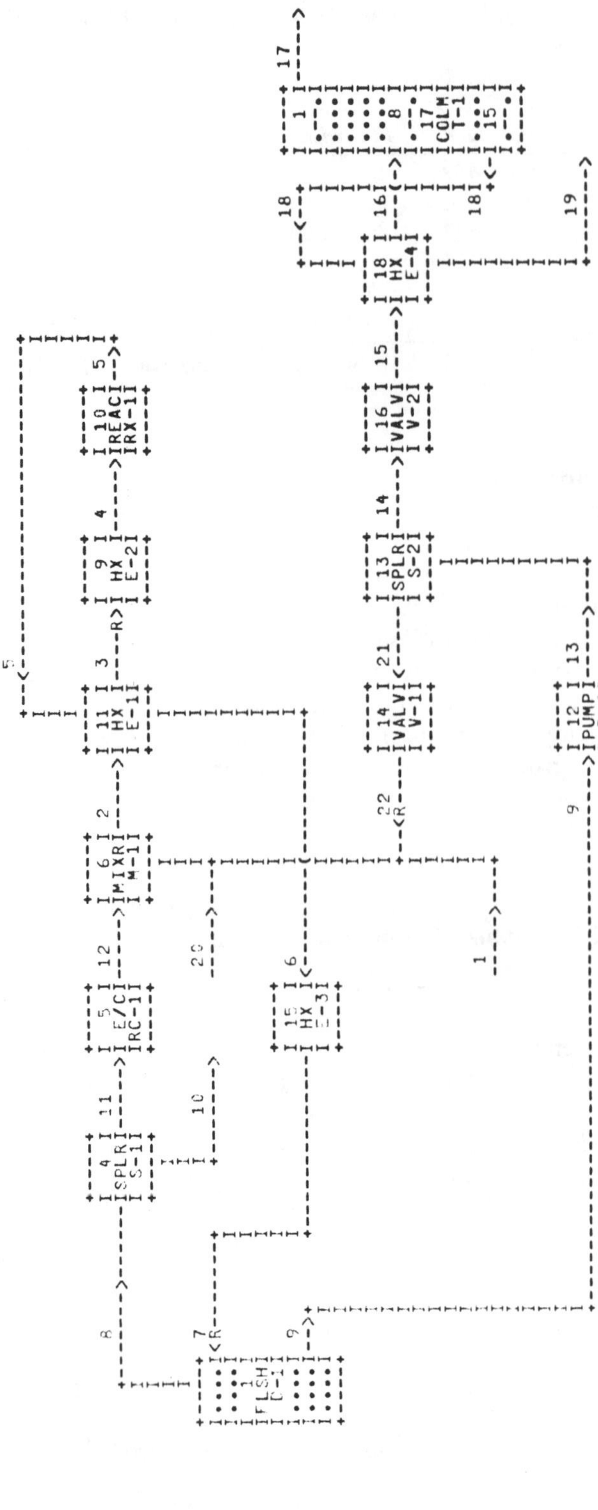

Figure EX-7.2 Block diagram generated by PROCESS™
(Reproduced by permission, Simulation Science, Inc., Fullerton, California.)

ing quickly and analytically the amount of purge required so that the split of flash drum vapors between purge and recycle can be set. The calculator equation for the purge can be readily derived by simultaneous solution of overall hydrogen and inert balance.

- Composition of STRM 8 same as STRM 10
- Composition of STRM 9 same as STRM 22

Hydrogen balance

$$(Y_H)_{20}F_{20} + (Y_H)_9F_{22} = (n_H)_9 + (Y_H)_8F_{PU} + (n_{BZ})_1 3X_A \quad \text{(EX-7.3)}$$

Inert balance

$$(Y_I)_{20}F_{20} + (Y_I)_9F_{22} = (n_I)_9 + (Y_I)_8F_{PU} \quad \text{(EX-7.4)}$$

From Eq. EX-7.3

$$F_{20} = \frac{(n_4)_9}{(Y_H)_{20}} + \frac{(Y_H)_8}{(Y_H)_{20}} F_{PU} + \frac{(n_{BZ})_1 3X_A}{(Y_H)_{20}} - \frac{(Y_H)_9F_{22}}{(Y_H)_{20}}$$

From Eq. EX-7.4

$$F_{20} = \frac{(n_I)_9}{(Y_I)_{20}} + \frac{(Y_I)_8F_{PU}}{(Y_I)_{20}} - \frac{(Y_I)_9}{(Y_I)_{20}} F_{22}$$

$$F_{PU}\left[\frac{(Y_I)_8}{(Y_I)_{20}} - \frac{(Y_H)_8}{(Y_H)_{20}}\right] = \frac{(n_H)_9}{(Y_H)_{20}} + \frac{(n_{BZ})_1 3X_A}{(Y_H)_{20}} + \frac{(Y_I)_9F_{22}}{(Y_I)_{20}} - \frac{(n_I)_9}{(Y_I)_{20}} - \frac{(Y_H)_9F_{22}}{(Y_H)_{20}} \quad \text{(EX-7.5)}$$

Multiplying Eq. Ex-7.5 by $(Y_H)_{20}$ gives

$$F_{PU}\left[\frac{(Y_H)_{20}(Y_I)_8}{(Y_I)_{20}} - (Y_H)_8\right] = (n_H)_9 + (n_{BZ})_1 3X_A + \frac{(Y_H)_{20}(Y_I)_9F_{22}}{(Y_I)_{20}} - \frac{(n_I)_9}{(Y_I)_{20}} - (Y_H)_9F_{22} \quad \text{(EX-7.6)}$$

where subscripts H, BZ and I refer to hydrogen, benzene, and inerts (C_1 to C_3), respectively; number subscripts refer to the entering or leaving streams so designated and PU designates the purge. The symbol n refers to the actual number of moles of the indicated component; F refers to the total molar flow of indicated streams. Conversion of the limiting reactant, benzene, is indicated by the symbol X_A, and Y represents mole fraction.

The final form for the case at 3:1 recycle flash liquid can be seen in Appendix B.1 under calculator. When there is no recycle flash liquid, stream 22 entries must be omitted.

The required values are retrieved from the program for each cycle, and the correct value for the purge rate necessary for a material balance for the updated conditions is calculated each cycle. The ratio of hydrogen to inerts in the hydrogen makeup is 5.667.

Application of Controllers (see Figs. EX–7.1 and EX–7.2)

In order to attain the desired conditions of 4:1 H_2/benzene and 20% inerts in stream 2, it is necessary to introduce calculation controllers. The first controller, CT-1, adjusts the H_2 feed to give the desired 4:1 ratio in stream 2, and CT-2 adjusts the recycle stream in order to meet the inerts content specification. Obviously, these changes affect other streams in the loop, and several iterations, assisted when necessary by Wegstein's convergence accelerator, are required for convergence of stream 7. In an actual plant, on-stream analyzers and controllers may or may not be used. When not used, the unit is started up at total recycle, and purging is begun when spot analysis of the total feed indicates a desired inerts specification. The approximate H_2/benzene ratio can be adjusted by flow control on the makeup hydrogen. Once desired conditions are reached, set points are maintained over a long period of time. The calculational controllers, however, are necessary for efficient solution of the loop.

RECYCLE LOOP RESULTS

Output from the SIMSCI PROCESS™ program [2] for this case consists of complete material balances, exchanger duties, flash calculations, compressor work, tower calculations, and state of fluid throughout the process. For reactor design purposes all that is needed is the composition in mole fraction at the reactor inlet, which is reproduced in Table EX–7.1 for the case of no-recycle flash liquid and 3:1 flash liquid recycle. As for all computer calculations, the results should be checked for reasonableness and consistency with the imposed constraints, the following observations are pertinent.

Product Specification (99.8 wt% C_6H_{12} and 0.05 wt% C_6H_6)

99.95 and 0.0476 for no flash-bottoms recycle

99.98 and 0.013 for 3:1 flash-bottoms recycle

This result is reasonable since the recycle of flash bottoms cause the resulting bottoms to have a higher ratio of C_6H_{12}/C_6H_6 than the no-recycle case.

Table EX-7.1 Reactor Feed for 4:1 H_2/BZ and 20% Inerts

	Mole Fraction: No Recycle Cyclohexane	*Mole Fraction: 3:1 Flash Bottoms-to-Benzene Ratio*
H_2	0.63526	0.40331
C_1	0.16176	0.15826
C_2	0.02548	0.02698
C_3	0.01133	0.01402
C_6H_6	0.15885	0.10068
C_6H_{12}	0.00732	0.29675
	1.00000	1.00000
Dew point @ 10 atm	368°K	403.4°K

Overall Material Balance Professional simulators usually provide an overall mass material balance for each component in terms of the sum of the feed and reaction minus the product for which case the result should be zero for each component. Obviously such is never completely achieved due to roundoff errors and practical convergence criteria. In the two cases calculated the percentage deviation was largest for major recycle components: H_2 in the case of the no flash-bottoms recycle (−0.1%) and cyclohexane in the case of 3:1 flash-bottoms recycle (0.25%). Neither error is unacceptable.

Separations A shortcut method using hand calculations and API K values confirmed reasonableness of the distillation calculations.

Dew Points One dew point was quickly checked by using API K_v values and found satisfactory.

Overview of Results All streams were examined for reasonableness or correctness (e.g., 20% inerts and 4:1 H_2/benzene in stream 2) and found satisfactory

REACTOR CALCULATIONS

A single multitubular reactor and multiple-stage adiabatic reactors can be quickly evaluated using the inlet composition in Table EX–7.1 and the conditions and constraints summarized in Table 5.1 for a one-dimensional, ideal plug-flow, pseudohomogeneous fixed-bed reactor. A multipurpose interactive fixed-bed design program (PFRSIM), which uses the Livermore solver for ordinary differential equations (LSODE), provided a rapid means for studying both types of reactors [3]. Figure EX–7.3 illustrates the input screens. See Appendixes B.2 and B.3 for further details.

Multitubular Reactor

The multitubular reactor, although a nonadiablatic reactor, is often called an *isothermal* reactor, since the goal is to operate under conditions as close to isothermal as possible in order to attain some crucial selectivity in a multireaction system or to avoid exceeding, as in this study, some maximum temperature. Isothermicity is seldom realized, for the initial portion of the reaction is usually rapid; and the linear rate of heat transfer is no match for the exponential increase of the rate of heat production by the reaction with increasing temperature.

As pointed out in Chapter 5, the nonadiabatic model is the least reliable because of the wall effects in the small-diameter tubes and the inaccuracies in the heat-transfer correlations. The most reliable design is accomplished by scale-up from a single-tube pilot unit of the same length and with the same mass velocity and operating conditions as the planned large unit. A model, however, has great value in pursuing ideas for alternative operating scenarios, including various catalyst options both in the planning stages of development work and during regular plant operation. Prior to detailed pilot-plant or plant experience, generalized heat-transfer correlations must be used as in this example, but these can be checked and adjusted in pilot-plant or plant, as demonstrated in Example 11.

A fouling factor of 0.001 hr ft^2°F/BTU (7.37 sec cm^2°C/cal) was used in calculating the overall heat-transfer coefficient in this example to ensure a conservative result based on long-term operation.

Design Strategy for Nonadiabatic Reactor

1. Execute PROCESS™ simulation to obtain results without flash-bottoms recycle. Determine dew points of reactor inlet at 10 atm and reactor outlet at 9 atm. Do not operate below these temperatures for this vapor-phase design. Use PROCESS™ to determine dew point. The user inputs the component compositions and the pressure along with an estimate of the dew point.
2. Determine crucial variables that set temperature maximum, tube length, and number.
3. Try different combinations of the crucial variables first with catalyst A and then with catalysts B and BL in order to establish an acceptable design consistent with the applicable constraints.
4. Determine the operating flexibility afforded by the proposed designs. As the catalyst deactivates, the hot spot will move down the tube and necessitate an increase in temperature to maintain production. Cases at higher temperatures and shorter length should be calculated to determine the feasibility of the proposed design for such eventualities.

```
 ENTER THE NAME OF YOUR DATA FILE.
C6H6DAT
 ENTER THE REACTOR TYPE       ,TYPE
 [ 1 = AXIAL, 2 = RADIAL ]
1
 ENTER THE HEAT TRANSFER MODE,MODE
 [ 1 = ISOTHERMAL, 2 = ADIABATIC, 3 = NONADIABATIC ]
3
 ENTER THE INSIDE DIAMETER    ,DI      IN CM
2.0574
 ENTER THE WALL THICKNESS     ,XW      IN CM
.2413
 ENTER THE JACKET TEMPERATURE,TJ       IN DEG. K
430
 ENTER THE MASS VELOCITY      ,G       IN G/(CM^2*S)
.45
 ENTER THE INLET TEMPERATURE  ,TI      IN DEG. K
430
 ENTER THE INLET PRESSURE     ,PI      IN ATM
10
 ENTER THE INLET COMPOSITION  ,NI      IN GMOLS/GMOL LR
 . . . FOR H2
3.9991
 . . . FOR C6H6
1.0
 . . . FOR CYCL
.0461
 . . . FOR CH4
1.0183
 . . . FOR C2H6
.1604
 . . . FOR C3H8
.0713
 ENTER THE STEP SIZE          ,ZSTEP   IN CM
.1
 ENTER THE PRINTING INTERVAL  ,ZPRINT  IN CM
30
 ENTER THE BED LIMIT          ,ZLMT    IN CM
550
 ENTER THE CONVERSION LIMIT   ,XLMT
.9995
 ENTER THE YIELD LIMIT        ,YLMT
.9995
 ENTER THE BED VOID FRACTION  ,EPS
.36
 ENTER THE CATALYST ACTIVITY  ,ACT
 . . . FOR REACTION 1
.2
 ENTER AN IDENTIFICATION NOTE FOR THIS RUN (10 CHAR MAX).
RUN 1
 ENTER THE LEVEL OF OUTPUT FOR THE REPORT.
       [1] HEADING + SUMMARY
       [2] HEADING + TABLE
       [3] HEADING + TABLE + SUMMARY

3
 DO YOU WANT TO SAVE THE TEMPERATURE PROFILE RESULTS IN A FILE FOR PLOTTING?
N
```

Figure EX-7.3 Input screens for interactive, one-dimensional, pseudohomogeneous, ideal plug-flow reactor simulation (PFRSIM)

Note options: adiabatic, nonadiabatic, isothermal, and axial or radial. See Appendix B.2 for more information about the program.

Crucial Variables

A series of runs can be made rather rapidly to establish the crucial variables. The interactive program permits changing one or more variables without reentering all the input data. A typical printout is given in Table EX–7.2, and a summary of preliminary runs using catalyst A is given in Table EX–7.3. Clearly, based on the results in Table EX–7.3, jacket temperature and mass velocity are the most sensitive variables. The inlet temperature has essentially no effect other than slightly changing the position of the hot spot. The jacket temperature plays a major role since it is very close to the reaction temperature for a major portion of the reactor and, in particular, the low rate region.

Table EX-7.2 Typical Printout for Nonadiabatic Reactor Catalyst BL


```
I.   REACTOR CONFIGURATION

     REACTOR TYPE          , TYPE    = AXIAL
     HEAT TRANSFER MODE, MODE    = NONADIABATIC
     INSIDE DIAMETER       , DI      =      2.0574 CM
     WALL THICKNESS        , XW      =       .2413 CM
     JACKET TEMPERATURE, TJ      =    430.0000 DEG. K

II.  FEED CONDITIONS

     MASS VELOCITY         , G       =       .4500 G/(CM^2*S)
     INLET TEMPERATURE     , TI      =    430.0000 DEG. K
     INLET PRESSURE        , PI      =     10.0000 ATM
     MOLECULAR WEIGHT      , MWF     =     18.1661G/GMOL
     INLET COMPOSITION
          COMPONENT          NI           XI
          ---------          --           --
             H2            3.9991       .63526
             C6H6          1.0000       .15885
             CYCL           .0461       .00732
             CH4           1.0183       .16176
             C2H6           .1604       .02548
             C3H8           .0713       .01133

III.  PRINTING AND EXIT CRITERIA

     STEP SIZE             , ZSTEP   =       .1000 CM
     PRINTING INTERVAL     , ZPRINT  =     30.0000 CM
     BED LIMIT             , ZLMT    =    550.0000 CM
     CONVERSION LIMIT      , XLMT    =       .9995
     YIELD LIMIT           , YLMT    =       .9995
```

```
IV.   CATALYST DESCRIPTION

      PARTICLE DIAMETER , DP        =       .3175 CM
      BED DENSITY       , RHOB      =      1.2820 G/CM^3
      PARTICLE DENSITY  , RHOP      =      1.8920 G/CM^3
      BED VOID FRACTION , EPS       =       .3600
      CATALYST ACTIVITY , ACT FOR...
                          R# 1      =       .2000
```

Z	T	P	X	Y	R(1)	R(2)
.0	430.0	10.000	.0000	.0000	.15280E-04	
1.4	434.7	9.995	.0076	.0076	.18292E-04	
2.9	439.7	9.991	.0174	.0174	.21835E-04	
4.4	444.6	9.986	.0291	.0291	.25783E-04	
5.9	449.6	9.981	.0427	.0427	.30240E-04	
7.3	454.5	9.976	.0576	.0576	.34964E-04	
8.6	459.3	9.972	.0734	.0734	.39925E-04	
9.9	464.6	9.967	.0915	.0915	.45482E-04	
11.1	469.8	9.963	.1103	.1103	.51095E-04	
12.2	475.0	9.960	.1296	.1296	.56481E-04	
13.2	480.0	9.956	.1488	.1488	.61310E-04	
14.1	484.6	9.953	.1674	.1674	.65292E-04	
15.0	489.2	9.950	.1870	.1870	.68606E-04	
15.9	493.7	9.947	.2075	.2075	.70969E-04	
16.9	498.4	9.944	.2309	.2309	.72298E-04	
18.0	502.9	9.940	.2568	.2568	.72257E-04	
19.3	507.0	9.936	.2872	.2872	.70696E-04	
21.1	510.7	9.930	.3276	.3276	.67195E-04	
24.0	512.5	9.921	.3882	.3882	.61133E-04	
30.4	507.4	9.901	.5035	.5035	.49865E-04	
34.0	502.3	9.891	.5587	.5587	.44206E-04	
37.2	497.2	9.882	.6022	.6022	.39380E-04	
40.2	492.1	9.874	.6386	.6386	.35045E-04	
43.1	487.2	9.867	.6698	.6698	.31077E-04	
46.0	482.2	9.859	.6974	.6974	.27383E-04	
48.9	477.3	9.852	.7217	.7217	.24014E-04	
51.9	472.4	9.845	.7436	.7436	.20907E-04	
55.0	467.7	9.838	.7632	.7632	.18118E-04	
58.4	463.1	9.831	.7818	.7818	.15540E-04	
62.2	458.4	9.822	.7996	.7996	.13204E-04	
66.6	454.0	9.813	.8169	.8169	.11103E-04	
72.0	449.6	9.802	.8347	.8347	.92054E-05	
79.2	445.5	9.787	.8541	.8541	.74629E-05	
89.8	441.7	9.766	.8768	.8768	.58262E-05	
107.8	438.1	9.731	.9057	.9057	.41750E-05	
137.8	435.1	9.674	.9381	.9381	.26101E-05	
167.8	433.3	9.619	.9588	.9588	.16947E-05	
197.8	432.2	9.564	.9723	.9723	.11202E-05	
227.8	431.5	9.510	.9813	.9813	.74820E-06	
257.8	431.0	9.456	.9874	.9874	.50294E-06	
287.8	430.7	9.402	.9914	.9914	.33952E-06	
317.8	430.5	9.347	.9942	.9942	.22985E-06	
347.8	430.3	9.293	.9960	.9960	.15590E-06	
377.8	430.2	9.238	.9973	.9973	.10590E-06	

Table EX-7.2 (*continued*)

407.8	430.1	9.183	.9982	.9982	.71999E-07
437.8	430.1	9.128	.9988	.9988	.48991E-07
467.8	430.1	9.073	.9991	.9991	.33372E-07
497.8	430.0	9.017	.9994	.9994	.22756E-07
502.2	430.0	9.008	.9995	.9995	.21513E-07

```
1                                                              RUN 1

                        SUMMARY OF RESULTS

   I.  OUTLET CONDITIONS

       Z  =   502.2 CM
       T  =   430.0 K
       P  =   9.008 ATM
       X  =   .9995
       Y  =   .9995
       MW =   34.69 G/GMOL

       COMP       N(J)        X(J)
       -------------------------
       H2        1.0007       .3035
       C6H6       .0005       .0002
       CYCL      1.0456       .3171
       CH4       1.0183       .3089
       C2H6       .1604       .0487
       C3H8       .0713       .0216

   II.  HEAT TRANSFER INFORMATION

       TMIN =   430.0 K  AT  Z =         .0 CM
       TMAX =   512.5 K  AT  Z =       23.9 CM
       UI =   .19342E-01  CAL/CM^2*K*SEC
       Q   =  -.21151E+04  CAL/SEC

DO YOU WANT TO CHANGE ANY FEED OR REACTOR
CONDITIONS AND EXECUTE THE PROGRAM AGAIN?
```

- Composition at inlet may be given as mole fraction or moles/mole of limiting reactant. Whichever is given, the other is calculated automatically and both are part of the output.
- If the question at the end is answered "yes," only those variables the user wishes to change need to be entered interactively. This procedure makes possible rapid study of the effect of various variables.

Table EX-7.3 Summary of Runs Nonadiabatic Reactor Catalyst A

Case No.	*G* g/cm^2 *sec*	*Inlet Temp, °K*	*Jacket Temp, °K*	*Maximum Temp, °K*	*Z cm*	*ΔP atm*
1	0.45	400	400	564.7	190.1	0.35
2	0.35	400	400	569.7	135.8	0.16
3	0.6	400	400	558.7	275.4	0.89
4	0.6	390	390	553	439.1	1.38
5	0.5	390	390	557.9	344	0.74
6	0.5	385	385	554.1	447	1.0
7*	0.4	375	375	539.2	600+	1.0

- 4:1 H_2/benzene
- No recycle cyclohexane
- 10 atm @ inlet
- Dew point: Inlet = 368°K @ 10 atm
  Outlet = 389.8 @ 9 atm

*Below outlet dew point.

Operation with High-Activity Catalyst (Catalyst A)

With the combined constraints of maxima in ΔP and temperature of 1 atm and 525°K, respectively, it would be difficult to operate the reactor with catalyst A, as can be seen from Table EX–7.3. Two alternates for the nonadiabatic case remain, use of a catalyst with a lower activity such as catalyst B or recycling a portion of the product cyclohexane in order to increase the heat capacity of the system. The lower-activity catalyst is preferred since the recycle case would require many more tubes to carry the increased flow.

Operation with Medium-Activity Catalyst (Catalysts B and BL)

Since the interactive program allows rapid changing of one or more parameters, it was possible to quickly determine the activity level, jacket, and inlet temperatures, and mass velocity required to meet the given constraints. Results of various trials are summarized in Table EX–7.4, where it can be seen that a satisfactory design can be obtained by using catalyst B with an inlet and jacket temperature of 420°K, a mass velocity of 0.45 g/cm^2 sec, and a length of

Table EX-7.4 Summary of Runs Nonadiabatic Reactor Catalyst B and Catalyst BL

Case No.	*G* *g/cm² sec*	*Inlet* *Temp, °K*	*Jacket* *Temp, °K*	*Maximum* *Temp, °K*	*Z cm*	*ΔP atm*
CATAYLST B (Factor = 0.25):						
1	0.6	440	440	525	418	1.54
2	0.4	430	430	530.6	336.7	0.52
3	0.45	430	430	526.5	385	0.76
4	0.48	430	430	524.1	414.3	0.93
5	0.4	420	420	521.4	459.2	0.69
6	0.45	420	420	515.3	529.2	1.0
CATALYST BL (Factor = 0.20):						
7	0.5	440	440	519.5	437.6	1.1
8	0.5	450	450	527.9	364.1	0.94
9	0.6	450	450	522.6	446.4	1.71
10	0.6	440	440	512.5	540	2.04
11	0.45	440	440	523.3	389	0.79
12	0.45	430	430	512.5	502.2	1.0

- 4:1 H_2/benzene
- No recycle cyclohexane
- 10 atm @ inlet
- Dew Point: Inlet = 368°K @ 10 atm
  Outlet = 389.8 @ 9 atm

529 cm (case 6). The design selected allows flexibility for raising temperatures as activity declines and the hot spot moves down the tube. Note in Table EX–7.4 that the temperature can be raised to 430°K without excessively exceeding the limiting maximum temperature of 525°K. The desired conversion under these conditions is reached after 385 cm of length of active catalyst. This means that if the temperature is raised to 430°K, the design selected can operate at full production and within specification when the first 144 cm of the bed is inactive. This design is a little tight, which means that even catalyst B, with an activity factor of 0.25, is a bit too active for the nonadiabatic case. Assuming a slightly less active catalyst called catalyst BL with an activity factor

0.20, yields a more flexible design requiring less catalyst, because a higher jacket temperature is possible, which raises the rate in the constant temperature region.

Thus the selected nonadiabatic design should be catalyst BL shown as case 12 in Table EX–7.4, with inlet and jacket temperatures at 430°K and a total catalyst length of 502 cm, which permits operation at 440°K when the first 113 cm of the bed is deactivated (see Fig. EX–7.4).

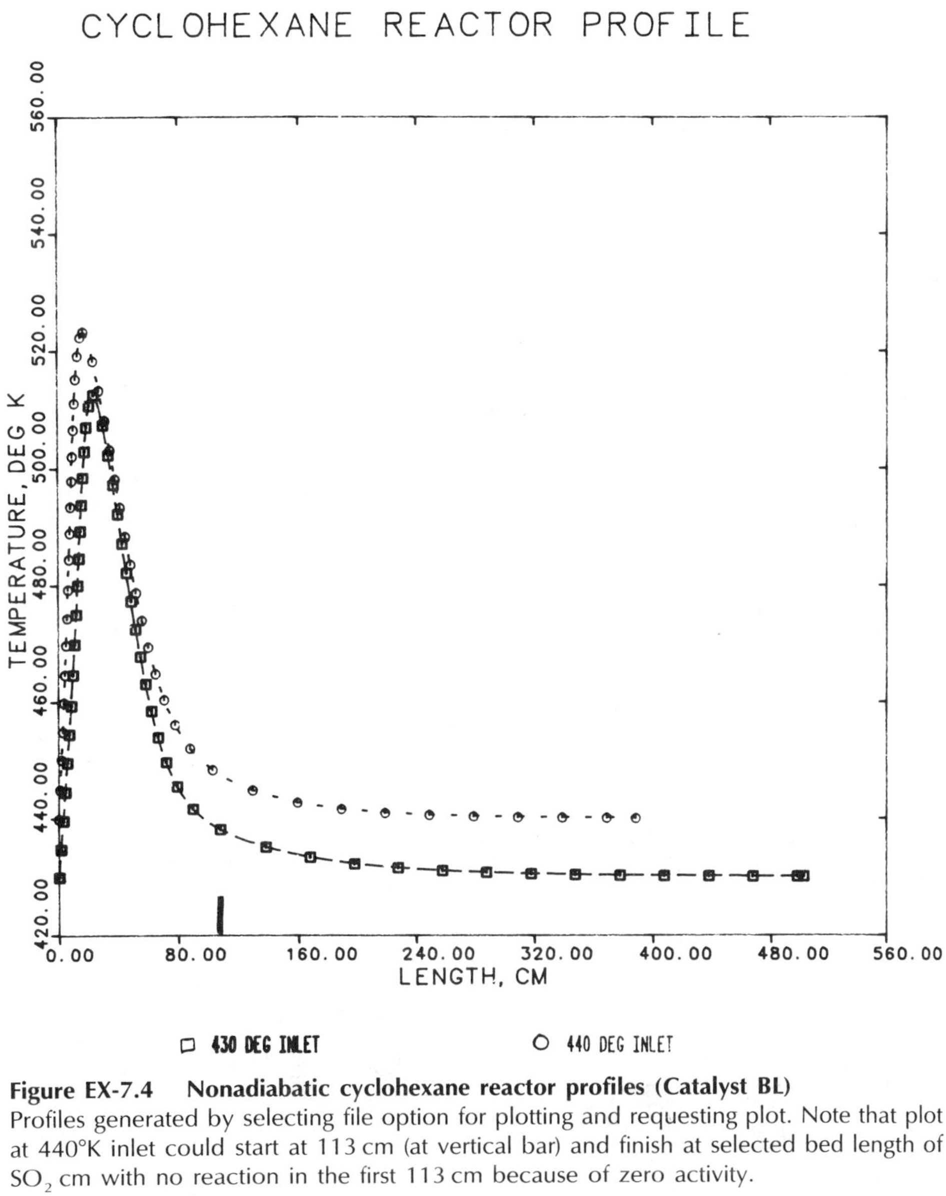

Figure EX-7.4 Nonadiabatic cyclohexane reactor profiles (Catalyst BL)
Profiles generated by selecting file option for plotting and requesting plot. Note that plot at 440°K inlet could start at 113 cm (at vertical bar) and finish at selected bed length of SO_2 cm with no reaction in the first 113 cm because of zero activity.

Estimate of Error Introduced by Ignoring Radial Gradient

The one-dimensional nonadiabatic model uses an inside heat-transfer coefficient based on a flat temperature profile. Equation 5.30 provides a convenient criterion for judging the acceptability of ignoring the radial temperature gradient.

For an error in the rate due to temperature 5% or less, the following applies:

$$\frac{|\Delta H|\hat{r}_1\rho_b D^2}{\lambda_r T_w} < \frac{1.6RT_w}{E}$$

A quick check of this criterion can be made by calling for rate and physical property data from the program for conditions at the highest rate, which are as follows:

$Z = 17$ cm

$P = 9.959$ atm

$\hat{r}_1 = 0.723865\text{E}-4$ @ 498.9°K and 0.757845E−4 @ 508.9°K

$T = 498.9$°K

$\mu = 0.0001578$ g/cm sec

$\lambda_f = 0.0002973$ cal/cm sec °K

$\mathbf{G} = 0.45$ g/cm^2 sec

$T_w = 430$°K

$X_A = 0.2333$

$\Delta H_1 = 51{,}620$ cal/mole

$c_p = 13.7626$ cal/g mole °C

$M_w = 20.44$, mixture

$\rho_b = 1.282$ g/cm^2

$D = 2.0574$ cm

$\varepsilon = 0.36$

From Table 4.2,

$$h_w = \frac{\lambda_f}{D_p}\, 0.18\, \text{Re}^{0.93}\text{Pr}^{0.33}$$

$$\text{Re} = \frac{(0.3175)(0.45)}{0.0001578} = 905.4 \qquad \text{Pr} = \frac{(13.763)(0.0001578)}{(20.44)(0.000297)} = 0.358$$

$$h_w = \frac{0.0002973}{0.3173}\, 0.18(905.4)^{0.93}(0.358)^{0.33} = 0.0675 \text{ cal/cm}^2 \text{ sec °K}$$

From Table 4.2,

$$\frac{h_w D_p}{\lambda_r} \frac{\varepsilon}{1-\varepsilon} = 0.27$$

$$\lambda_r = \left(\frac{h_w}{0.27}\right)\left(\frac{0.36}{0.64}\right) = \frac{(0.0675)(0.3175)}{0.27}\left(\frac{0.36}{0.64}\right) = 0.04466 \text{ cal/cm sec °K}$$

The ratio of the log of the rates at 498.9 and 508.9°K is related to the effective E:

$$\ln \frac{0.0000757845}{0.0000723864} = 0.045875 = \frac{E}{R}\left(\frac{1}{498.9} - \frac{1}{508.9}\right)$$

$$E = 2314 \text{ cal/mole}$$

Now applying the criterion, gives

$$\frac{(51620)(0.7238\text{E}-4)(1.282)(2.0574)^2}{(0.04466)(430)} < \frac{(1.6)(1.987)(430)}{2314}$$

$$1.056 < 0.632 \quad \text{for 5\% error}$$

$$1.056 < 1.264 \quad \text{for 10\% error}$$

Therefore the radial gradient can be ignored, since a 10% error in the region of the hot spot is not excessive.

Adiabatic Reactors in Series

In order to test the viability of using adiabatic reactors in series, it was necessary to increase the heat capacity of the system by using a substantial recycle of product cyclohexane. To remain in the gas phase, the inlet temperatures would have to be slightly above the dew point. The dew point for the inlet to the first bed at 10 atm was calculated to be 403.4°K for a 3:1 recycle ratio of flash liquid to benzene feed (see Table EX–7.1 for compositions).

Design Strategy for Adiabatic Reactor System

1. Execute PROCESS™ simulation to obtain results for 3:1 flash-bottoms recycle. Determine dew points at inlet of each reactor using appropriate pressure.
2. Determine crucial variables that set bed length and outlet temperature.
3. Select catalyst for each bed that proves best for a reasonable design.
4. Selected design should have enough capacity to handle some deactivation in the first bed. More significant deactivation can be handled by arranging pipe and valving so that the order of the reactors can be changed.

Use of Scaling Factors as a Design Aid

Adiabatic reactors present a fruitful area for applying scaling factors in design. Consider the mole balance (Eq. 5.33) for the limiting reactant A, in this instance benzene.

$$\frac{dn_{na}}{dZ} = \frac{M_F \rho_b}{\mathbf{G} Y_{A_0}} (-\hat{r}_A) \tag{EX-7.7}$$

$$\frac{dZ}{\mathbf{G}} = \frac{dn_{Aa}}{(M_F \rho_b / Y_{A_0})(-\hat{r}_A)} \tag{EX-7.8}$$

or

$$\frac{L}{\mathbf{G}} = \int_{(n_{Aa})_{In}}^{(n_{Aa})_{Out}} \frac{dn_{na}}{(M_F \rho_b / Y_{A_0})(-\hat{r}_A)} = \int_0^{X_A} \frac{dX_A}{(M_F \rho_b / Y_{A_0})} \hat{r}_A \tag{EX-7.9}$$

Now for the adiabatic case, temperature change is a function of X_A; therefore, rate is a function of X_A and initial temperature and composition. Hence, for given inlet conditions and X_A the RHS of Eq. EX-7.9 is a constant, and a single value of $L/\mathbf{G}$ applies. Once an inlet temperature and an allowable design outlet temperature or conversion are selected, the adiabatic program can be run to yield a length at the mass velocity selected. Other values of $\mathbf{G}$ and L may be selected that give the same $L/\mathbf{G}$ to yield the same bed profile.

There are definite constraints, however, on the values of L and $\mathbf{G}$. In addition to the constancy of $L/\mathbf{G}$, these are

1. $L/D_p > 100$ to avoid significant axial mixing.
2. Allowable ΔP per bed (in this instance 0.363 atm) must not be violated.
3. ΔT between catalyst and fluid must, if ignored, not cause more than a 5–10% error in the rate. (See Eqs. 5.8 to 5.10).

Another useful relation can be rationalized from the pressure-drop equation (Eq. 4.20), which at constant fluid properties can be written as

$$\frac{\Delta P}{L} \propto \mathbf{G}^2$$

or

$$\Delta P \propto \mathbf{G}^2 L \propto \mathbf{G}^2 \left(\frac{L}{\mathbf{G}}\right) \mathbf{G} \propto \left(\frac{L}{\mathbf{G}}\right) \mathbf{G}^3$$

Since $L/\mathbf{G}$ is a constant for set conditions,

$$\Delta P \propto \mathbf{G}^3 \tag{EX-7.10}$$

or

$$\frac{\Delta P_2}{\Delta P_1} = \left(\frac{\mathbf{G}_2}{\mathbf{G}_1}\right)^3 \tag{EX-7.11}$$

Details of Adiabatic Reactor Design

An inlet temperature slightly above the dew point was selected for each reactor. Since each of the reactors in the series is to be identical in size so that positions in the train can be rotated by appropriate valving, the selecton of bed depth and **G** for reactor 1 is the initial and crucial exercise. The $L/\mathbf{G}$ set for the first bed must be set so that adequate conversion occurs in the first bed, and subsequent beds will be able to complete the reaction with the same $L/\mathbf{G}$ values. Since control of inlet temperatures is highly restricted in order to stay in the gas phase, selection of different catalyst activity levels may have to be employed in the several beds.

Catalyst A in Bed 1 A single run was made with catalyst A, the most active catalyst. Because of the high activity, a high **G** of 1.0 was used to extend the bed so as to avoid an excessively shallow bed. Results are shown in Table EX–7.5. A reasonable design basis would be an outlet conversion of 44%, yielding an adiabatic temperature of 515.5°K, a reasonable distance from the maximum allowable temperature of 525°K.

By interpolation from Table EX–7.3, 44% conversion occurs at 515.5°K and 20.72 cm,

$$\therefore \text{ Required } \frac{L}{\mathbf{G}} = \frac{20.72}{1.0} = 20.72$$

The target outlet pressure is

$$10 - 0.363 = 9.637 \text{ atm}$$

Since $\Delta P \propto \mathbf{G}^2 \propto \mathbf{G}^2(L/\mathbf{G})\mathbf{G} \propto 20.72\mathbf{G}^3$, the pressure at 20.72 cm, from Table EX–7.5, is

$$9.798 - (0.011)(0.72) = 9.790 \text{ atm}$$

$$\Delta P = 10 - 9.790 = 0.210$$

$$\text{Desired } \mathbf{G}_2 = \mathbf{G}_1 \left(\frac{0.363}{0.210}\right)^{0.333} = (1.0)\left(\frac{0.363}{0.210}\right)^{0.3333} = 1.2$$

Thus

$$L = (20.72)(1.2) = 24.86 \text{ cm}$$

Clearly, catalyst A is not amenable to this system, since criterion 1, the miniumum bed-depth criterion, is violated.

$$\text{Minimum bed depth} = 100\ D_p = (100)(0.3175) = 31.8 \text{ cm}$$

Table EX-7.5 Trial Run, Adiabatic Bed #1 Catalyst A (Act = 1.0)

Z	T	P	X	Y	$R(1)$
.0	405.0	10.000	.0000	.0000	.17459E-04
1.0	407.4	9.990	.0088	.0088	.19127E-04
2.0	410.0	9.981	.0184	.0184	.21064E-04
3.0	412.9	9.971	.0290	.0290	.23329E-04
4.0	416.0	9.962	.0408	.0408	.25988E-04
5.0	419.6	9.952	.0540	.0540	.29121E-04
6.0	423.5	9.942	.0688	.0688	.32814E-04
7.0	427.9	9.932	.0855	.0855	.37148E-04
8.0	432.9	9.923	.1045	.1045	.42169E-04
8.9	437.9	9.914	.1237	.1237	.47243E-04
9.7	442.9	9.906	.1428	.1428	.52074E-04
10.4	447.5	9.899	.1609	.1609	.56363E-04
11.1	452.5	9.891	.1805	.1805	.60455E-04
11.7	457.0	9.885	.1984	.1984	.63554E-04
12.3	461.7	9.879	.2170	.2170	.66029E-04
12.9	466.5	9.873	.2362	.2362	.67655E-04
13.5	471.3	9.867	.2558	.2558	.68268E-04
14.1	476.2	9.861	.2754	.2754	.67803E-04
14.7	480.9	9.854	.2947	.2947	.66315E-04
15.3	485.4	9.848	.3134	.3134	.63959E-04
15.9	489.8	9.842	.3314	.3314	.60956E-04
16.6	494.5	9.834	.3512	.3512	.56949E-04
17.3	498.9	9.827	.3696	.3696	.52717E-04
18.1	503.5	9.819	.3888	.3888	.47925E-04
19.0	508.1	9.809	.4084	.4084	.42865E-04
20.0	512.6	9.798	.4277	.4277	.37832E-04
21.0	516.6	9.787	.4447	.4447	.33465E-04
22.0	520.1	9.777	.4599	.4599	.29720E-04
23.0	523.2	9.766	.4733	.4733	.26522E-04
24.0	525.9	9.755	.4853	.4853	.23786E-04
25.0	528.4	9.744	.4962	.4962	.21437E-04
25.4	529.3	9.740	.5002	.5002	.20591E-04

Notes:

- Although printing interval was set at 1.0 cm, program prints more often in region of rapid temperature change.
- This is a portion of the output of PFRSIM plug-flow reactor simulator written by Doyle E. Broom. Copyright 1988, 1987, 1986, 1985, 1984 by The University of Texas System. See Appendix B.2.

Catalyst B in Bed 1 A reasonable trial for catalyst B can be deduced from the result on catalyst A, since the rate equation changes from A to B by means of a multiplier (0.25), for which case the following applies for the target conditions:

$$\left(\frac{L}{\mathbf{G}}\right)_{\mathrm{B}} = \left(\frac{L}{\mathbf{G}}\right)_{\mathrm{A}} \frac{1}{0.25} = \frac{20.72}{0.25} = 82.88$$

Determine value of **G** to give desired ΔP:

$$\Delta P \propto \mathbf{G}^2 L$$

$$\mathbf{G}_B^2 L_B = \mathbf{G}_A^2 L_A \quad \text{or} \quad \mathbf{G}_B^3 \left(\frac{L}{\mathbf{G}}\right)_B = \mathbf{G}_A^3 \left(\frac{L}{\mathbf{G}}\right)_A$$

$$\mathbf{G}_B = 1.2 \left[\left(\frac{L}{\mathbf{G}}\right)_A \Big/ \left(\frac{L}{\mathbf{G}}\right)_B \right]^{1/3} = 0.756$$

Thus $L_2 = 62.7$ cm. This conclusion was checked by running the program, which gave good agreement of 62.9 cm at $X = 0.44$, and $T = 515.5$°K.

This bed depth, though meeting criterion 1, is shallow and makes an expensive distributor necessary to ensure even distribution of flow. A further, more serious, disadvantage for this medium-activity catalyst is that the relatively low $L/\mathbf{G}$, which must be the same in each bed, is not adequate in the two subsequent beds to achieve the desired conversion and meet all other constraints. This conclusion is demonstrated in Table EX–7.6, even with the change of allowing the maximum temperature to reach 525°K and using the most active catalyst A in beds 2 and 3. Of course, a fourth reactor could be specified, but the shallow beds already make this option unfavorable. The lower **G** of 0.689 was selected to produce a slightly lower ΔP for flexibility in operation.

Catalyst C in Bed 1 The low-activity catalyst requires a much higher $L/\mathbf{G}$ so that in maintaining equal $L/\mathbf{G}$ in each bed, beds 2 and 3 will have adequate catalyst to carry the reaction essentially to completion. With the same entry and conversion conditions, the required $L/\mathbf{G}$ can be calculated by using data on catalyst B for maximum temperature rise (Table EX–7.6): For bed 1,

$$\left(\frac{L}{\mathbf{G}}\right)_C = \left(\frac{L}{\mathbf{G}}\right)_B \frac{1}{0.1} = \left(\frac{65}{0.689}\right) \frac{1}{0.1} = 943$$

where 0.1 is the multiplier for converting catalyst B kinetics to catalyst C.

$$943\mathbf{G}^3 = (0.689)^3(94.3)$$

or

$$\mathbf{G} = 0.32 \quad \text{and} \quad L = (943)(0.32) = 301.8 \text{ cm}$$

Use 300 cm and set that as limit of the bed. Results are assembled in Table EX–7.7 along with results for beds 2 and 3, in which catalysts B and A were used, respectively, to better the target conversion in the final reactor. This

Table EX-7.6 Summary of Adiabatic Beds in Series (Catalyst B in Bed #1; Catalyst A in Beds #2 and 3)

All Reactors:
G = 0.689 g/cm^{2}sec
Length = 65 cm

Reactor No.	Inlet P, atm	Inlet Temp,°K	Outlet P, atm	Outlet Temp,°K	Catalyst
1	10	415	9.680	524.7	B
2	9.444	410	9.150	516.1	A
3	9.014	430	8.752	453.2	A

Compositions (Mole Fraction)

	Reactor #1 Inlet	Reactor #2 Inlet	Reactor #3 Inlet	Reactor #3 Outlet
H_2	0.40331	0.30217	0.18220	0.1520
C_6H_6	0.10068	0.06129	0.01450	0.0027
Cyclohexane	0.29675	0.40346	0.53015	0.5620
CH_4	0.15826	0.18508	0.21692	0.2249
C_2H_6	0.02698	0.03160	0.03700	0.0384
C_3H_8	0.01402	0.01640	0.01920	0.0199
MW		44	51.56	53.46

Excessive benzene in Reactor #3 discharge.

gives a safety factor that can be demonstrated by using the same reactor sequence but assuming that the last 60 cm (20%) of bed 1 has been deactivated. Table EX–7.8 indicates that the selected design will handle this situation readily.

This proposed design at $L/\mathbf{G}$ of 938 meets criteria 1 and 2, and the calculation that follows shows that it also meets criterion 3.

Table EX-7.7 Summary of Design Conditions for Adiabatic Beds in Series (Catalyst C in Bed #1; Catalyst B in Bed #2, and Catalyst A in Bed #3)

All Reactors:

$G = 0.32\ g/cm^2\ sec$

Length = 300 cm

Reactor No.	Inlet P, atm	Inlet Temp,°K	Outlet P, atm	Outlet Temp,°K	Dew Point @ Inlet,°K	Catalyst
1	10	405	9.665	524.2	403.4	C
2	9.529	410	9.220	524.1	408.9	B
3	9.084	425	8.818	444.7	415.5	A

Low ΔP heat exchangers specified.

Compositions (Mole Fraction)

	Reactor #1 Inlet	Reactor #2 Inlet	Reactor #3 Inlet	Reactor #3 Outlet
H_2	0.40331	0.30270	0.17122	0.1453
C_6H_6	0.10068	0.06140	0.01010	0.0000
Cyclohexane	0.29675	0.40300	0.54185	0.5692
CH_4	0.15826	0.78500	0.21992	0.2268
C_2H_6	0.02698	0.03150	0.03740	0.0386
C_3H_8	0.01402	0.01640	0.01950	0.0201
MW		43.97	52.25	53.88

Check of Criterion 3

The ΔT between catalyst and fluid can be estimated by using Eq. 5.8 for the heat-transfer coefficient between the catalyst and the fluid.

$$\frac{h_s D_p}{\lambda_f} = 2 + 1.1\,\mathrm{Pr}^{0.33}\mathrm{Re}^{0.6}$$

where

$$\lambda_f = \text{thermal conductivity.}$$

Table EX-7.8 Summary of Operating Adiabatic Reactors in Series (Catalyst C in Bed #1 with last 60 cm deactivated; Catalyst B in Bed #2, and Catalyst A in Bed #3)

All Reactors:
G = 0.32 g/cm² sec
Length = 300 cm

Reactor No.	Inlet P, atm	Inlet Temp,°K	Outlet P, atm	Outlet Temp,°K	Dew Point @ Inlet,°K	Catalyst
1	10	415	9.7	525.1	403.4	C
2	9.6	423	9.3	527.9	408	B
3	9.1	425	8.8	449.1	415.5	A

Compositions (Mole Fraction)

	Reactor #1 Inlet	Reactor #2 Inlet	Reactor #3 Inlet	Reactor #3 Outlet
H_2	0.40331	0.30105	0.1765	0.1450
C_6H_6	0.10068	0.0645	0.1230	0.0000
Cyclohexane	0.29675	0.3947	0.5362	0.5695
CH_4	0.15826	0.1829	0.2184	0.2268
C_2H_6	0.02698	0.0312	0.0373	0.0387
C_3H_8	0.01402	0.0162	0.0193	0.0200

Bed 2 exhibits the highest rate; therefore, it is the one to check. Values of rates in the range of interest were determined at constant composition to evaluate the effect of temperature on the rate

$$\frac{(\hat{r}_1)_2}{(\hat{r}_2)_1} = \frac{0.937 \times 10^{-5}}{0.879 \times 10^{-5}} = 1.07 \text{ for a } 5° \text{ change}$$

Thus an error of 7% in the rate would occur if the catalyst surface were 5° higher than the bulk temperature used for rate calculations in the model. Such an error is acceptable. Physical properties were obtained by using a special printout feature of the program.

$$\mathrm{Pr} = \frac{c_p \mu}{\lambda_f} = \frac{(23.2)(0.000149)}{(43.6)(0.000192)} = 0.413$$

$$\mathrm{Re} = \frac{D_p \mathbf{G}}{\mu} = \frac{(0.3175)(0.32)}{(0.000149)} = 682$$

$$\frac{h_S D_p}{\lambda_f} = 2 + 1.1\ (0.413)^{0.33}(682)^{0.6} = 43.1$$

$$h_s = \frac{(43.1)(0.000192)}{0.3175} = 0.026$$

$$h_s a_m(\Delta T) = (-\Delta H)\hat{r}$$

$$\Delta T = \frac{(51{,}620)(0.58\mathrm{E}-5)}{(8.6)(0.026)} = 1.3°$$

This ΔT is much less than the acceptable error; hence criterion 3 is satisfied.

REACTOR CONFIGURATIONS

The process design of the reactor configurations consists of the following simple calculations.

1. Required product rate (basis: 50 MM gal./hr cyclohexane; 95% operating factor)

$$\frac{(50 \times 10^6\ \mathrm{gal.})(6.522\ \mathrm{lb/gal.})(454\ \mathrm{g/lb})}{(365)(0.95)(24)(3600)} = 4941.7\ \mathrm{g/sec}$$

2. Required feed rate based on loop calculations

Operation	*F_T/F_{PT} from Plant Simulation*
No product CH recycle	11432.29/8058.86 = 1.4186
3:1 CH to BZ feed	37366.69/8103 = 4.6115

where F_T is total mass of feed to reactor system and F_{PT} is total mass of pure cyclohexane product from stabilizer.

Multitubular Reactor Configuration

The required number of tubes is calculated as follows:

$$\text{No. of tubes} = \frac{(4941.7)(F_T F_{PT})}{\mathbf{G}\,\pi D^2/4}$$

No Flash-Bottoms Recycle

$$\text{No. of tubes} = \frac{(4941.7)(1.4186)}{(0.45)[\pi(2.0574)^2/4]} = 4686 \text{ tubes}$$

Approximate Shell Diameter and Length Tubes layout is triangular. Outside diameter of tubes is 2.54 cm.

$$p = 1.25 D_O$$

$$n_t = \frac{3n_D^2 + 1}{4}$$

$$D_I = p(n_D + 1)$$

where D_O is OD of tube. D_I is ID of shell, n_t is total number of tubes, n_D is number of tubes at bundle diameter, and p is the pitch.

$$n_D = \left(\frac{4n_t - 1}{3}\right)^{1/2} = \left[\frac{4(4686) - 1}{3}\right]^{1/2} = 79$$

$$p = 1.25(2.54) = 3.175 \text{ cm}$$

$$D_I = 3.175(79 + 1) = 248 \text{ cm or } 97.6 \text{ in.}$$

The cooling fluid would be boiling boiler feedwater, which by thermosyphon effect would be carried to a steam drum for separation of saturated steam for superheating in a furnace-stack exchanger.

Shell Thickness (*ASME Code*)

$$t_v = \frac{PR_I}{\mathbf{SE} - 0.6P} + C$$

where t is thickness in inches; P is pressure in psig; R_I is shell inside radius; **S** is allowable stress (13800 psi for carbon steel at 650°F and below from Table 6.2); **E** is joint efficiency (use 1, 0); C is corrosion allowance in inches (use 0.125 in.).

$$\text{Design temperature} = \text{maximum jacket temperature } (440°\text{K}) + 50°\text{F}$$
$$= 332 + 50 = 382°\text{F}$$

Pressure of boiling feed @ 383°F water is 200.5 psia (185.8 psig) from steam table.

$$\text{Design pressure} = (200.5)(1.2) = 240.6 \text{ psi or } 226 \text{ psig}$$

Therefore,

$$t_v = \frac{(226)(97.6/2)}{13800 - (0.6)(226)} + 0.125 = 0.932 \text{ in.}$$

Summary Tubes: 4686 one-inch OD tubes packed with catalyst BL to a height of 502.2 cm (16.48 ft), with space of approximately 4″ at each end for bed-holding inserts, total required length of tube is 17′2″.

Shell

ID = 97.6 in., OD = 99.48 in.

Length = 17′2″ tangent to tangent

Ellipsoidal heads: must be removable for loading and unloading.

Adiabatic Reactor Configurations (3:1 CH/BZ)

$G = 0.32$ g/cm^2 sec

No. of reactors = 3

Operating temperature = 525°K (485°F); operating pressure = 10 atm (132.3 psig)

Design temperature = 535°F; design pressure = 159 psig

Diameter:

$$D = \left[\frac{(4941.7)(4.6115)}{(0.32)(\pi/4)}\right]^{0.5} = 301.12 \text{ or } 9.88 \text{ ft } (9'\ 10.5'')\text{ID}$$

Thickness (based on ASTM A 204 GrA 0.5 Mo steel) with $S = 16300$ psi:

$$t = \frac{(159)(9.88)(12)(0.5)}{16300 - (0.6)(159)} = 0.582 \text{ in.}$$

Outside diameter: 9.977 ft (304 cm)

Height:

Catalyst bed: 300 cm (118.12″)

Inert support: 30.5 cm (12″)

Inert top layer: 30.5 cm (12″)(plus 1″ balls to fill head)

Tangent to tangent: 361 cm or 11′10″

See Fig. 1.1 for typical arrangement and Fig. 5.6 for suggested distributor, which must be designed carefully for this large-diameter reactor.

FINAL REACTOR SELECTION

The operating flexibility of the adiabatic system is severely constrained by the restriction of gas-phase operation. Since the control of the reactors is ultimately based on the inlet temperature and different-activity catalysts in each bed, successful operation is tenuous. There is no sure way to set the activities or control changes in them. Thus all that remains for control is the inlet temperature, which is limited by a minimum corresponding to the dew point temperature. If the first-bed catalyst has a higher than expected activity, the outlet temperature will exceed the allowable maximum. This observation leads to the logical conclusion that a mixed-phase operation should be designed.

A mixed-phase operation would have the following advantages.

- Wide flexibility in operation can be realized, since inlet temperature could be set so as not to exceed the allowable outlet for any activity catalyst.
- A high-activity catalyst can be selected so that low-temperature operation far from the range of significant side reactions will be possible.
- Mixed-phase feed will produce cooling of the reaction mix as flashing of the benzene and cyclohexane liquids occurs along the bed. Although heat exchanges between beds will still be necessary for bed inlet temperature control, the duties of these exchangers will be much lower than for the all-gas case.

The nonadiabatic all-vapor-phase reactor is viable when a medium-activity catalyst, as specified, is used. Flexibility in the jacket temperature through control of the exit steam pressure at the steam disengaging drum provides a convenient means for controlling conversion. The design jacket temperature is 68° above the inlet dew point so that, if a batch of catalyst with higher initial activity is installed, lower-temperature operation is possible. Also an increase in temperature by 10° can permit continued full operation with on-specification product when 22% of the catalyst has been deactivated. The reactor feed system should be manifolded for the use of cold hydrogen when needed to reduce reaction temperature rapidly. In such cases the hydrocarbon feed is reduced or eliminated until the temperature surge is corrected.

Of the two vapor-phase designs, the nonadiabatic case is the preferred one. It provides much greater operating flexibility than does the adiabatic design. The most flexible design would be adiabatic reactors in series operating in the mixed-phase regime. Such operation should be tested in a pilot-reactor system under various inlet conditions to demonstrate operating protocol, lack of extreme reactor sensitivity, and the performance of very high activity catalyst at low temperature where side-reaction products will be negligible. A stronger support than Kieselguhr, such as alumina, will be required to prevent catalyst breakage as liquid cyclohexane evaporates within the catalyst particles during use.

CONCLUSIONS

The design model was shown to be very useful in studying reactor alternates. The insights obtained through these calculations led to a better idea involving mixed-phase operation that would necessitate further development work. A rather fundamental axiom has been demonstrated by these calculations. In particular, for reversible exothermic reaction systems and for reaction systems with undesired side reactions, the best design usually employs a high-activity catalyst and low-temperature operation.

REFERENCES

1. Nagat, S., Hashimoto, K., Taniyama, I., Nishida, H. and Iwana, S., *Kagaku Kogaku*, (abridged Eng. Edition) *2*(1) 18 (1964).
2. PROCESS™, Simulation Sciences, Inc., Fullerton, California, 1987.
3. Broom, Doyle, PFRSIM^{CR} in "Thermal Cracking of a Light Coal Pyrolysis Tar and Related Compounds," Ph.D. Dissertation, University of Texas, Austin, Texas, 1989.

EXAMPLE 8

Use of a Reactor Model as an Aid to Catalyst Selection: Styrene Production with Radial-Flow Reactor

OBJECTIVE

This example demonstrates the utility of a reactor model in comparing catalyst performance and determining the economic credits for changing to a more selective but more costly catalyst. A two-stage, radial-flow, adiabatic reactor model is used, which is a common type for ethylbenzene dehydrogenation at low pressures where ΔP must be low.

PROBLEM STATEMENT

The next turnaround for the styrene monomer plant is scheduled 12 months from now. A new high-selectivity catalyst (A92) for the ethylbenzene dehydrogenation step is being marketed by the supplier of our current catalyst (A2), and it is necessary to make a judgment concerning the advantages, if any, of changing to the new catalyst. Research and Development, using bench-scale test reactors, has obtained parameter values for the new catalyst using the current rate forms.

The ethylbenzene production unit is operating at peak capacity, so any increase in yield must be accompanied by an increase in selectivity, since conversion must stay constant under the ethylbenzene feed limiting condition. Compare the new catalyst performance with that of the existing catalyst as rapidly as possible so that the plans for further evaluations can be developed if necessary.

REACTION DATA

Chemistry, kinetics, and catalyst data are summarized in this section.

Main Reactions

$$C_6H_5CH_2CH_3 \leftrightarrow C_6H_5CH=CH_2 + H_2 \tag{1}$$

ethylbenzene styrene

$$C_6H_5CH_2CH_3 \rightarrow C_6H_6 + C_2H_4 \tag{2}$$

ethylbenzene benzene

$$C_6H_5CH_2CH_3 + H_2 \rightarrow C_6H_5CH_3 + CH_4 \tag{3}$$

ethylbenzene toluene

Other Reactions

$$C_2H_4 + 2\,H_2 \rightarrow 2\,CH_4 \tag{4}$$

$$CH_4 + 2\,H_2O \rightarrow CO_2 + 4\,H_2 \tag{5}$$

Small amounts of heavier tars are also formed.

Kinetics

The following published rate forms for reactions (1), (2) and (3) [1] with the parameter values given below apply for the two catalysts. Plant observations suggest that 80% of the C_2H_4 is rapidly converted to CH_4, and 76% of the total methane is converted by reaction (5) to CO_2.

$$\hat{r}_1 = \frac{k_{p_1}K_E(P_E - P_H P_S/K_1)}{1 + K_E P_E + K_S P_S} \tag{EX-8.1}$$

$$\hat{r}_2 = \frac{k_{p_2}K_E P_E}{1 + K_E P_E + K_S P_S} \tag{EX-8.2}$$

$$\hat{r}_3 = \frac{k_{p_3}K_E P_E}{1 + K_E P_E + K_S P_S} \tag{EX-8.3}$$

$$\hat{r}_4 = 0.8\hat{r}_2 \tag{EX-8.4}$$

$$\hat{r}_5 = 0.76[\hat{r}_3 + 2(0.8\hat{r}_2)] \tag{EX-8.5}$$

where subscripts E, H, and S refer to ethylbenzene, hydrogen, and styrene, respectively; partial pressures are in atm, and rates are given in moles/g cat. sec.

	A		*E*	
Parameter	*Catalyst A2*	*Catalyst A92*	*Catalyst A2*	*Catalyst A92*
$k_{p_1} = A_1 \exp(-E_1/RT)$	6.04E2	3.706E7	29,268	48,761
$k_{p_2} = A_2 \exp(-E_2/RT)$	2.4835E6	3.431E6	50,800	50,800
$k_{p_3} = A_3 \exp(-E_3/RT)$	1.2E − 1	8.568E − 3	21,800	21,800

Both catalysts:

$$K_E = 1.3363\text{E}4 \exp[(3.68195\text{E}4 - 54.05T)/RT]$$

$$K_S = 1.0730\text{E}5 \exp[(3.68195\text{E}4 - 54.05T)/RT]$$

Equilibrium constant:

$$K_1 = \exp(16.12 - 15350/T)$$

Note: The rate constant values represent estimated average values for the first six months of use. Actually, as activity declines, temperature must be raised so that the inlet temperatures that will be selected are average values for the period.

Other Data

Mixture viscosity	$\mu = 2 \times 10^{-4} + 1.825 \times 10^{-7}(T - 600)$
Inlet conditions (both catalysts)	$\mathbf{G} = 0.106\ \text{g/cm}^2\ \text{sec}$ $P = 1.5$ atm first bed, ΔP between beds = 0.05 atm
Constant total conversion	$X_A = 0.617$ or 61.7%
Inlet temperature (same for both beds)	$T = 862°\text{K}$ average for first 6-month period as observed with catalyst A2, for which selectivity has averaged 95%

REACTOR DESCRIPTION

Type: Radial (two identical reactors in series with intermediate reheating to original inlet temperature)

Vessel ID	13.5 ft (411.5 cm)
Basket ID	11.18 ft (340.8 cm)
Distributor OD	5 ft (152.4 cm)
Length of bed (axial direction)	26.67 ft (812.9 cm)

Catalyst Data

	Existing (A2)	*Proposed (A92)*
Chemical content	Fe_2O_3, K_2CO_3 promoted with Ce_2O_3 and Mo_2O_3	Fe_2O_3, K_2CO_3 promoted with Cr_2O_3, Ce_2O_3, Mo_2O_3, and $CeCO_3$
Extrudate	1/8″ × 1/4″ (3 × 6 mm)	1/8″ × 1/4″ (3 × 6 mm)
Equivalent spherical diameter, D_p	0.381/cm	0.381/cm
Bulk density	1.282 g/cm^3	1.282 g/cm^3
Particle density	2.137 g/cm^3	2.137 g/cm^3
Bed void fraction	0.4	0.4
Cost	\$175/ft^3 (\$6180.07 m^3)	\$250/ft^3 (\$8828.67/m^3)
Catalyst life	18 months	Not known; assume same as A2
Activity average		
0–6 months	1.0	Assume same as A2
6–12 months	0.8	
12–18 months	0.6	

Feed and Product Values

MW	*Product*	*\$/lb*	*\$/lb mole*	*\$/kg mole*
106.17	Ethylbenzene	0.25	26.543	58.465
104.15	Styrene	0.32	33.328	73.410
78.14	Benzene	0.16	12.498	27.529
92.142	Toluene	0.10	9.214	20.300
28.054	Ethylene		0.95	2.093
16.043	Methane		0.57	1.256
44.010	Carbon dioxide		0	0
2.002	Hydrogen		0.19	0.419
	Fuel gas (plant price) \$1.50/MM BTU (\$5.952/kcal)			

CALCULATION PROCEDURE

The following steps were used in implementing the radial reactor option for PFRSIM program, which, as explained on p. 136, simply uses the radial distance of the reactor as the Z coordinate. In this case the inlet position is at the outside radius (76.2 cm) of the inner pipe distributor (see Fig. 1.2), and the end of the bed is the inside radius of the catalyst basket (170.4 cm). Although the inlet mass velocity **G** is set at 0.106 g/cm^2 sec for all cases, the value of **G** changes with position because of the changing cross-sectional area.

$$\mathbf{G} = \frac{m_f}{2\pi r_t L}$$

where m_f is the total mass flowing per second and L is the axial length of the beds.

1. Use the model on the A2 catalyst to calculate average performance for the first six months. Compare with that reported. If satisfactory, go to step 2.
2. Insert parameter values for A92 catalyst and repeat calculations for the two beds. Try different inlet temperatures until conversion as observed for the A2 catalyst is attained.
3. Now insert activity factors for reaction (1) only for each catalyst and calculate successively the 6–12 and 12–18 month performances again.

ECONOMICS

A rapid procedure [2] for performing economic analyses on the reaction system involves attaching the values given above to each component of the several stoichiometric equations and calculating cost per mole of the key component reacted for each reaction. This procedure is analogous to that of determining heats of reaction from heats of formation of the reactants and products.

Reaction 1

$$C_6H_5C_2H_5 \leftrightarrow C_6H_5CHCH_2 + H_2$$

$$-26.543 + 33.328 + 0.19 = \$6.98/\text{lb mole} \ (\$15.37/\text{kg mole})$$

Reaction 2

$$C_6H_5C_2H_5 \rightarrow C_6H_6 + C_2H_4$$

$$-26.543 + 12.498 + 0.95 = \$-13.10/\text{lb mole} \ (\$-28.86/\text{kg mole})$$

Reaction 3

$$C_6H_5C_2H_5 + H_2 \rightarrow C_6H_5CH_3 + CH_4$$

$$-26.543 - 0.19 + 9.214 + 0.57 = \$-16.95/\text{lb mole} \ (\$-37.33/\text{kg mole})$$

Reaction 4

$$C_2H_4 + 2\,H_2 \rightarrow 2\,CH_4$$

$$-0.95 - (2)(0.19) + 2(0.57) = \$-0.19 \text{ lb mole} \ (\$-0.42/\text{kg mole})$$

Reaction 5

$$CH_4 + 2\,H_2O \rightarrow CO_2 + 4\,H_2$$
$$-0.57 - 2(0) + 0 + 4(0.19) = \$0.19/\text{lb mole}\ (\$0.42/\text{kg mole})$$

Since steam and CO_2 are not conveniently recoverable in the process, both are given values of zero. The condensed steam can be recovered after treatment, but that requires maintaining the same exit conversion in each case by increasing the temperature an appropriate amount, which is found by trying different temperatures.

The results of model calculations are summarized in Table EX–8.1, and a typical output for bed 1 is given in Table EX–8.2.

Energy Cost Increment

Catalyst A92 requires additional heat input based on difference in temperature for two cases (875 − 862 = 13):

$$[M_{\text{EB}}(c_p)_{\text{EB}} + M_{\text{W}}(c_p)_{\text{W}}]13 = \frac{(148{,}518)(1000)}{3}\,[(c_p)_{\text{EB}} + 2(c_p)_{\text{W}}]13$$

$$= \frac{1.48518 \times 10^8}{3}\,(0.66 + (2)(0.53))13 = 1.107 \times 10^9\ \text{cal/hr}$$

or

$$1.107 \times 10^9/0.7 = 1.581 \times 10^9\ \text{cal/hr for furnance efficiency of 70\%}$$

Some of this heat can be recovered, and this is a worst-case analysis.

Catalyst Cost Increment (Beginning of 18-Month Period)

A92	\$250 ft^3
A2	\$175 ft^3

Catalyst volume (2 beds) $(11.18^2 - 5^2)(\pi/4)(26.67)(2) = 4189\ \text{ft}^3$

$$\Delta_{\text{CAT}} = -(250 - 173)(4189) = \$314{,}174$$

A summary of cash flow for each six-month period is given in Table EX–8.3.

Table EX-8.1 Summary of Catalyst Performance

Period months	Catalyst	Inlet Temp. °K	Selectivity %	PRODUCT (moles/mole of EB fed to Bed #1)								
				EB	*STY*	H_2	*BEN*	*TOL*	C_2H_4	CH_4	H_2O	CO_2
0-6	A2	862	95.17	0.3834	0.5868	0.6591	0.0095	0.0251	0.0019	0.0086	11.7262	0.0269
(α = 1.0)	A92	875	96.03	0.3825	0.5930	0.6707	0.0226	0.0019	0.0045	0.0091	11.7221	0.0209
6 - 12	A2	864	94.56	0.3846	0.5819	0.6638	0.0113	0.0222	0.0022	0.0096	11.7189	0.0306
(α = 0.8)	A92	879	95.3	0.3818	0.5892	0.6818	0.0270	0.0020	0.0054	0.0109	11.7110	0.0345
12 - 18	A2	870	93.54	0.3787	0.5809	0.6817	0.0149	0.0256	0.0031	0.0119	11.7049	0.0376
(α = 0.6)	A92	886	94.05	0.3790	0.5841	0.7016	0.0345	0.0023	0.0069	0.0139	11.6924	0.0438

Notes: Activity factors shown are for reaction #1 only. Conversion in all cases 61.7 to 62%.

Table EX-8.2 Printout for Bed #1 Catalyst A2

```
                              STYRENE REACTOR


I. FEED CONDITIONS

     REACTOR TYPE          , TYPE     =        2.0000 (1 OR 2)
     BED LENGTH            , L        =      812.9000 CM
     REACTOR MODE          , MODE     =        2.0000 (1, 2, OR 3)
     MASS VELOCITY         , G        =         .1060 G/(CM**2*S)
     INLET TEMPERATURE,      TI       =      862.0000 DEG. K
     JACKET TEMPERATUR,      TJ       =         .0000 DEG. K
     INLET PRESSURE        , PI       =        1.5000 ATM
     INLET COMPOSITION
          COMPONENT             NI             XI
          ---------             --             --
             EBE              1.0000         .07825
             STY               .0000         .00000
             H2                .0000         .00000
             BEN               .0000         .00000
             TOL               .0000         .00000
             C2H4              .0000         .00000
             CH4               .0000         .00000
             H2O             11.7800         .92175
             CO2               .0000         .00000
     INSIDE DIAMETER       , DI       =      152.4000 CM
     WALL THICKNESS        , XW       =         .1000 CM
     STEP SIZE             , ZSTEP    =        5.0000 CM
     PRINTING INTERVAL,      ZPRINT   =       10.0000 CM
     BED LIMIT             , ZLMT     =      170.4000 CM
     CONVERSION LIMIT      , XLMT     =         .9000 (DIMENSIONLESS)
     YIELD LIMIT           , YLMT     =         .9000 G-MOLES PR/G-MOL


II. CATALYST DESCRIPTION


     PARTICLE DIAMETER, DP =     .3810 CM
     BED VOID FRACTION, EPS =     .40000
     BED DENSITY, RHOB = 1.28200 G/CM**3



                           SUMMARY OF RESULTS

         Z =    171.2 CM
         T =   791.6 K
         P =   1.381 ATM
         X =   .4262
         Y =   .4095
         MW =   24.08
         OUTLET COMPOSITION


         COMP        N(J)        X(J)

         EBE         .5738       .0434
         STY         .4095       .0310
         H2          .4491       .0340
         BEN         .0046       .0004
         TOL         .0120       .0009
         C2H4        .0009       .0001
         CH4         .0047       .0004
         H2O       11.7505       .8888
         CO2         .0147       .0011
         TMIN =    791.6 K AT Z =       171.2 CM
         TMAX =    862.0 K AT Z =        76.2 CM
```

Table EX-8.2 *(continued)*

STYRENE REACTOR

Z	T	P	X	Y	R(1)	R(2)
76.2	862.0	1.500	.0000	.0000	.19221E-04	.27426E-06
86.2	822.9	1.488	.2367	.2301	.25857E-05	.22740E-07
96.2	811.0	1.475	.3083	.2994	.12726E-05	.11127E-07
106.2	804.4	1.463	.3480	.3375	.76833E-06	.74898E-08
116.2	800.2	1.450	.3733	.3616	.50570E-06	.58153E-08
126.2	797.4	1.438	.3905	.3778	.34895E-06	.48916E-08
136.2	795.3	1.425	.4029	.3892	.24807E-06	.43252E-08
146.2	793.9	1.413	.4119	.3974	.18008E-06	.39529E-08
156.2	792.7	1.400	.4187	.4033	.13285E-06	.36956E-08
166.2	791.9	1.387	.4240	.4078	.99387E-07	.35108E-08
171.2	791.6	1.381	.4262	.4095	.86389E-07	.34374E-08

Software PFRSIM written by Doyle E. Boom copyright by The University of Texas System, 1988, 1987, 1986, 1985, 1984. Reproduced by permission. See Appendix B. 2 for details. Radial, adiabatic option used with inside bed radius of 76.2 cm and outside bed radius of 171.2.

Table EX-8.3 Summary of Incremental Cash Flow

Period (months)	*Δ Inlet Temp. A92 vs A2*	*Δ \$ Energy per Hour*	*Δ \$ Products*	*Net Δ \$/Hr*	*Net Revenue Δ \$/Mo*
0 - 6	13	- 9.41	186.82	177.41	127735
6 - 12	15	-11.76	191.49	179.73	129407
12 - 18	15	-11.76	163.46	151.70	109224

Δ = A92 Case - A2 Case

Product Values (First Six Months)

Basis: 1 kg mole EB fed, 100% recovery of products.

$$V_{PR} = \underbrace{(\text{moles styrene})(15.37)}_{\text{reaction 1}} + \underbrace{(\text{moles benzene})(-28.86)}_{\text{reaction 2}}$$

$$+ \underbrace{(\text{moles toluene})(-37.33)}_{\text{reaction 3}}$$

$$+ \underbrace{(\text{moles benzene-moles } C_2H_4)(-0.42)}_{\text{reaction 4}} + \underbrace{(\text{moles } CO_2)(0.42)}_{\text{reaction 5}}$$

where V_{PR} is product value in \$/kg mole total EB fed, and moles indicated are moles appearing in product stream.

Catalyst A2 (from Table EX–8.2)

$$\begin{aligned} V_{PR} &= (0.5868)(15.37) + (0.0095)(-28.86) + (0.0201)(-37.33) \\ &\quad + (0.0095 - 0.0019)(-0.42) + (0.0269)(0.42) \\ &= \$8.00/\text{kg mole total EB fed} \end{aligned}$$

Catalyst A92 (from Table EX–8.2)

$$\begin{aligned} V_{PR} &= (0.5930)(15.37) + (0.0226)(-28.86) + (0.0019)(-37.33) \\ &\quad + (0.0226 - 0.0045)(-0.42) + (0.0290)(0.42) \\ &= \$8.40/\text{kg mole total EB fed} \end{aligned}$$

$$\Delta\ \text{Value} = 8.40 - 8.00 = \$0.40/\text{kg mole total EB}$$

Total feed flow: (2:1 H_2O/EB mass ratio)

$$\frac{(0.106\ \text{g/sec})(3600\ \text{sec/hr})(\pi)(152.4)(812.9)}{1000} = 148{,}518\ \text{kg/hr}$$

for a 2:1 mass ratio of steam to ethylbenzene of

$$\frac{148{,}518}{(3)(106)} = 467.04\ \text{kg moles EB/hr}$$

$$\text{Total}\ \Delta\ \text{value} = (0.40)(467.04) = \$186.82/\text{hr}$$

Calculation of Payout Time

A quick assessment frequently used for an early indication of project promise is the payout time, which is usually calculated without any corrections for taxes.

$$\text{Payout time} = \frac{\Delta\text{cost of cat.}}{\text{Net}\ \Delta\$\ \text{cash flow/time}}$$

Using first 6 months (see Table EX-8.3):

$$\text{Payout time} = \frac{314{,}174}{127{,}735/30} = 74\ \text{days}$$

This is a very attractive preliminary assessment of the advantage of changing to catalyst A92.

Calculation of Discounted Rate of Return

The rate of return is a frequently used measure that allows comparison with other projects and can be useful if used with due regard for the major uncertainties associated with predicting performance. In order to place all such evaluations on a comparable basis, cash received in each period (one month in this case) is discounted to the beginning of the period, and then, in turn, each subsequent six-month period so calculated must be discounted to the beginning of the project.

$$P_w = \sum_{1}^{n} R_n \frac{1}{(1+i)^n} \tag{EX-8.1}$$

where P_w is the present worth, R_n is the cash flow in period n, and i is the normal or periodic rate of return.

The series of Eq. EX-8.1 can be simplified for each six-month period, since R_n is considered to be a constant value for each month in the six-month period based on the original assumption of average conditions.

$$P_w = \bar{R}_n \sum_{n=1}^{n} \frac{1}{(1+i)^n} = \bar{R}_n \frac{(1+i)^n - 1}{i(1+i)^n} \tag{EX-8.2}$$

where P_w is the present worth of the investment and $\bar{R}_n$ is the uniform per month cash flow for a given period. For the three six-month operating periods for which n is in months,

$$P_w = \bar{R}_1 \frac{(1+i)^6 - 1}{i(1+i)^6} + \bar{R}_2 \frac{(1+i)^6 - 1}{i(1+i)^6} \left[\frac{1}{(1+i)^6}\right] + \bar{R}_3 \frac{(1+i)^6 - 1}{i(1+i)^6} \left[\frac{1}{(1+i)^{12}}\right] \tag{EX-8.3}$$

The bracketed terms in each case correct that six-month portion of the present worth from its beginning period to the present time (time = 0). Equation EX-8.3 is often solved for i by trial and error, but use of a nonlinear equation solver, such as IMSL NEQNF, is much quicker and is easy to use when a convenient interactive driver, such as used in Example 3, is applied. All that needs to be placed into the driver is Eq. EX-8.3 rearranged as follows:

$$\begin{aligned} F(1) = {} & \mathrm{PAR}(1) * (1 + X(1)) * * 6 - 1)/X(1) * (1 + X(1)) * * 6 \\ & + \mathrm{PAR}(2) * (1 + X(1)) * * 6 - 1)/X(1) * (1 + X(1)) * * 12 \\ & + \mathrm{PAR}(3) * (1 + X(1) * * 6 - 1)/X(1) * (1 + X(1)) * * 18 - \mathrm{PAR}(4) \end{aligned}$$

Where $X(1)$ is i; PAR(1), PAR(2), and PAR(3) are $\bar{R}_1$, $\bar{R}_2$, and $\bar{R}_3$, respectively (see Table EX-8.3); PAR(4) is the present worth, in this case the Δdollar required for the new catalyst.

The solution of this equation, with a present worth of the original Δcatalyst cost of \$314,174, PAR(4), and R values as given in Table EX–8.3, is 0.454. Hence,

Monthly rate of return = 45.4%

Nominal rate of return per year = (45.4)(12) = 545%

Effective rate of return per year = $100\,[(1 + 0.454)^{12} - 1] = 8828\%$

By any measure the proposed catalyst change is attractive. Because of the extremely high return, the simple payout time is a satisfactory measure.

UNCERTAINTIES AND FINAL DECISION

Before a final decision is reached, careful examination of the uncertainties that will be faced is necessary. These uncertainties include the following:

1. *Market strength for styrene.* A tight market has been assumed, such that all increases in styrene yield will be absorbed by the market without decline of selling price.
2. *Feed and product values.* The spread between styrene and ethylbenzene is crucial. It can increase or decrease. Energy costs, low in this case, could increase.
3. *Catalyst life and activity.* They may be different than assumed. Calculations were based on the assumption of identical activity characteristics for the new catalyst (A92), which have been well documented for the existing catalyst (A2).

Market strength requires constant attention related in this case to customers of downstream products as well as development of new products. Feed and product differentials are often governed by macroeconomic forces, but well-wrought contracts and consumer satisfaction can play a significant role. Within the technical areas, the uncertainty concerning the long-term performance of the new catalyst is a crucial issue. Obviously, if the catalyst is totally new, it may only have been subjected to life tests of 30 to 60 days; and some legitimate concern about long-term performance is certainly appropriate. As an example, if the new catalyst had to be changed in six months, cash flow for the year would decline by \$1,706,336 (\$659,087 for seven days lost production and \$1,047,250 for catalyst replacement). This decline exceeds the added cash flow for the new catalyst A92 compared with A2 (127,735 × 12 = 1,532,820).

Since the next turnaround is 12 months away, it is feasible to begin a life test on the new catalyst by using a pilot reactor installed adjacent to the existing reactor if a connection in the feedline exists so that a slip stream from the feed to the plant can be used. Such a test would surely provide the background for a confident decision. One should also remember that the technical representative for the catalyst manufacturer is a reliable and competent source of information on the supplier's catalyst, and that representative should be contacted for details of the latest tests and experiences with the new catalyst.

CONCLUSIONS

The value of reasonable models in the decision-making process associated with reactor operation has been demonstrated by this example. Though useful in identifying possibilities for improved performance, many other important issues must be considered in the decision process that require a thorough knowledge of the market and at times actual pilot-plant work with the catalyst.

REFERENCES

1. Modell, D.J., in "Chemical Engineering Computing," Vol. 1, pp. 100–113, AIChE, New York, 1972.
2. DeHey, J., Personal communication, March 1988.

EXAMPLE 9

Use of Reactor Model in Troubleshooting: Catalytic Reforming

OBJECTIVE

This example has dual objectives:

1. Demonstrate the utility of a reactor model in troubleshooting
2. Demonstrate kinetic modeling of multiple-reaction systems by using lumping of compounds into classes involving many reactants and products

Since many successful processes were originally scaled up from pilot-plant results, reactor models are used most often for reactor fault diagnosis (troubleshooting) or process improvement studies. This situation is particulary peculiar to commodity industries. When profit margins decline because of competition or radical increases in raw material and/or energy costs, a reliable model is a most valuable aid in the continuing task of reducing operating costs and overcoming operating problems. One such process, which is a classic example of competent experimental research and development, is catalytic reforming.

BACKGROUND ON CATALYTIC REFORMING

Catalytic reforming involves the conversion of a paraffinic naphtha to a large proportion of aromatics, which can be used as high-octane base stock for unleaded gasoline or as raw material for organic chemicals and polymers. The feed is a complex mixture of compounds in the range of C_6–C_{10}, and the product stream may contain as many as 300 compounds. Modeling requires some attempt at simplification by selecting pure compounds and lumping others into a mixture of isomers of the same carbon number. Proprietary models designed on this basis will typically have from 13 to 22 compound lumps and involve various reactions yielding up to 35 rate equations.

These models have been successfully applied rather continuously in the task of improving the return on units operated by a given company at multiple locations. The once burdensome task of model discrimination and parameter evaluation has been made much easier by custom software such as SIMUSOLV™ (see Example 11), but the experimental effort necessary to obtain reliable and adequate data remains formidable for such complex systems as catalytic reforming. In some cases such effort may be difficult to justify.

Models appearing in the open literature for catalytic reforming follow procedures similar to those for the proprietary models, but they are further simplified by reducing the number of lumped pseduocomponents. The experimental data must be well planned, but can be less voluminous because of the fewer parameters involved.

One such approach in the open literature [1] employs five rate equations based on five general reactions and three pseudocomponents in the feed, each having properties defined by an average pure component representative of each class represented. This technique is applicable to any type of complex system and will give good results when the applicable rate parameters are obtained from extensive experimental and plant data.

The following general reactions were proposed:

Napthene dehydrogenation

$$C_nH_{2n} \leftrightarrow C_{2n-6} + 3\,H_2 \tag{1}$$

Paraffin cyclization

$$C_nH_{2n+2} \leftrightarrow C_nH_{2n} + H_2 \tag{2}$$

Paraffin hydrocracking

$$C_nH_{2n+2} + \left(\frac{n-3}{3}\right)H_2 \rightarrow \frac{n}{15}\,(CH_4 + C_2H_6 + C_3H_8 + C_4H_{10} + C_5H_{12}) \tag{3}$$

Napthene hydrocracking

$$C_nH_{2n} + \frac{n}{3}\,H_2 \rightarrow \frac{n}{15}\,(CH_4 + C_2H_6 + C_3H_8 + C_4H_{10} + C_5H_{12}) \tag{4}$$

Pentane hydrocracking

$$C_5H_{12} + H_2 \rightarrow \tfrac{1}{2}(CH_4 + C_2H_6 + C_3H_8 + C_4H_{10}) \tag{5}$$

Reaction (1) is endothermic and tends to be the dominant reaction in the first reactor, since the exothermic hydrocracking reactions do not reach a point of significant conversion until higher temperatures are sustained in the later reactors.

Although hydrogen affects the forward reactions of reactions (1) and (2) adversely, it is used as a means for inhibiting coke formation that deactivates the catalyst.

The subscript n on the various compounds refers to a carbon number selected, by the following equation, to represent a pseudocomponent representative of the feedstock of interest:

$$M_F = n_P M_P + n_N M_N + n_A M_A$$

$$M_F = n_P(12n + 2n + 2) + n_N(12n + 2n) + n_A(12n + 2n - 6)$$

Hence,

$$n = \tfrac{1}{14}(M_F - 2n_P + 6n_A) \qquad \text{(EX-9.1)}$$

where n_P, n_N, and n_A refer to moles paraffin, naphthene, and aromatic, respectively, per mole of hydrocarbon feed; M_P, M_N, and M_A are the molecular weights of, respectively, the paraffins, naphthenes, and aromatics and n is the average carbon number.

PROBLEM STATEMENT

Difficulty is being experienced with reactor 1 in a train of four reactors of a catalytic-reforming unit designed to process 30,000 BPD of light naphtha. In order to increase the capacity of the unit by 10%, the inlet pressure was lowered from 300 psia to 200 psia, and the H_2/oil mole ratio was lowered from 5:1 to 3:1. The first bed is most sensitive to pressure because of the importance of reaction (1) in it. The before and after results are given below:

a. Compare the aromatics flow rate in the product from the first bed for both cases (moles/cm^2 sec).

Reactor 1 (observations)

	Previous Results	*Results after Change in P*
Mass velocity, **G**	2750 lb/hr ft^2 0.3733 g/sec cm^2	3025 lb/hr ft^2 0.4106 g/sec cm^2
Temp. In (°F)	920 (766.67°K)	920 (766.67°K)
Inlet pressure (psia)	300 (20.4 atm)	200 (13.605 atm)
Outlet pressure (psia)	279 (19.0 atm)	175 (11.9 atm)
H_2/oil ratio (moles)	5:1	3:1
MW, total product	22	28
% aromatics in product, Bed 1	3.7	3.5

Reactor dimensions : 16 ft ID × 10 ft bed length

b. Using the reactor model, determine if the result observed corresponds with that predicted by the model. If the two do not agree, offer one or more hypotheses based on problems with the reactor.

Reaction Data

Catalyst (Pt-Re/Al_2O_3)
Comp. 0.3% Pt, 0.3% Re; size 0.18 cm × 0.54 cm extrudate

$$D_P = \frac{6}{a_S} = \frac{6d_c}{4 + 2d_c/h_c}$$

Bulk density	0.66 g/cm^3
Bed void fraction	0.31
Particle density	1.12 g/cm^3

Feed
Naphtha: 59.2° API hydrofined naphtha

Feed analysis

	Liquid Volume%	*Mole%*
Naphthenes	30.3	32.0
Paraffins	61.1	57.3
Aromatics	8.6	10.6
Molecular weight	112	

Use C8's as model components based on Eq. EX-9.1

Type	Compound
Naphthene	*trans*-1, 4 Dimethylcyclohexane
Paraffin	2-Methylheptane
Aromatic	*m*-Xylene

Recycle gas stream

Component	Mole%
H_2	68.8
CH_4	7.1
C_2H_6	9.2
C_3H_8	9.0
C_4H_{10}	4.4
C_5H_{12}	1.5
	100.0

Rate equations (1)

$$A = (1 + K_P P_P + K_N P_N + K_A P_A)3600$$

$$\hat{r}_1 = \frac{k_{p_1}(K_N P_N - K_A P_A P_H^3 / K_1}{A}$$

$$\hat{r}_2 = \frac{k_{p_2}(K_P P_P - K_N P_N P_H / K_2)}{A}$$

$$\hat{r}_3 = k_{p_3} K_P P_P / A$$

$$\hat{r}_4 = k_{p_4} K_N P_N / A$$

$$\hat{r}_5 = k_{p_5} K_P P_{c_5} / A$$

K and $k = \exp(A' - B/T)$ $T[=]°\mathrm{K}$ $\hat{r}[=]$ moles/g cat. sec

Constant	*A′*	*B*	*Constant*	*Adsorption K's (Temp Independent)*
k_{p1}	24.5	19 533	K_N	0.49
k_{p2}	24.45	20 222	K_P	0.02
k_{p3}	32.50	25 522	K_A	2.03
k_{p4}	29.53	25 256		
k_{p5}	38.10	30 511		
K_1	45.22	24 361		
K_2	13.45	2 778		

where subscripts A, N, P refer, respectively, to aromatics, naphthenes, and paraffins.

LOGICAL PROCEDURE

In many situations of the type addressed in this example, a model is already in the computer storage, and the user can use the program to quickly compare the before and after cases. First the plug-flow, adiabatic model is run, using the before conditions to be certain that it predicts that performance satisfactorily. Then the new conditions are used, and the results may be compared with confidence with the plant observations.

In this case the general-purpose program described in Example 7 was used, and the rate and property data given above were inserted in the template. Results of the comparative runs are shown in Table EX–9.1, and comparisons of observed and calculated results are shown in Table EX–9.2.

Aromatics flow is calculated as follows:
Predicted new condition:

$$\text{Aromatic rate} = \frac{G(Y_{\text{ARM}})}{\text{MW}} = \frac{(0.4106)(0.0503)}{(28.39)} = 7.275 \times 10^{-4} \text{ g moles/cm}^2 \text{ sec}$$

Table EX–9.1 Calculated Product Streams for First Bed of Catalytic Reformer (previous conditions and new conditions)

Previous Conditions			*New Conditions*		
$Z = 304.8$ CM			$Z = 304.8$ CM		
$T = 709.2$ K			$T = 704.0$ K		
$P =$ 19.221 ATM			$P = 11.935$ ATM		
$X =$ .6154			$X =$ .5775		
$Y =$ .6514			$Y =$ .6036		
$Mh =$ 21.95			$Mh =$ 28.39		
OUTLET COMPOSITION			OUTLET COMPOSITION		
COMP	*N(J)*	*X(J)*	*COMP*	*N(J)*	*X(J)*
NAP	.3846	.0146	NAP	.4225	.0227
PAR	1.6366	.0623	PAR	1.6696	.0899
ARM	.9826	.0374	ARM	.9348	.0503
H2	17.4137	.6624	H2	11.0512	.5947
CH4	1.6777	.0638	CH4	1.0192	.0549
C2H6	2.1546	.0820	C2H6	1.3017	.0701
C3H8	.5737	.0218	C3H8	1.2780	.0688
C4	1.0646	.0405	C4	.6514	.0351
C5	.3989	.0152	C5	.2528	.0136

Calculations by PFRSIM. See Appendix B.2 for details.

Table EX-9.2 Comparisons of Aromatics Flow from First Bed

Condition	*Aromatics Flow, gmoles/cm²sec*
A. Observed Results at Old Condition	6.3×10^{-4}
B. Calculated Results at Old Condition	6.36×10^{-4}
C. Observed Results at New Condition	5.133×10^{-4}
D. Calculated Results at New Condition	7.275×10^{-9}

RESULTS

From Table EX–9.2 it is clear that the model provides a reasonable estimate of performance for the old condition and gives one confidence in its predictive ability at other conditions not out of the range of its known applicability. If the new case is in such a range of conditions, the operating results are significantly below the predicted results. Possible causes include

1. Flow bypassing in bed due to partial plugging
2. Excessive temperature during regeneration prior to starting at new conditions.

EXAMPLE 10

Expert System for Reactor Diagnostics: Steam Reforming

OBJECTIVE

In this example the development of a computer-based expert system is presented with the objective of illustrating a logical procedure and indicating its value as an aid to competent human judgment.

PROCEDURE

As discussed in Chapter 8 the development of an expert system involves four steps.

1. Assembling a narrative account of known problems and the associated symptoms, causes, and recommended actions
2. Preparing from the narrative material an organized table of symptoms, causes, and recommended courses of action
3. Developing and entering into the expert system software data base all symptoms (abnormal conditions) and distinguishing features, all known causes, a series of applicable rules, and recommended courses of action
4. Testing the finished system

ASSEMBLING A NARRATIVE ACCOUNT

Table EX–10.1 presents an organized group of notes based on information from the literature, catalyst suppliers, and unit operators and technical personnel. Because of the proprietary nature of the operating lore, this latter contribution is not as significant as it would be for a system being developed for plant use. In such cases the input of company operators, technical service engineers, and R & D personnel would be a major component of the narrative material. In any event, the goal is to write down all the information on problems that have been experienced and solved in the past. Much of this type of information gathering involves study of old files, reading relevant literature, and interviewing experienced people. The notes taken must then be organized in categories as shown in Table EX–10.1. In so doing, the knowledge of several sources may be required to complete particular problems, and rechecking of the details proves an invaluable aid to producing accurate descriptions and analyses.

Table EX–10.1 Methane Steam Reforming Operating Lore and Experiences Combined with Research Findings

I. Conditions observed during start-up period

A. Fresh or regenerated catalyst contains nickel as NiO, which must be reduced to Ni in order to produce an active catalyst. Since the reforming reaction produces H_2, it is common practice to operate initially at a reduced feed rate and high H_2O/CH_4 (7 to 10) until adequate reduction is attained and thus full activity. If hydrogen is available in storage, a small added stream of H_2 will facilitate the reduction and reduce start-up time. Without added H_2, even unreduced catalyst will produce some reforming when the flow rate is reduced. The higher steam rate serves to accelerate the reaction, but the H_2 partial pressure at the inlet region will be lower than further along in the tube. There is a possibility, therefore, if the reduction phase of start-up is not sufficient in length or if the steam rate is not high enough to cause adequate hydrogen production and regular operating conditions are established prematurely, that the first portion of the catalyst in the tubes will be less active. With the heat flux for design production rates, the less-active region will not be able to absorb the total heat transferred as endothermic reaction heat and a modest amount of sensible heat to maintain a desired temperature profile. Accordingly, the temperature of the gas and the tubewall will increase.

If the increase is significant and long-lasting, carbon and coke may form, the ΔP will increase, and coke buildup will continue to the point where shutdown and regeneration will be necessary. Usually, however, the condition should be caught early so that complete reduction can be effected and full activity attained.

Observed conditions

- Increased tubewall temperature [hot tube area (hot bands) near inlet of all tubes]
- Increased ΔP; may increase with time if higher hydrocarbons are present
- Lower production (more CH_4 at outlet) or increased temperature required to maintain production

B. Poor loading procedure can cause significantly different bed porosities in various tubes so that poor distribution to the various tubes will occur upon

Table EX-10.1 (*continued*)

start-up. Some tubes will carry more flow than normal and others much less. Average pressure drop may be higher or lower than normal, depending on the mix between tightly and loosely packed beds. The tubes experiencing lower flows may become significantly hotter. The positions of such tubes can be random or isolated to one section, depending on the packing practices. Continued operation can lead to coke formation and increasing ΔP.

Observed conditions

- Variation in tube temperatures between tubes
- Abnormal ΔP may increase with time on stream, and some tubes become very hot

II. Steady operation

A. Carbon formation (whisker)

In methane steam reforming the usual route to carbon formation on the catalyst is through methane cracking on active nickel

$$CH_4 \rightleftarrows C + 2\,H_2 \qquad \text{(endothermic)}$$

This process produces whiskers or filaments of carbon the same diameter as the Ni particle. The unusual result is that the nickel remains on the outer tip of the whisker as the carbon grows in length.

Based on the operating quotient $P^2_{H_2}/P_{CH_4}$ and the equilibrium ratio at the temperature at a point in the tube, the reaction becomes favorable when $P^2_{H_2}/P_{CH_4} < K$. This situation is most likely to occur in a region 15 to 30% of the distance from the inlet, where high tube temperatures and low hydrogen concentrations are possible. The tubes of top-fired furnaces will show a hot band in the region somewhere within the first 15–30% of the tube. For side-fired furnaces a lesser temperature increase would be seen more generally over the entire tube usually as hot areas in the form of stripes as on a tiger's tail.

Low H_2/CH_4 ratios are encouraged by low activity and a low steam-to-methane ratio, both of which can occur in regions of low-activity catalyst at the top portion of the tube. Sulfur poisoning can cause lowering of activity in this region in a unit that was previously performing satisfactorily. Excessive heat flux in any region due to a mistake can also lead to a temperature at which $K > P_{H_2}/P_{CH_4}$.

During the initial stages of whisker formation, an increased steam-to-methane ratio can reverse the process and cause the whiskers to disappear. In fact, the whisker growth may have been caused by a partial failure in steam flow, and restoration of the usual flow would remove the whiskers.

Usually the more severe problem is due to sulfur poisoning, which may be gradual. In this case the carbon grows slowly and, as it ages, collapses onto the catalyst. This aged carbon is not reactive and will not disappear by increasing the steam ratios. Whiskers, as they grow, become quite strong and ultimately cause catalyst breakdown, which increases pressure drop.

Observed conditions

- Hot bands or stripes.
- Sudden or gradual increase in ΔP.
- Associated with low H_2/CH_4 and low H_2O/CH_4.
- Activity may not be greatly affected; but does occur in regions of catalyst already lowered in activity such as by sulfur poisoning.
- Does not normally occur unless excessively low steam ratios are used.

B. Carbon formation (pyrolytic carbon)
Pyrolytic carbon or coke can be formed by a thermal noncatalytic process or a combination of both thermal and catalytic routes. Carbon precursors are often formed thermally and further reacted on catalyst and tube surfaces by a catalytic step to the final coke product, a highly hydrogen-deficient polymer. The coke can cover the catalyst, reducing its accessibility and, thus, activity. It also can partially fill in the voids between the catalyst particles, producing increased pressure drop and necessitating higher operating pressure to maintain throughout. This increased pressure in turn favors the secondary reactions that form coke precursors. Thus in a relatively short time, if conditions are severe enough, large amounts of coke can form.

In the case of methane, very high temperatures are required to produce pyrolytic carbon. But usually natural-gas streams will contain some higher hydrocarbons. It is also possible due to upsets in treatment facilities for slugs of higher hydrocarbons to appear in the methane feed. In such cases, in regions of even modestly high temperature and lower steam ratios, coke can form rapidly. The lower activity in the region causes higher tubewall temperatures, since the endothermic heat of the reaction is not adequate to consume a major portion of the heat from the tubes. Again, this situation further accelerates coke formation. Both the increased ΔP and hotter tubes will ultimately require action to remove the coke or replace the catalyst.

Again, sulfur poisoning also makes this kind of carbon formation more likely in the sulfur-deactivated region.

Observed conditions

- Increased ΔP
- Hot bands for top-fired or hot stripes along tubes for side-fired tubes where coke formation has occurred
- Significant activity decline
- Often occurs along with and is encouraged by sulfur poisoning
- Usually associated with hydrocarbons heavier than methane
- Increased CH_4 in product

C. Increased pressure drop can also be caused by the deposit of boiler feedwater solids often in the tubes closest to the inlet portion of the steam header. Thus tubes in this region will become hotter than others all along the length.

Table EX-10.1 *(continued)*

Observed conditions

- ΔP gradually increasing with time or sometimes more suddenly due to a slug of solids
- Ultimately, hotter tubes result in the region closest to the inlet

D. Sulfur poisoning

Hydrogen sulfide (H_2S) and organic sulfur compounds are strongly adsorbed on active nickel at reforming conditions. Steam-reforming units are preceded by a hot (350–400°C) ZnO bed, which removes H_2S, COS, and mercaptans (RSH). More stable compounds, such as thiophenes, are not removed effectively. If they are present, they must be reacted over a Co/Mo catalyst that converts them to H_2S, which then can be removed by the ZnO.

$$ZnO + H_2S \rightarrow ZnS + H_2O$$

The ZnO unit is usually adequate for natural-gas feeds, but upsets can occur that will allow sulfur compounds to enter the steam reformer. For example, loss of steam to preheater upstream of the ZnO unit can reduce temperature of the bed to the point that the adsorptions and reactions over ZnO will not occur, and sulfur compounds will be admitted to the steam-reforming catalyst. All organic sulfur compounds reaching the Ni catalyst are converted to H_2S, and it, in turn, is chemisorbed. On other occasions a switch in feedstock may present a natural gas with much higher sulfur content. These situations can occur suddenly and then be corrected, but it will take several months at normal operating conditions for the H_2S to be totally desorbed from the system. It adheres strongly to both the catalyst and the tube walls.

Since chemisorbed H_2S deactivates the Ni particulate to which it is adsorbed and prevents its use for catalyzing the main reaction, the activity declines with each H_2S molecule adsorbed. The lower-activity catalyst can then lead to other problems as described previously.

Observed conditions

- Increased sulfur content in outlet from ZnO unit
- Hot bands or stripes in top one-third of tube due to deactivated catalyst
- Increased CH_4 content in product
- Discharge from ZnO unit contains or did contain sulfur compounds above normal

E. Arsenic poisoning

Arsenic mimics the effect of sulfur on the catalyst. Arsenic compounds can be present in some natural-gas streams or carried over in recycle gases from some solvent CO_2 removal systems. Also can be present in ZnO adsorbent.

Observed conditions

- Same as for sulfur, only sulfur compounds in significant amounts will not be detectable

III. Regeneration and shutdown

A. Sulfur removal

Adsorbed sulfur can be removed by steaming at an outlet temperature 1200°F or higher by using 5–10 lb steam per pound of carbon [1]. Steaming must be done at low pressure (100 psia or less). High pressures cause the formation of NiO which agglomerates more readily than nickel:

$$Ni + H_2O \rightleftarrows NiO + H_2$$

The result is a catalyst with less Ni surface and thus less activity.

B. Combined sulfur and carbon removal

Usually both carbon and sulfur are present, and a combined steaming at 723°K inlet and steam-air treatment is recommended. Carbon can also be removed by steaming alone ($C + H_2O \rightleftarrows CO + H_2$), but the temperature must be higher (up to 970°K inlet temperature, depending on the age of the carbon). It is much faster to steam for sulfur removal and then add air slowly up to 5% while continuing the steaming. Air lowers the ignition point of the coke to around 720°K.

Observed conditions

- Some sulfur on the catalyst goes to $NiSO_4$, which can reform H_2S when the catalyst is reduced prior to start-up. At modest H_2 compositions below the range of 25–100 H_2O/H_2 [1], the $NiSO_4$ can be reduced to the NiO and the remaining H_2S purged. This procedure is used after the steam-air treatment.
- The initial steaming is continued until the SO_2 and H_2S analysis in the exit stream is at a low value that does not change any further with time.

$$Ni\,[\text{ads S}] + H_2O \rightleftarrows NiO + H_2S$$

$$H_2S + 2\,H_2O \rightleftarrows SO_2 + H_2$$

$$Ni + H_2O \rightleftarrows NiO + H_2$$

Keep pressure below 100 psia to avoid agglomerating the NiO.

- If catalyst is promoted with Na_2O or K_2O, the oxides will react with SO_2 produced during steaming to yield the corresponding sulfates. Upon reduction of NiO to Ni prior to start-up, the Na_2SO_4 and K_2SO_4 release H_2S and repoison the nickel; if Na and K content are high enough, no net removal of sulfur will occur.
- If it is necessary to keep furnace and catalyst warm for an extended period, use steam at pressures below 100 psia to avoid nickel agglomeration.

PREPARING AN ORGANIZED TABULATION

The next step involves the essential conversion of the narrative material into a structured list of symptoms, possible causes, and courses of action, as shown in Table EX–10.2. The symptoms are best restricted to those conditions most frequently and easily monitored by operators. Such conditions become symptoms when they evidence abnormal behavior. Characteristically, these symptoms are few, and, invariably, additional information, called in this table distinguishing features, is necessary to complete the diagnosis. Table EX–10.2 also shows the recommended means for confirming each diagnosis.

CONVERTING INFORMATION IN TABULATION TO STRUCTURED FORM

If a process engineer is assigned to a new unit and has prepared narrative notes and a table such as Table EX–10.2, personal use of this information in solving plant problems would require no further effort. But if there is need to make this expert information known to others, including other engineers and operators, further effort is necessary to construct a more rigid structure, which for most efficient use can be made part of an expert system. This task is accomplished by arranging the information in Table EX–10.2 into a form suitable for entry into expert system software, as shown in the diagram of Fig. EX–10.1. This diagram, although it contains all the necessary elements for problem diagnosis and suggested action, is not easy or rapid to use, even though the numbers on the connecting lines are helpful. It is for this reason that expert system programs have been developed for ease of entry of problem parameters, causes, and recommended actions so that the ultimate user will have an iterative tool for rapid use. The program selected for this example is Texas Instrument's Personal Consultant™ Plus [2], a highly user-friendly software package that makes full use of the utility of the LISP language while providing a menu-driven overlay that does not require the user to work with the LISP language. The structure of the program is shown in Fig. EX–10.2, using the names invoked by the program for the relevant items essential in causal reasoning.

PARAMETERS are the variables used by the IF/THEN RULES to formulate questions, assign values to parameters, infer conclusions, and communicate results to the user. They include operating time, abnormal conditions, problem diagnosis terms, suggested action, textual terms, and material offering explanations.

In addition to the material assembled in Fig. EX–10.1, textual material can be used to provide more detailed explanations of the causes and actions deduced. Also degrees of certainty for alternate causes can also be included. Figure EX–10.3 provides an example of frames from a single session with the finished program and illustrates its ease of use and the rapid resolution of a problem based on the expertise of people who have experienced such problems before.

3. Inadequate regeneration, leaving unburned coke on the catalyst
4. Upset in upstream desulfurization unit, causing higher H_2S content in liquid feed
5. Flow maldistribution at inlet

Item 1 does not seem likely, since observed pressure drops are not greatly different from the predicted values. Item 5 is not likely, since the new condition is at a higher mass flowrate and lower pressure, which should improve distribution. This reasoning leaves items 2, 3, and 4, each of which would cause catalyst deactivation by, respectively, Pt agglomeration, fouling, and poisoning. At this point the model's usefulness has been served, and more information must be obtained from the unit data log and laboratory tests to determine the most probable cause.

CONCLUSIONS

Models can serve useful purposes in troubleshooting early in the analysis of the problem by setting clearly the magnitude of the presumed difficulty. Other procedures of analysis of past data and laboratory work must follow in order to determine the causal relation. This example has also illustrated the utility of the lumping procedure in developing rate forms for complex reaction systems.

REFERENCE

1. Lazenby, T.M., Waterfield, D.N., Bath, M., and Rubie, P.E., *I. Chem. E. Symposium Series*, No. 35, Inst. Chem. Engrs. (London), 1972.

Table EX–10.2 Methane Steam-Reforming Diagnostics

	Symptoms	*Diagnosis*: Distinguishing Features	*Diagnosis*: Possible Causes	Confirmation	Action
A.	*Occurrences at or soon after start-up*				
A-1	• Tubewall temperature toward top of tube is higher than average T_w and increases with time • Pressure drop may increase with time • Above normal CH_4 content in product stream	ΔP increase will most likely occur with higher hydrocarbons in the feed.	• Lower activity near inlet due to inadequate H_2O/CH_4 ratio at start-up (much less than 7–10) or inadequate reduction period, causing high temperatures and coke formation in region.	Check steam and CH_4 flow records and analyze feed for sulfur content.	Switch to higher H_2O/CH_4 ratio and effect complete reduction.
			• High sulfur in feed due to faulty operation of hot ZnO unit (see item 6), causing same situation.		Correct operation at ZnO unit.
A-2	• Some tubes significantly hotter than others; ΔP across unit may be higher or lower than normal, but may increase with time on stream; will be seen soon after start-up.	Distribution of hot tubes may be random or grouped, depending on loading technique in use at time of loading each tube.	Faulty loading of tubes leading to poor distribution	Can only confirm by ΔP test across individual tubes after shutdown. Tests will show wide variations in some tubes from norm.	Shutdown and check ΔP across offending tubes. Repack those that deviate significantly from the average ΔP.

Table EX–10.2 *(continued)*

Symptoms	*Diagnosis* — *Distinguishing Features*	*Diagnosis* — *Possible Causes*	*Confirmation*	*Action*
B. *Occurrences during steady-state operation*				
B-1. • Hot bands or stripes; and, after prolonged operation, increased ΔP; activity may not be affected, in which case outlet CH_4 content is normal	• Can occur without heavier hydrocarbons in the feed • Exaggerated effect when preceded by sulfur poisoning	• Low H_2/CH_4 ratio and low H_2O/CH_4, indicating regions of low activity, which correspond to regions of high tubewall temperatures and which, together with low H_2, favors carbon whisker formation. • Excessive heat flux can also cause whisker formation.	Increase steam rate and effect should disappear in most cases unless C is aged. Examination of catalyst shows whisker carbon formation.	If in initial stages, increase H_2O/CH_4 ratio to reverse process. If condition is long lived, must regenerate (see item B.2).
B-2. • Hot bands or stripes in upper portion of tube or over entire tube • Increased ΔP; increasing and becoming very high at times • Above normal CH_4 content in product steam.	• Often occurs after sulfur poisoning due to high T_w • Can get worse rapidly with time	Deactivation caused by pyrolytic coke formation due to heavier hydrocarbons in feed or slug of same.	If caught early, operation with higher H_2O/CH_4 ratio may arrest further formation.	If excessive, shutdown and steam and air-steam regenerate.

B-3. • Increased ΔP and hot tubes in region near steam-header inlet	If sufficient tubes are plugged, CH_4 content of product will be higher.	Deposit of boiler feedwater solids in affected tubes	Confirmed upset in steam system At shutdown, inorganic solids will be found in affected tubes.	If tubes become too hot, must shut down, unload, and recharge affected tubes.
B-4. • Hot bands or stripes usually in top one-third of tube due to deactivation • Increased CH_4 content in product • Can be accompanied by coke formation in which case ΔP increases	• Increased S content in discharge from ZnO unit • If condition continues, deactivation can become so severe and ΔP so high, if coking occurs, that immediate shutdown will be required	Sulfur poisoning caused by H_2S breakthrough from ZnO unit or thiophene in feed that is not removed by ZnO unit.	Analysis of feed will show significant thiophene or H_2S content.	• Steam and air-steam regeneration • Correct operation of ZnO unit and determine sulfur compound types in feed. If thiophenes present, change feed source or correct operation of hydrodesulfurization unit if part of system.
B-5. • Hot banks or stripes usually in top one-third of tube due to deactivation • Increased CH_4 content in product • Can be accompanied by coke formation in which case ΔP increases	• If condition continues, deactivation can become so severe and ΔP so high, if coking occurs, that immediate shutdown will be required	Arsenic poisoning of catalyst	Analysis of feed will show significant As content.	• Regenerate • Determine entry point of As (feed or treatment solvent) and remedy.

Table EX–10.2 *(continued)*

Symptoms	*Diagnosis* — *Distinguishing Features*	*Diagnosis* — *Possible Causes*	*Confirmation*	*Action*
C. *Occurrences after regeneration*				
C-1. • Activity decline after sulfur and/or carbon removal as indicated by higher CH_4 content of product during operation	• Steaming was conducted at pressures exceeding 100 psia.	High-pressure steaming (above 100 psia) causes agglomeration of NiO reducing, thereby, the surface areas of the final active Ni and effectively reducing catalyst activity.	Low Ni surface area on a representative sample removed from tubes.	If severe, must replace the catalyst.
C-2. • No improvement in activity after steaming and steam-air regeneration as indicated by high CH_4 content in product during operation.		Catalyst has Na or K promoter or other alkali metal contaminants that form sulfates that produce H_2S in reducing atmosphere during start-up.	Catalyst removed from tubes shows significant amounts of alkali metals. Also sulfur content is high.	Replace catalyst.
C-3. • Performance is not restored by steaming and regeneration as indicated by high CH_4 content in product.	Regeneration was done after a prolonged operating period at high temperature.	Agglomeration of nickel on catalyst with attendant loss in active Ni area and thus activity.	Ni surface area of the catalyst that was removed from tubes is low.	Replace catalyst.

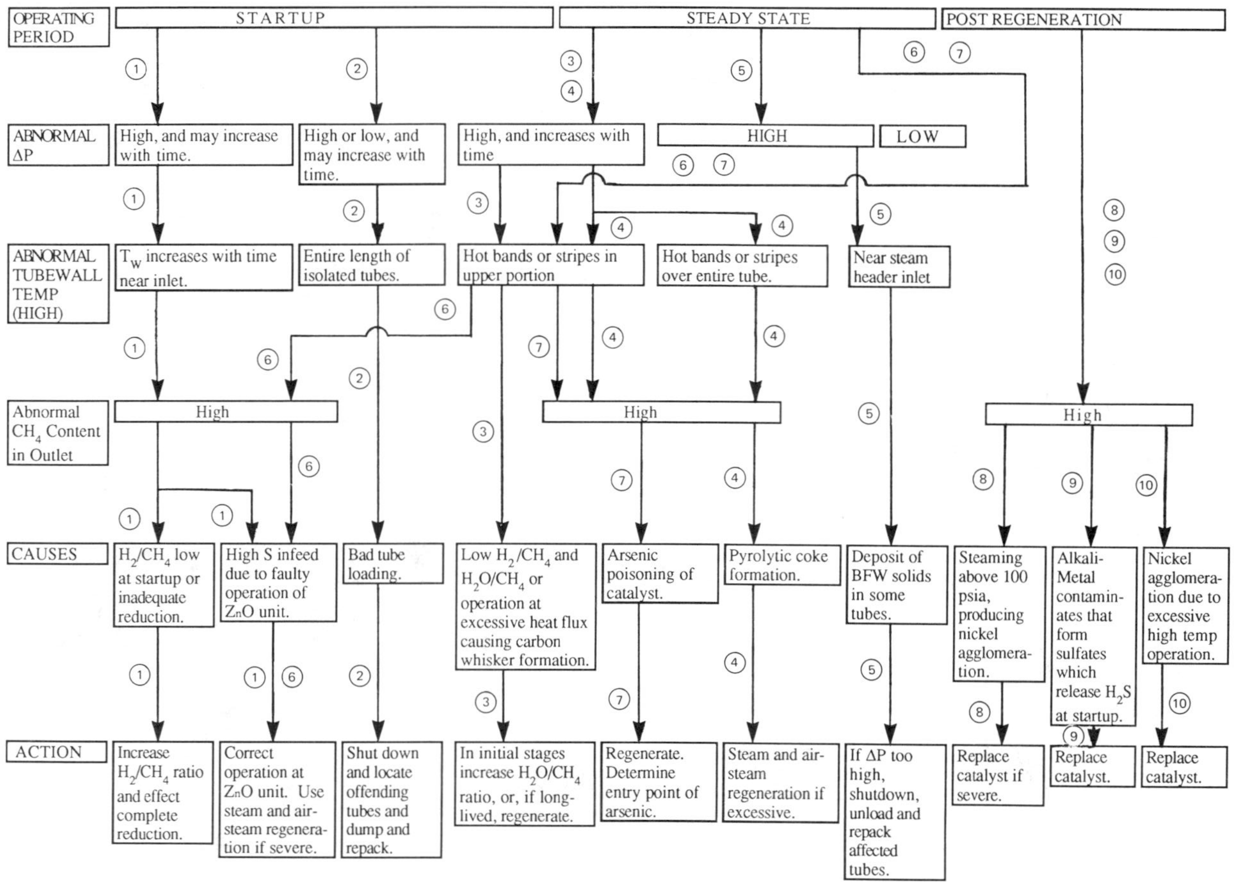

Figure EX-10.1 Diagnostic diagram
Note similarity to diagnostic charts in automobile shop manuals that have been in use for many years. The "yes" or "no" options are in this problem replaced by numbers because of the complexity of the system.

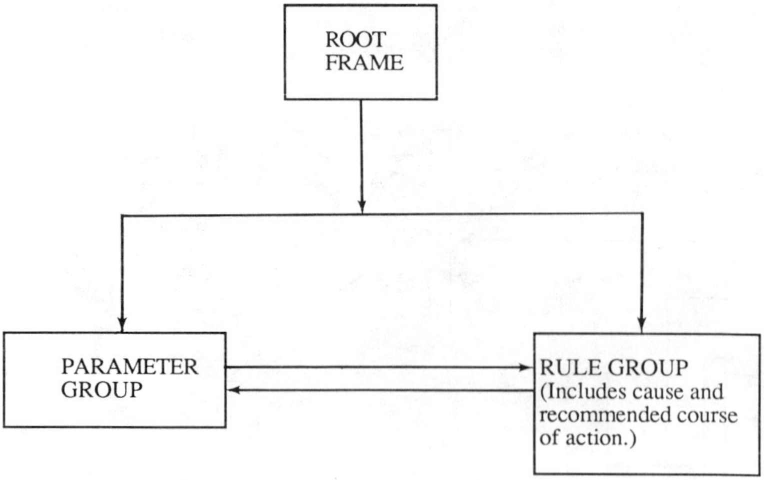

Figure EX-10.2 Simplified structure of Personal Consultant Plus™.
(Material extracted from Personal Consultant Plus™ Manual with permission of the publisher, copyright 1987, 1988, Texas Instruments, Inc. All rights reserved.)

```
                    Steam Reforming Diagnosis

Current objective:
This is an expert system to analyze steam reforming operating problems and
suggest possible actions. You will be prompted for information about
operating times, temperatures, analyses and so on. If information is not
available to you, enter what you think is a reasonable answer and then try to
obtain the needed data from log sheets, questions to colleagues, etc.
 ** End - RETURN/ENTER to continue
```

```
Activities:
 • CONSULT
   DEVELOP
   BUILD
```

Figure EX-10.3 Example frames from a single session using Personal Consultant Plus™ for synthesis gas plant diagnosis
Dot indicates item-entered. (Reproduced by permission, copyright 1987, 1988, Texas Instruments, Inc. All rights reserved.)

```
What is the current operating period?

 • STARTUP
   STEADY-STATE
   POST-REGENERATION
```

```
Is the TUBE-WALL-TEMPERATURE abnormal?

 • YES
   NO
```

```
Is the PRESSURE-DROP abnormal?

 • YES
   NO
```

```
What is the pressure drop?

   LOW
 • HIGH
```

```
Is the EXIT-METHANE-COMPOSITION abnormal?

   YES
 • NO
```

```
Where in the tube bank are the hot spots?

   NEAR-INLET-OF-TUBES
 • ENTIRE-LENGTH-OF-ISOLATED-TUBES
   BANDS-OR-STRIPES
   NEAR-STEAM-HEADER-INLET
```

```
Tubes must be carefully loaded in a consistent manner so that each will have
bed void fractions within about 10% as indicated by individual tube pressure
drop tests after loading. If such procedures are not followed and tubes with
widely deviating pressure drops are permitted,such tubes will carry more or
less flow than the others depending on the sign of the deviation. If the drop
is higher in a tube,less fluid will pass through it and less endothermic heat
will be required. Heat transfer will be less and wall temperature higher.
 ** End - RETURN/ENTER to continue
```

```
Conclusions:
The cause of the problem being diagnosed is as follows:  BAD-TUBE-LOADING

The action to be taken to correct the problem is as follows:
SHUTDOWN-CHECK-REPACK-FAULTY-TUBES


 ** End - RETURN/ENTER to continue
```

Figure EX-10.3 *(continued)*

CONCLUSIONS

Although an individual engineer may, through his own memory of previous experiences, rapidly solve process operating problems, an expert system carefully developed by people experienced with the operating unit in question can be of tremendous value to newly employed or newly assigned engineers. In fact, even the experienced engineer will find that such an expert system guards against the failure to remember certain crucial issues, particularly when under the stress caused by a serious operating problem.

REFERENCES

1. Rostrup-Nelsen, J.R., "Catalytic Steam Reforming," in *Catalysis-Science and Technology*, Vol. 5, Ed., J.R. Anderson and M. Boudart, Springer-Verlag, New York, 1983.
2. Personal Consultant Plus, Texas Instruments, Inc., Dallas, 1986.

EXAMPLE

11

Model Development from Recycle Microreactor and Pilot-Plant Data

OBJECTIVE

This example illustrates procedures for developing a model for a nonadiabatic reactor using rates obtained directly from a recycle microreactor and the overall heat-transfer coefficient from pilot-plant data. Use of the convenient data analysis program SIMUSOLV™ [1] is also demonstrated.

BACKGROUND

For most reaction systems it has been shown that a ratio of recycle to feed of 25 or higher is adequate to ensure continuous stirred reactor performance [2, 3], such that the rate may be directly obtained from inlet and outlet compositions in the absence of transport gradients

$$\hat{r} = \frac{X_1}{W/F_A} \qquad \text{(EX-11.1)}$$

where $\hat{r}_1$ is the rate of reaction 1; X_1 is the extent of reaction 1 per mole of A fed; W is the mass of catalyst; and F_A is the feed rate of A in moles/time. Two kinds of recycle reactors are shown in Figs. EX–11.1 and EX–11.2: external recycle and internal recycle. The latter, a Berty reactor, uses what is more appropriately called an internal fan for recirculating the gaseous reactants over the bed while the externally recycled reactor employs a compressor that can produce the desired recycle flow even at 1 atm [4]. Because of the pumping characteristics of the internal blower, it is not possible to obtain the desired mixing at 1 atm pressure where the density of the gas is low. At 2 atm or above, however, the Berty reactor will also readily provide the desired recycle rate and flow regime that will yield kinetic data free of temperature and concentration gradients [5].

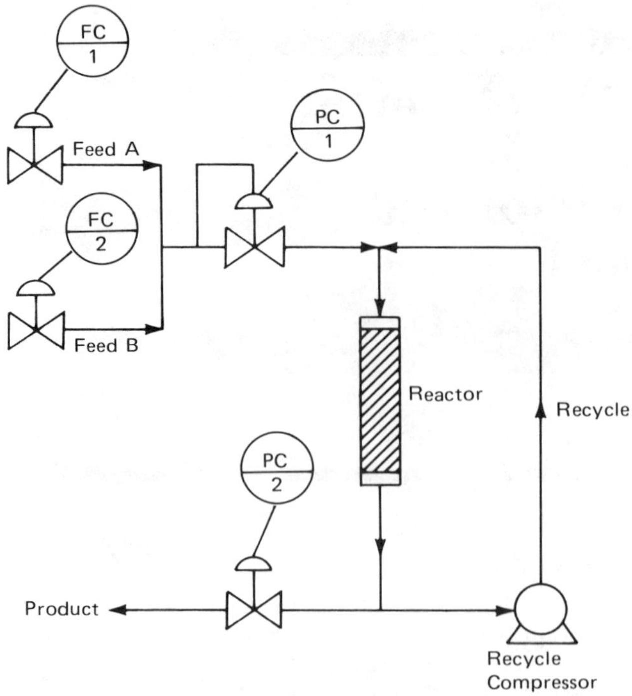

Figure EX-11.1 External-recycle reactor
Flow rates of feeds A and B are controlled by FC-1 and FC-2 and the back pressure of the mixture set by PC-1. The mixture is introduced into the recycle stream and fed to the reactor. A product stream is removed through the reactor pressure-control valve, and the rest of the stream recycled by means of the recycle compressor. Compressor and piping must have low dead volume. System should be tested for homogeneous (noncatalytic) reaction.

There has been much valuable effort in the development of recycle reactors for laboratory studies and criteria for ensuring valid gradient-free measurements [1]. The internally stirred Berty-type reactor has become the most popular because of its ease of operation and maintenance, and various ready-built units are available. These reactors are very useful for rapid, exothermic reaction systems that usually require nonadiabatic reactors for proper control of temperature and side reactions in commercial units. Such reactors are most often best designed by scale-up of a single-tube reactor to a multitubular unit as described on page 141, but searching for optimum conditions and implementation of predictive model control requires a reasonable model with minimum complexity. By employing the recycle and thus CSTR bench-scale reactor, one can obtain very good values of kinetic constants, and the pilot reactor may then be used to determine good values of heat-transfer coefficients as described below.

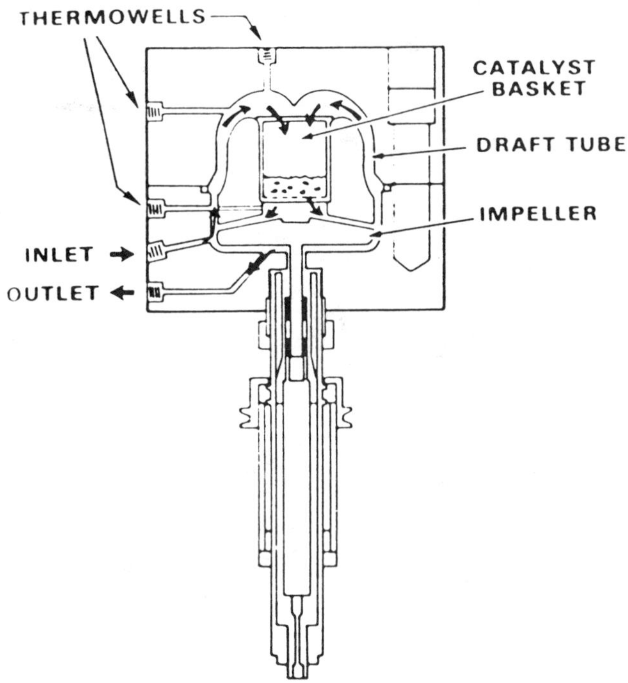

Figure EX-11.2 Internal recycle reactor (Berty reactor)
Catalyst is held in a stationary basket; and a turbine impeller driven by an Autoclave Engineers MagaDrive, a magnetically actuated rotary drive. Thermowell reaches into catalyst bed. High recycle rates ensure complete back-mixing, which produces uniform temperature and composition. (Manufactured by Autoclave Engineers, Erie, Pennsylvania. Illustration reproduced by permission: J.A. Mahoney, *J. Catal. 32*, 247 (1974).)

PROBLEM STATEMENT

Formaldehyde is produced by the partial oxidation of methanol using a Mo/Fe catalyst. A side reaction of oxidation of formaldehyde to CO and H_2O is controlled somewhat by the choice of catalyst and prevention of excessively high temperatures.

$$CH_3OH + \tfrac{1}{2}\, O_2 \rightarrow CH_2O + H_2O \qquad (1)$$

$$CH_2O + \tfrac{1}{2}\, O_2 \rightarrow CO + H_2O \qquad (2)$$

A number of rate forms have been proposed, among which the following has been presented for reactions (1) and (2) on powdered catalyst [6]:

Table EX–11.1 Recycle Reactor Data

PROC D1

DATA

DUMMY	*T*	*PMEOH*	*PFORM*	*PH20*	*R1*	*R2*
1	543	0.0793143	0.0445586	0.0463409	8.9062E-06	1.5625E-07
2	543	0.0766408	0.0454498	0.0472321	8.6458E-06	1.5625E-07
3	543	0.0614908	0.0704026	0.0744128	6.7708E-06	1.9792E-07
4	543	0.0583717	0.0663923	0.0708481	6.6146E-06	2.0833E-07
5	543	0.0993656	0.0289631	0.0302998	1.1198E-05	1.9271E-07
6	543	0.0966921	0.0280719	0.0289631	1.0677E-05	1.7187E-07
7	543	0.0891172	0.0392116	0.0405483	1.0156E-05	1.6667E-07
8	543	0.0917907	0.0409939	0.0423307	1.0208E-05	1.4062E-07
9	543	0.0877804	0.0499056	0.051688	9.8958E-06	1.7708E-07
10	543	0.108277	0.0151499	0.0151499	1.1875E-05	1.0937E-07
11	602	0.0988002	0.0464361	0.0474241	3.1406E-05	3.9583E-07
12	602	0.016302	0.11856	0.128934	7.1875E-06	5.2604E-07
13	602	0.0345801	0.108186	0.11856	1.2708E-05	5.2604E-07
14	602	0.0350741	0.124488	0.137332	1.3125E-05	6.4583E-07
15	602	0.0237121	0.0953422	0.105222	9.8958E-06	4.8958E-07
16	602	0.0251941	0.0968242	0.106704	9.8958E-06	4.6875E-07
17	602	0.0568101	0.0943542	0.10127	2.1406E-05	7.6042E-07
18	602	0.0829922	0.0597741	0.0627382	2.7396E-05	6.0417E-07
19	602	0.0232181	0.168454	0.193648	1.0781E-05	0.00000075
END						
END						

Note: Data from Ref. 7 were entered into this data file using units of °K, atm, and moles/g cat. sec, respectively, for T, P, and the observed rate.

Table EX–11.2 Formaldehyde Pilot-Plant Temperature Profiles (varying operating conditions as noted)

```
PROC S1                          PROC S2
  SET TJ = 506                     SET TJ = 528.0
  SET ZEND = 66.0                  SET ZEND = 75
  SET G = .132                     SET G = .189
END                              END
PROC D1                          PROC D2
DATA                               DATA
```

Z	T	Z	T
0.0	572.999	0.00	573.000
5.28	614.170	6.00	608.741
10.56	616.440	10.50	613.869
15.18	606.032	15.00	611.314
20.46	579.890	20.25	601.851
25.08	555.162	25.50	585.841
30.36	534.984	30.00	570.854
34.98	524.127	36.00	554.958
40.26	516.879	40.50	546.893
42.24	515.102	45.00	541.460

```
PROC S3
  SET TJ = 545.0
  SET ZEND = 92
  SET G = .265
END

PROC D3
DATA
```

Z	T
0.00	573.001
4.60	598.054
10.12	610.084
13.80	611.323
20.24	608.002
24.84	602.577
30.36	593.164
35.88	582.304
40.48	573.959
45.08	567.135
50.60	561.085
55.20	557.532
60.72	554.577
65.31	552.872
70.83	551.437
76.36	550.437
81.00	549.818

$$\hat{r}_1 = \frac{k_{p_1} P_M}{1 + a_1 P_M + a_2 P_W} \tag{EX-11.2}$$

$$\hat{r}_2 = \frac{k_{p_2} P_F}{1 + b_1 P_M + b_2 P_W} \tag{EX-11.3}$$

At commercial operating conditions a_1 is much larger than a_2 while b_1 and b_2 are nearly equal. Reaction (2) becomes significant only after higher conversions are reached, at which point the water term dominates. Thus, approximate forms may be postulated to correlate rates for a single catalyst size as follows:

$$\hat{r}_1 = \frac{k_{p_1} P_M}{1 + a P_M} \tag{EX-11.4}$$

$$\hat{r}_2 = \frac{k_{p_2} P_M}{1 + b P_W} \tag{EX-11.5}$$

a. Using the rate data given in Table EX–11.1 that were obtained at 1 atm pressure on 2 × 3 mm extrudates of promoted Fe_2O_3/MoO_3 in an externally recycled reactor [7], determine the best values of k_{p_1}, k_{p_2}, a, and b.
b. Using the single-tube, pilot-plant data given in Table EX–11.2, determine a multiplying factor for the general overall heat-transfer equation that will produce a reasonable fit of the calculated curve with the observed temperatures.

DETERMINATION OF KINETIC CONSTANTS (Part a)

The data in Table EX–11.1 were obtained by varying W/F_A and the composition of the feed so as to ensure data over a wide range of partial pressures. The rates were calculated directly from measured compositions of feed and product, using Eq. EX-11.1 for each reaction.

The rate equations EX-11.4 and EX-11.5 can readily be placed in a form linear in the coefficients by writing the inverse forms.

$$\frac{1}{\hat{r}_1} = \frac{a}{k_{p_1}} + \frac{1}{k_{p_1}}\left(\frac{1}{P_M}\right) \tag{EX-11.6}$$

$$\frac{P_F}{\hat{r}_2} = \frac{1}{k_{p_2}} + \frac{b}{k_{p_2}} P_W \tag{EX-11.7}$$

The values of k_{p_1}, k_{p_2}, a, and b can be determined by the method of least squares for the two temperatures studied, as summarized in Figs. EX–11.3 and

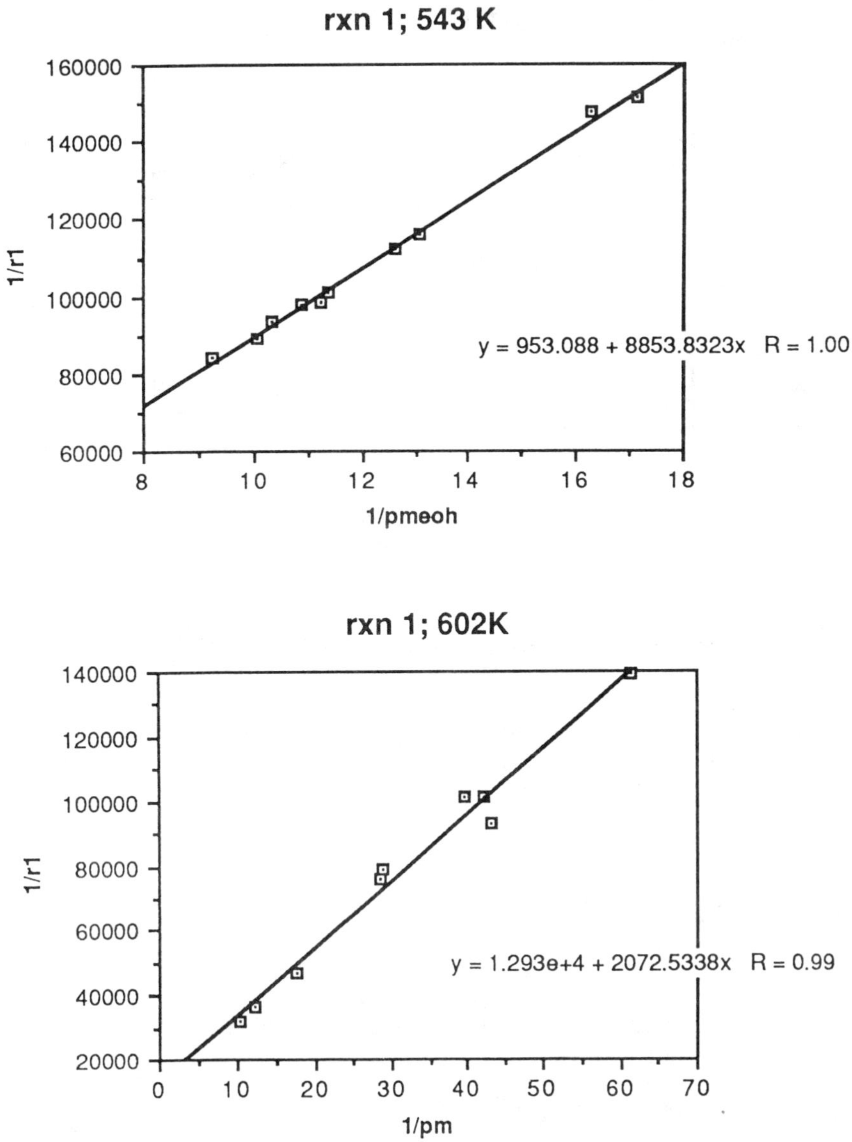

Figure EX-11.3 Reaction 1 least-squares fit using Cricket Graph™ Software.

EX–11.4. There are many handy programs for this purpose, including those for pocket calculators. The software CRICKET GRAPH™ [8] was used in this case to draw the least-squares line and plot the data points. Using the two sets of values at the respective temperatures of 543°K and 602°K, values of E/R, E, and the preexponential factor A, were determined as follows.

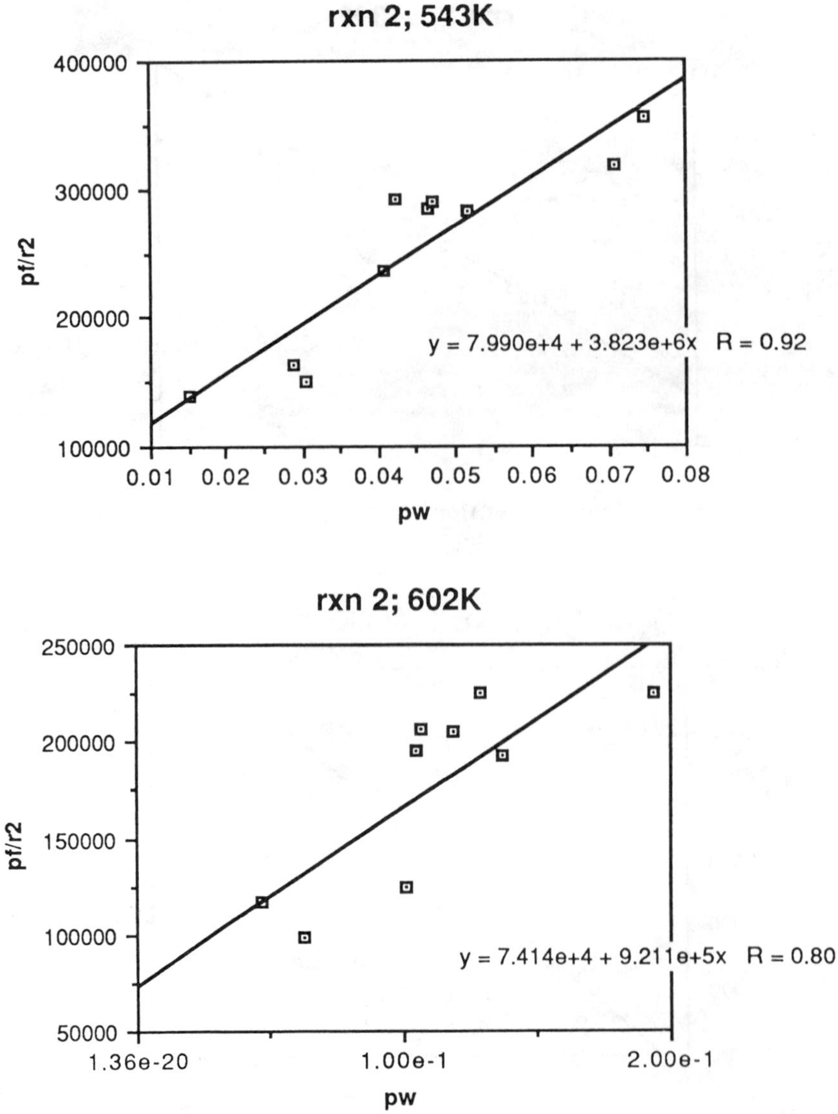

Figure EX-11.4 Reaction 2 least-squares fit using Cricket Graph™ Software.

For reaction (1),

$$k_{p_1} = 1.1294\text{E} - 4 \,@\, 543°\text{K} \text{ and } 4.825\text{E} - 4 \,@\, 602°\text{K}$$

$$a = 0.10765 \,@\, 543°\text{K} \text{ and } 6.2387 \,@\, 602°\text{K}$$

$$\ln \frac{1.129\text{E} - 4}{4.82501\text{E} - 4} = \frac{-E}{R}\left(\frac{1}{543} - \frac{1}{602}\right)$$

$$\frac{E_1}{R} = 8072.5$$

$$E_1 = 16040$$

$$\ln 1.129\text{E} - 4 = \ln \text{A} - \frac{8072.5}{543}$$

$$A_1 = 318.3$$

or

$$k_{p_1} = 318.3 \exp\left(\frac{-16040}{RT}\right)$$

Similarly

$$a = 1.057\text{E}17 \exp\left(\frac{-44.816}{RT}\right)$$

For reaction (2),

$$k_{p_2} = 1.252\text{E}{-5} \,@\, 543°\text{K and } 1.349\text{E} - 5 \,@\, 602°\text{K}$$

$$b = 47.847 \,@\, 543°\text{K and } 12.427 \,@\, 602°\text{K}$$

From which,

$$k_{p_2} = 2.685\text{E}{-5} \exp\left(\frac{-823.74}{RT}\right)$$

$$b = 4.9407\text{E}{-5} \exp\left[-\left(\frac{-14884}{RT}\right)\right]$$

These values minimize the sum of the squares of the differences between observed and calculated values of $1/\hat{r}_1$ and $P_F/\hat{r}_2$, respectively, and can serve as given or as first approximations for parameter optimization, in which case the resulting coefficients minimize the square of the errors between calculated and observed rates. The Dow software, SIMUSOLV™ [1] was designed for this latter purpose. Since algebraic rate equations are being used rather than differential equations, the input shown in Fig. EX–11.5 is different from that given in the heat-transfer coefficient determination given below in part b.

Results from this analysis are summarized in Table EX–11.3. In this case linear regression gave a good fit, as evidenced by the close agreement between the linear regression and SIMUSOLV™ optimization. In most cases parameters obtained directly by linear regression make good initial estimates for the more statistically correct SIMUSOLV™ optimization. In this example the assumption of normally distributed absolute error in $1/\hat{r}_1$ and $P_F/\hat{r}_2$ used for the method for least squares permits a reasonable linear fit. Such is not always the case.

```
$ ty linfor.csl
PROGRAM - LINEAR FIT OF FORMALDEHYDE RATE DATA

   INITIAL

      VARIABLE  DUMMY  = 0.

      CONSTANT TSTOP   = 19.,TO=543,PMEOH=0,PFORM=0,PH2O=0,...
               R=1.987
      CONSTANT A1=0,A2=0,A3=0,A4=0,E1=0,E2=0,EA=0,EB=0

       CINT   = 1.0

   END

   DYNAMIC
      T=LINEAR(T)
      PMEOH=LINEAR(PMEOH)
      PFORM=LINEAR(PFORM)
      PH2O=LINEAR(PH2O)

      K1=A1*EXP(-E1/R/T)
      K2=A2*EXP(-E2/R/T)
      A=A3*EXP(-EA/R/T)
      B=A4*EXP(-EB/R/T)

      R1=K1*PMEOH/(1.+A*PMEOH)
      R2=K2*PFORM/(1.+B*PH2O)

      TERMT( DUMMY .GE. TSTOP )     $ 'DEFINE TERMINATION CONDITION'

      DERIVATIVE

      END

   END

   TERMINAL
   END

END
```

Figure EX-11.5 SIMUSOLV input code for determining best values of rate constants. (Copyright 1988, Dow Chemical Company, Midland, Michigan.)

The plots given in Figs. EX–11.6 and EX–11.7 show the calculated rates at each experimental point as part of the line. Symbols represent the measured values. The results for reaction (1) are good, but there is considerable scatter for reaction (2), as shown in Fig. EX–11.7, for the higher temperature data. Clearly, this region should be further studied by experiment along with at least one additional temperature. More variation in water partial pressures is also needed (see Table EX–11.2).

The power of the SIMUSOLV^{TM} routine rests in the ease by which the user can interact with ideas about the reaction system and rapidly receive outcomes in terms of graphs, as illustrated in Figs. EX–11.6 and EX–11.7. The calcula-

Table EX-11.3 Comparision of Kinetic Terms by Linear Regression (Least Squares) and SIMUSOLV Parameter Optimization

	Linear Least Squares	*SIMUSOLV Parameter Optimization*
E_1	16040 cal mole	16033
A_1	318.3	318.3
E_a	44816	44806
A_a	1.057E17	1.057E17
E_2	823.74	823.21
A_2	2.685E − 5	2.6873E − 5
E_b	14884	14888
A_b	4.9407E − 5	4.9404E − 5

tional burden, including linking SIMUSOLV™ with optimization routines and graphical procedures, is eliminated, and the user is free to view results quickly, thereby stimulating the mind to reach conclusions or develop new ideas about the reaction system.

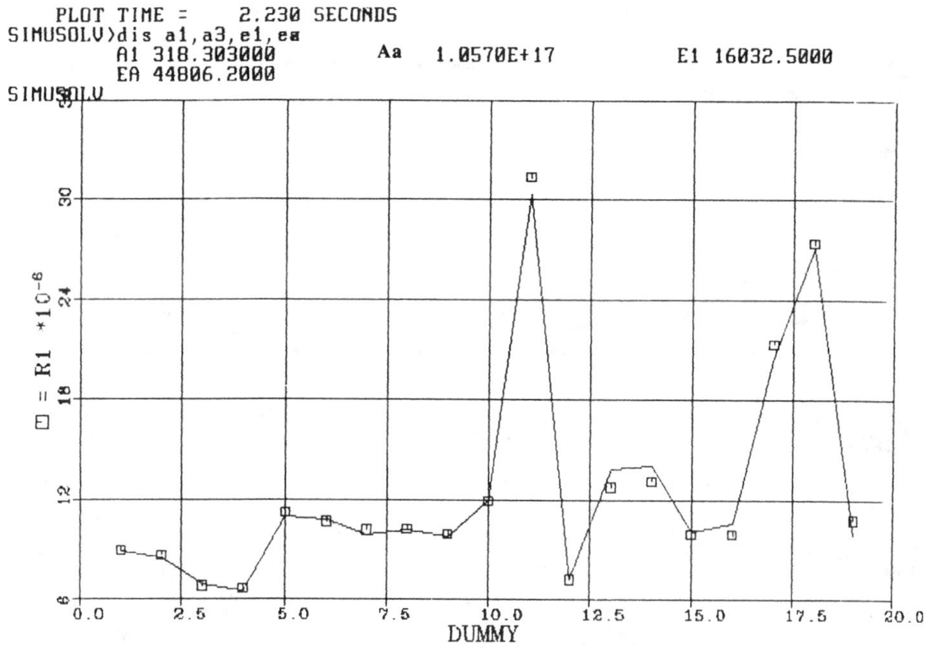

Figure EX-11.6 Comparison of measured and calculated rates for reaction 1 based on optimized values of activation energies and preexponential factors.
Calculations and plot executed by SIMUSOLV™. Dummy variable = observation number. (Copyright 1988, Dow Chemical Company, Midland, Michigan.)

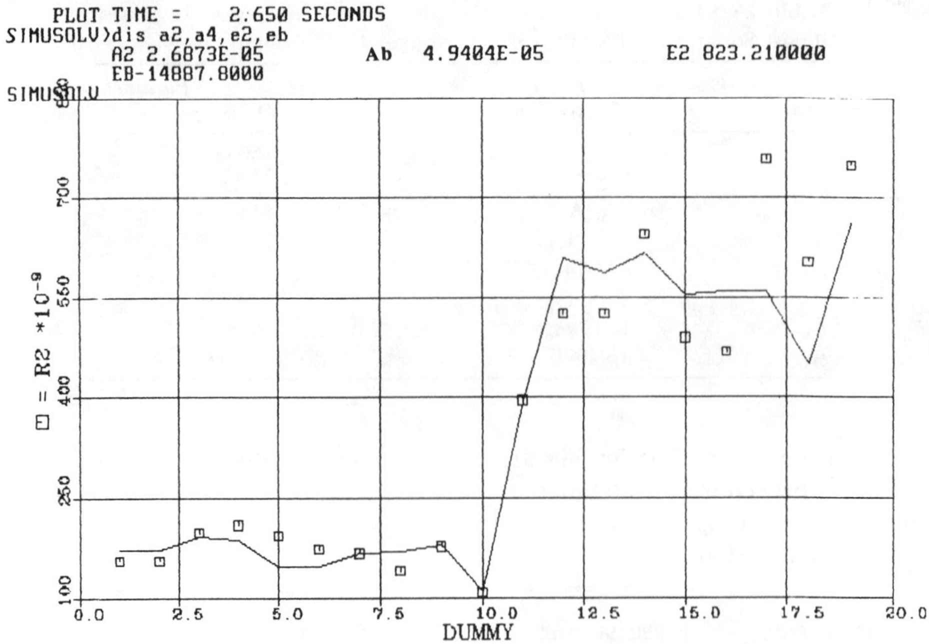

Figure EX-11.7 Comparison of measured and calculated rates for reaction 2 based on optimized values of activation energies and preexponential factors.
Calculations and plot executed using SIMUSOLV™. Dummy variable = observation number. (Copyright 1988, Dow Chemical Company, Midland, Michigan.)

INSIDE HEAT-TRANSFER COEFFICIENT FROM PILOT-PLANT DATA (Part b)

The actual commercial reactor for this exothermic reaction system will be a multitubular reactor; and the model, to be complete, must include not only the rate equations and parameter values but also differential equations for the nonadiabatic case. In this example the one-dimensional, pseudohomogeneous model is selected with heat transfer expressed in terms of an overall heat-transfer coefficient U, based on the average radial temperature T_M at any axial position of interest. This temperature has been shown to be at $R_b/\sqrt{2}$ or 0.707 times the inside radius of the tube based on a parabolic radial profile [9, 10].

Accurately measuring the temperature in the 1-in. OD tube is difficult. The usual thermocouple well inserted the length of the bed produces an atypical flow pattern and creates temperature errors because of the sizable amount of heat conducted along the well. Very thin bare thermocouples inserted radially through special pressure-tight fittings may be set at $0.707R_b$ for obtaining reasonable measured values of the mean temperature at various axial locations. Inserting any temperature device into the bed changes the flow pattern in the vicinity, but this method is less disadvantageous than the axial well.

The differential equations for the one-dimensional model together with physical properties, rate expressions, parameter values previously determined, and the general correlation for the inside heat-transfer correlation along with a design value for the outside film coefficient were inserted into the SIMUSOLV^{TM} program, as shown in Fig. EX–11.8. The multiplying factor COR applied to the equation for the inside film coefficient is used to correct

```
PROGRAM  'FORMALDEHYDE SYTHESIS--nonisothermal reactor'
     INITIAL
     CONSTANT RG1=82.06,RG2=1.987
     CONSTANT RP=.175,TO=573,TJ=0,G=0,PO=1.6
     CONSTANT NMEOHO=1.0,NO2O=1.1,NCH2OO=0.0065,NH2OO=0.1467,...
              NCOO=0.0563,NN2O=8.7905,NTO=11.100
     CONSTANT X10=0,X20=0,YIELDO=0,...
              ZEND=0,ZO=0,COR=.1
     CONSTANT RHOB=.84970,DI=2.0574,XW=0.1,...
              DP=.25,EPS=0.496,G=0,RHOP=1.892,INTMIN=0.1

'MOLECULAR WEIGHTS'
      CONSTANT MW1=32.042,MW2=31.999,MW3=30.026,MW4=18.016,...
               MW5=28.010,MW6=28.013

'HEAT CAPACITY CONSTANTS & HEATS OF FORMATION'
 CONSTANT A1=5.052,B1=1.694E-2,C1=6.179E-6,D1=-6.811E-9,HF1=-48.08E3,...
          A2=6.713,B2=-.879E-6,C2=4.170E-6,D2=-2.544E-9,HF2=0.000,...
          A3=5.607,B3=7.540E-3,C3=7.130E-6,D3=-5.494E-9,HF3=-27.70E3,...
          A4=7.701,B4=4.595E-4,C4=2.421E-6,D4=-0.859E-9,HF4=-57.80E3,...
          A5=7.373,B5=-0.307E-2,C5=6.562E-6,D5=-3.037E-9,HF5=-26.42E3,...
          A6=7.440,B6=0.324E-2,C6=6.400E-6,D6=-2.790E-9,HF6=0.000

      YMEOHO=NMEOHO/NTO
      YO2O=NO2O/NTO
      YCH2OO=NCH2OO/NTO
      YH2OO=NH2OO/NTO
      YCOO=NCOO/NTO
      YN2O=NN2O/NTO

      MWFEED=YMEOHO*MW1+YO2O*MW2+YCH2OO*MW3+YH2OO*MW4+...
             YCOO*MW5+YN2O*MW6

'CONVERSION FACTOR FOR Z'
     INLET=MWFEED*RHOB/G/YMEOHO



       VARIABLE Z=0.0
       TSTOP=ZEND

       C=50

       CINT=TSTOP/C

     END
```

Figure EX-11.8 SIMUSOLV^{TM} program insert for pilot-plant reactor.
(Copyright 1988, Dow Chemical Company, Midland, Michigan.)

```
  DYNAMIC
      DERIVATIVE
     YMEOH=NMEOH/NT
     YO2=NO2/NT
     YCH2O=NCH2O/NT
     YH2O=NH2O/NT
     YCO=NCO/NT
     YN2=NN2/NT
     MWMIX=YMEOH*MW1+YO2*MW2+YCH2O*MW3+YH2O*MW4+YCO*MW5+YN2*MW6
     PM=YMEOH*P
     PF=YCH2O*P
     PW=YH2O*P
'KINETICS'
'INSERT YOUR NUMBERS HERE'
     KP1=
     KP2=
     A=
     B=
     R1=KP1*PM/(1.+A*PM)
     R2=KP2*PF/(1.+B*PW)
     DX1DZ=R1*INLET
     DX2DZ=R2*INLET
     X1=INTEG(DX1DZ,X10)
     X2=INTEG(DX2DZ,X20)
     NMEOH=1-X1
     NO2=NO20-.5*X1-.5*X2
     NCH2O=NCH2OO+X1-X2
     NCO=NCOO+X2
     NH2O=NH2OO+X1+X2
     NN2=NN20
     NT=NMEOH+NO2+NCH2O+NCO+NH2O+NN2

     YIELD=(NCH2O-NCH2OO)/NMEOHO
'TEMPERATURE CHANGE'
     CP1=A1+B1*T+C1*T**2+D1*T**3
     CP2=A2+B2*T+C2*T**2+D2*T**3
     CP3=A3+B3*T+C3*T**2+D3*T**3
     CP4=A4+B4*T+C4*T**2+D4*T**3
     CP5=A5+B5*T+C5*T**2+D5*T**3
     CP6=A6+B6*T+C6*T**2+D6*T**3
'CP OF THE MIXTURE'
     CPM=YMEOH*CP1+YO2*CP2+YCH2O*CP3+YH2O*CP4+YCO*CP5...
        +YN2*CP6
     MUM=4.2886E-7*T+4.916E-5
     KFM=8.26752E-5+1.6535E-7*(T-400)
     RHOM=P*MWMIX/(RG1*T)

     T1=T-298.0
     T2=(T**2-298.0**2)/2.0
     T3=(T**3-298.0**3)/3.0
     T4=(T**4-298.0**4)/4.0
'DETERMINE HEATS OF REACTION'
     DH1=HF1+A1*T1+B1*T2+C1*T3+D1*T4
     DH2=HF2+A2*T1+B2*T2+C2*T3+D2*T4
     DH3=HF3+A3*T1+B3*T2+C3*T3+D3*T4
     DH4=HF4+A4*T1+B4*T2+C4*T3+D4*T4
     DH5=HF5+A5*T1+B5*T2+C5*T3+D5*T4
     DH6=HF6+A6*T1+B6*T2+C6*T3+D6*T4
```

Figure EX-11.8 *(continued)*

```
      HRXN1=DH3+DH4-.5*DH2-DH1
      HRXN2=DH4+DH5-.5*DH2-DH3
      RXNHT=(HRXN1*R1+HRXN2*R2)*RHOB

'HEAT TRANSFER TERM'
      DO=DI+2.0*XW
      RE=DP*G/MUM
      HO=600.0/7373.26
      HI=COR*KFM/DI*1.26*(RE**.95)*EXP(-6*DP/DI)
      UZ=1.0/(1.0/HI+DI/(DO*HO))
      HXFR=4.0*UZ*(TJ-T)/DI
      SENSHT=NT*CPM*G*YMEOHO/MWFEED

      DTDZ=(-RXNHT+HXFR)/SENSHT
      T=INTEG(DTDZ,TO)

'CHANGE IN PRESSURE'
      CFACT=1.0/1013200.0
      FK=1.80+180.0*(1.0-EPS)*MUM/(DP*G)
      F=-FK*G**2*(1.0-EPS)/(DP*RHOM*EPS**3)
      DPDZ=F*CFACT
      P=INTEG(DPDZ,PO)


      END
     TERMT(YIELD.GT.0.90)
     TERMT(Z.GT.TSTOP)
   END
```

Figure EX-11.8 *(continued)*

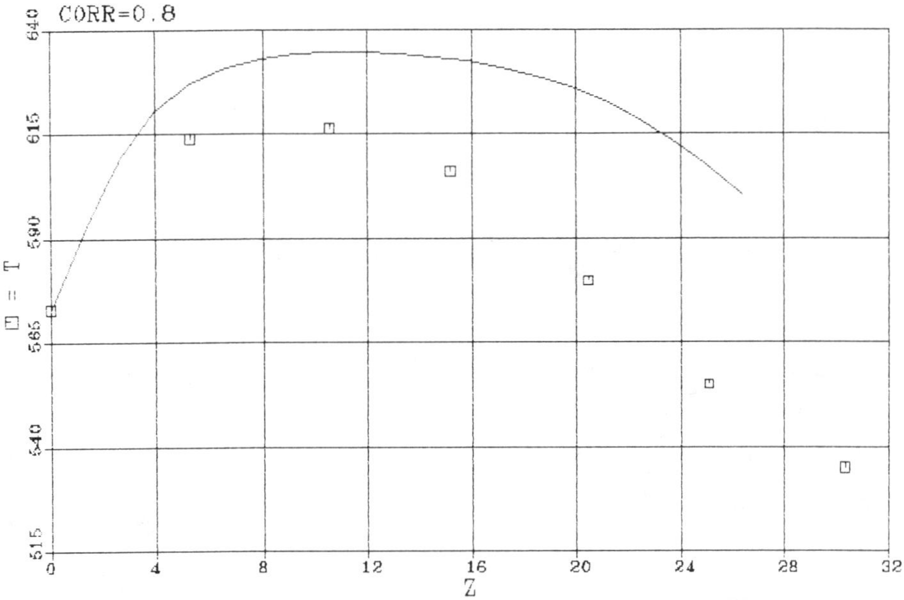

Figure EX-11.9 First estimate of correction factor for heat-transfer coefficient using SIMUSOLV™.

(Copyright 1988, Dow Chemical Company, Midland, Michigan.)

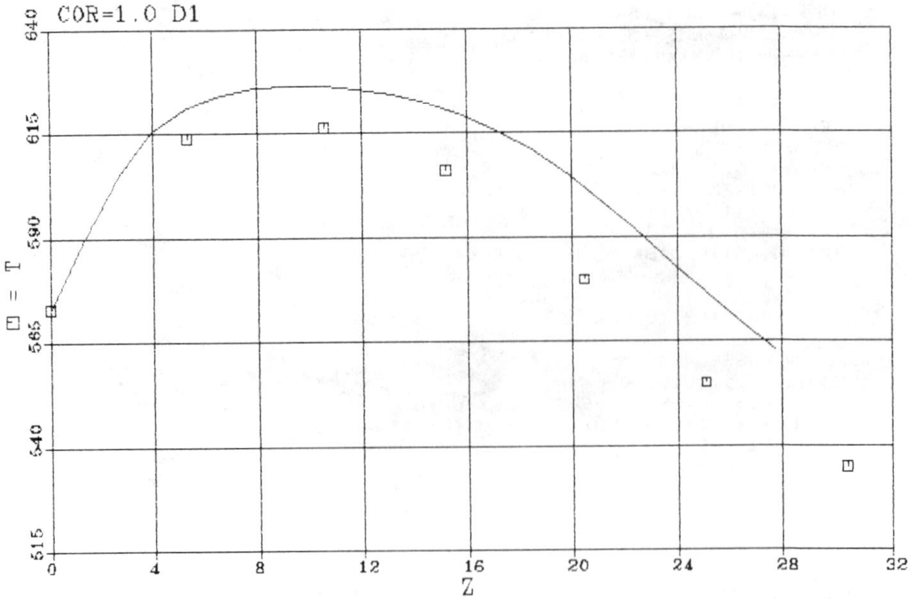

Figure EX-11.10 Second estimate of correction factor for heat transfer coefficient using SIMUSOLV™.
(Copyright 1988, Dow Chemical Company, Midland, Michigan.)

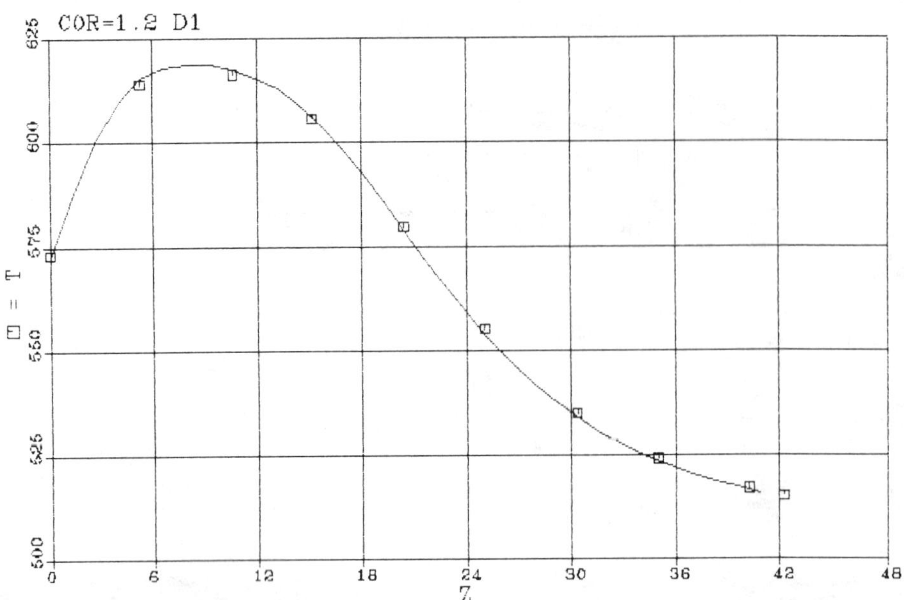

Figure EX-11.11 Final estimate of correction factor for heat-transfer coefficient using SIMUSOLV™.
(Copyright 1988, Dow Chemical Company, Midland, Michigan.)

for differences between the actual value and that calculated from the generalized equation. The procedure for using SIMUSOLV™ in evaluating COR is a straightforward iterative approach that displays a plot of the calculated temperature profile for the conditions of each run, upon which is also plotted the observed temperatures. The proper multiplier is determined by setting successive estimates of COR followed by appearance of plots, which quickly leads the user to a satisfactory fit of calculated and measured values. By using measured values at several mass velocities and comparing with calculated values, confidence in the selected value of the multiplier can be enhanced. The plots in Figs. EX–11.9 to 11 show the sequence of estimates and results for the first set of data for which the best fit of 1.2 was found for the multiplier COR. The other two sets of data (Table EX–11.2) also gave good fits with this same value.

VERIFICATION OF THE MODEL

The pilot-plant reactor can be used for verifying the model by either operating the full-length bed with intermediate sample probes or by using shorter beds with the remainder of the tube, preferably the upper portion, filled with inerts. If different lengths of shorter beds and the same mass velocity as for the full bed are used, outlet compositions will correspond to regions of rapid composition changes where the rates are high. In this manner the model can be well tested. If, instead, the actual full-length bed were used with no intermediate sampling, the outlet compositions would conform to equilibrium values or high conversions, where rates are not changing rapidly with position and temperature is essentially constant. Such data are of little value in testing the efficacy of a model.

REASONABLENESS OF THE ONE-DIMENSIONAL MODEL

Using the criterion given in Eq. 5.31, as demonstrated in Example 7, and the maximum rates near the hot spot, the one-dimensional model is satisfactory for this system within an acceptable error of 10 percent.

REFERENCES

1. SIMUSOLV, The Dow Chemical Company, Midland, Michigan, 1987.
2. Carberry, J.J., "Chemical and Catalytic Reactor Engineering," McGraw-Hill, New York, 1976.

3. Serrano, C., and Carberry, J.J. *Appl. Catal.*, *19*, 119 (1985).
4. Berty, J.M., *Chem. Eng. Prog.*, *70*(5) 78 (1974).
5. Carberry, J.J. *Ind. Eng. Chem. Res.*, *26*, 2557 (1987).
6. Bibin, V.N., and Popov, B.I., *Kinetika i Kataliz*, *10*(6), 1326 (1968).
7. Boskov, V.S., Bibin, V.N., Malinovskaya, O.A., and Popov, B.I., *Kinetika i Kataliz*. *13*(5), 1318 (1972).
8. "Cricket Graph™", Cricket Software, Inc., Philadelphia, Penn., 1985.
9. Beck, Jr., J., and Singer, E., *Chem. Eng. Prog.*, *47*, 534 (1951).
10. Quinton, J.H., and Storrow, J.A., *Chem. Eng. Sci.*, *5*, 245 (1956).

APPENDIX

A.1

Interactive Driver Routine for IMSL Nonlinear Algebraic Equation Solver NEQNF

User fills in required information in program, including the functions to be solved and then changes parameter. Values entered interactively as desired. Written by Doyle E. Broom. Revised by H.F. Rase to adapt to new issue of IMSL MATH/LIBRARY.

```
      PROGRAM NONLIN(INPUT,OUTPUT,TAPE5=INPUT,TAPE6=OUTPUT)
      INTEGER N,NP,ITMAX
C     INSERT VALUES OF N,NUMBER OF VARIABLES, AND NP, NUMBER OF
C     ADJUSTABLE PARAMETERS.
      PARAMETER (N= ,NP= )
      COMMON/A/PAR(NP)
      REAL PAR,ERREL,FCN,FNORM,X(N),XG(N)
      EXTERNAL FCN
      DATA ITMAX/200/
      WRITE(5,*)'ENTER THE VALUES OF THE PARAMETERS.'
      DO 10 I=1,NP
      WRITE(5,*)'ENTER PAR(',I,')'
      READ(5,*)PAR(I)
   10 CONTINUE
      WRITE(5,*)'ENTER THE ACCEPTABLE RELATIVE ERROR.'
      READ(5,*)ERREL
   15 WRITE(5,*)'ENTER THE INITIAL GUESSES FOR THE ',N,' VARIABLES.'
      DO 20 I=1,N
      WRITE(5,*)'ENTER XG(',I,')'
      READ(5,*)XG(I)
   20 CONTINUE
      CALL NEQNF(FCN,ERREL,N,ITMAX.XG,X,FNORM)
            WRITE(5,*)'THE SOLUTIONS ARE:'
            WRITE(6,*)'THE SOLUTIONS ARE:'
            DO 30 I=1,N
            WRITE(5,*)'X(',I,') = ',X(I)
            WRITE(6,*)'X(',I,') = ',X(I)
            WRITE(5,*)'WITH FNORM=', FNORM
            WRITE(6,*)'WITH FRNOM=', FNORM
   30       CONTINUE
      WRITE(5,*)'IF FNORM IS NOT CLOSE TO 0,START',
     $          'OVER AGAIN WITH BETTER GUESSES.'
      STOP
      END
      SUBROUTINE FCN(X,F,N)
      INTEGER N
C     INSERT VALUE OF NP, THE NUMBER OF ADJUSTABLE PARMETERS.
      PARAMETER (NP= )
      COMMON/A/PAR(NP)
      REAL X(N),F(N)
C     INSERT FUNCTIONS: F(1),F(2),ETC
```

APPENDIX A.2

Interactive Entries from Screen for Equilibrium and Adiabatic Plotting Program

Driver routine written by A.W. Alsop. Two data files used, one having heat-capacity equation constants and the other having a value of ln K determined from Stull at each of three temperatures in the region of interest.

```
.ADIMAC,TKPDAT,C6H6DAT
 FILE ADILGO ALREADY ASSIGNED.
 -(DELTA H)/R=      25873.54658
 (DELTA S)/R=         -46.60373
 GIVE THE MIN. AND MAX. TEMP. THAT YOU WISH TO 'GRAPH OVER
 IN DEG. K, EX. 300.0,600.0
400,600
 SELECT MODE FOR THE GRAPH GENERATION;
 1 - VARY FEED COMPOSITIONS
 2 - VARY SYSTEM PRESSURE
2
 HOW MANY CASES DO YOU WISH TO EXAMINE (5 MAX.)
4
 FOR CASE  1 ENTER PRESSURE IN ATM.
1
 FOR CASE  2 ENTER PRESSURE IN ATM.
5
 FOR CASE  3 ENTER PRESSURE IN ATM.
10
 FOR CASE  4 ENTER PRESSURE IN ATM.
15
 ENTER FEED MOLAR FLOW RATE OR MOLE FRACTION
 DO NOT CHANGE THE LIMITING REACTANT HERE!
  H2   =
4
  C6H6 =
1
  CYCL =
0
  CH4  =
0
  C2H6 =
0
  C3H8 =
0
  WHAT DO YOU WANT TO LABEL THIS PLOT?
CYCLOHEXANE EQUILIBRIUM
 NUMBER OF SQUARE INCHES PLOTTED           1.01999999999999  E+002
 NUMBER OF PLOTTER BUFFERS WRITTEN         000000000000015
 STOP
      .588 TM SECONDS EXECUTION TIME
```

APPENDIX

B.1

Input Listing for Cyclohexane Plant Recycle Using SIMSCI Process™

(Reproduced by permission: Simulation Sciences.)

```
VERSION 2.01
*************

        TM
PROCESS    INPUT LISTING - PAGE  1

 GENERAL DATA SECTION
 TITLE   USER = PLH, *
         PROBLEM = REACTCY, *
         DATE = JANUARY 1988,*
         PROJECT = 373 PROB #1
  DIME   METRIC ,TIME=HR                 , *
                 WT=KG                   , *
                 TEMPERATURE=K           , *
                 PRESSURE=PSIA           , *
                 ENERGY=KCAL             , *
                 WORK=KW                 , *
                 LIQVOLUME=CUM           , *
                 VAPVOLUME=CUM           , *
                 VISCOSITY=CP            , *
                 CONDUCTIVITY=KCH        , *
                 SURFACE=DYNE            , *
                 XDENSITY=DENSITY
TOLER STREAM=0.002
SEQUENCE   D-1, CAL1, SET1, S-1, RC-1, M-1, CT-1, CT-2,E-2,*
          RX-1,E-1,P-1,S-2,V-1, E-3, V-2,T-1,E-4
PRINT MBAL, INPUT=NONE, FRACTION
PLOT NETWORK,FORMAT=WIDE,NOFAIL
COMPONENT DATA SECTION
  LIBID 1,H2 / 2 , C1 / 3 , C2 / 4 , C3 / *
        5, C6H6 / 6 , CH
THERMODYNAMIC DATA SECTION
  METHODS SYSTEM=SRK , DENSITY(L)= LK , TRANSPORT=LIBRARY
STREAM DATA SECTION
  PROPERTIES  STRM            = 1           , *
              TEMPERATURE     =  311        , *
              PRESSURE        =  162        , *
              PHASE           =     L       , *
              RATE(M)         = 100         , *
              COMPOSITION(M)  = 5,99.96288 /6,0.03712
  PROPERTIES  STRM            = 20          , *
              TEMPERATURE     =  311        , *
              PRESSURE        =  162        , *
              PHASE           = V           , *
              COMPOSITION(M)  = 1 ,85           / 2 , 12        / *
                                3 , 2           / 4 , 1    ,*
              RATE(M)         = 300  $ ESTIMATE TO BE CORRECTED BY CT-1.
   PROPERTIES STRM           = 22,*
              TEMPERATURE     = 317,*
              PRESSURE        = 162,*
              REFS            = 9,*
              RATE(M)         = 300
  $ ESTIMATE OF STREAM 22 GIVEN FOR CALCULATOR INITIAL CALC.
  $ TEMPERATURE OF STREAM 3 GIVEN SO E-1 CAN BE SOLVED LATER.
  PROPERTIES  STRM            = 3           , *
              TEMPERATURE     = 385         ,*
              PRESSURE        = 152            , *
              REFS            = 2
  $ STARTING POINT: STREAM 7. ESTIMATED FLOW AND COMPOSITION

        TM
PROCESS    INPUT LISTING - PAGE  2

  PROPERTIES  STRM            = 7           , *
              TEMPERATURE     = 317           , *
              PRESSURE        = 122         , *
              COMPOSITION(M)  = 14.53/21.54/4.60/2.6/0.0/56.73,*
              RATE(M)         = 687.5
  $ESTIMATE OF STRM 16 FEED TO COLUMN
    PROPERTIES  STRM=16,TEMP=350,PRESS=200,REFSTRM=15
```

```
UNIT OPERATIONS SECTION
 MIXER UID=M-1
   FEEDS    1, 20, 12, 22
   PRODUCT  V=2
 CONTROLLER UID=CT-1
 SPEC STRM=2, COMP=1, REFC=5,  RATE=4,  $H2/C6H6 RATIO IN STRM 2
 PARA STRM=20  $ ADJUST H2 MAKEUP
 HX UID=E-1     $ REASONABLE TEMPERATURE FOR STREAM 3 SELECTED.
   COLD FEED=2, V=3 ,DP=10
   HOT FEED=5 ,V=6 ,DP=5
   SPEC  COLD,TEMP=385
 HX UID=E-2
   COLD FEED=3 ,V=4, DP=5
   UTILITY STEAM, PSAT=300
   SPEC  COLD, TEMP=425
 REACTOR UID=RX-1  $ PROVIDES MOLE BALANCE FOR DESIRED CONVERSION.
   FEED  4
   PROD  V=5
   OPER  PHAS=V,DP=15, TEMP=425
   CALC  CONVERSION, NOHBAL
   STOIC 1,-3/ 5,-1/ 6,1
   CONVERSION   0.9995
   BASE COMP=5
 HX UID=E-3
   HOT  FEED=6   ,M=7 ,DP=5
   UTILITY WATER, TIN=300 ,TOUT=317
   SPEC  HOT, TEMP=317
 FLASH UID=D-1
   FEED  7
   PROD  L= 9   ,V=8
   ADIA  PRES=122
 CALCULATOR UID=CAL1
   RETRIEVE P(1), STRM=9, COMP=1 ,RATE(M)
   RETRIEVE P(2), STRM=9, COMP=2,4, RATE(M)
   RETRIEVE P(3), STRM=1, COMP=5, RATE(M)
   RETRIEVE P(4), STRM=9, COMP=2,4, FRACTION(M)
   RETRIEVE P(5), STRM=22, RATE(M)
   RETRIEVE P(6), STRM=8, COMP=2,4 FRACTION(M)
   RETREIVE P(7), STRM=8, COMP=1, FRACTION(M)
   RETREIVE P(8), STRM=9, COMP=1, FRACTION(M)
   RESULT     1 ,    PURGE
   PROCEDURE
   $ PURGE RATE FROM SOLUTION OF HYDROGEN AND INERT BALANCES.
R(1)=(P(1)+P(3)*3*0.9995+5.667*P(4)*P(5)-5.667*P(2)-P(8)*P(5))/(5.667*P(6)-P(7))
   RETURN
 SET UID=SET1
          TM
    PROCESS    INPUT LISTING - PAGE  3

       SPEC UNIT=S-1    ,SPEC=1 ,IDNO=1   ,REFUNIT=CAL1,REFIDNO=1
     SPLITTER UID=S-1 NAME= H2 PURGE
       FEED 8
       PROD V=10  ,V=11
       SPEC STRM=10, RATE=1
     COMPRESSOR UID=RC-1
       FEED 11
       PROD V=12
       OPER  POUT=162             ,EFF=75
     CONTROLLER UID=CT-2  $ ADJUST INERTS TO 20% OF TOTAL FEED(STRM 2).
       PARAMETER STRM=11
       SPEC STRM=2, COMP=2,4, FRAC(M)=0.20
     PUMP UID=P-1, NAME=TOWERFEED
       FEED 9
       PRODUCT L=13
       OPER PRESS=250
    HX UID=E-4
      COLD FEED=15,M=16,DP=20
      HOT FEED=18, L=19,DP=20
      SPEC COLD, TEMP=350
    COLUMN  UID=T-1,NAME=STRIPPER
       FEED 16,8
       PROD  OVHD=17,10,  BTMS=18
       PARA TRAY=15
       HEAT 1,1/2,15
       PSPEC  TOP=200
       COND TYPE=1, PRESS=195
       SPEC STRM=18, COMP=6, RATE(M)=.999, REFS=15
       SPEC  REFLUX=3
       VARI HEAT=1,2
       ESTI MODEL=1,  REFLUX=30
    SPLITTER UID=S-2
      FEED 13
      PRODUCT L=21,L=14
      SPEC STRM=21,RATE=300
    VALVE UID=V-1
      FEED 21
      PRODUCT L=22
      OPER PRESS=162
    VALVE UID=V-2
      FEED=14
      PRODUCT L=15
      OPER PRESS=220
    RECYCLE DATA
      ACCEL TYPE=WEGS, START=3
       LOOP NO=1, START=D-1, END=E-3,TRIAL=20
        TM
 PROCESS   SIMULATION PROGRAM - VERSION 2.01
        TM
 PROCESS   OUTPUT DIRECTORY -  373 PROB #1     REACTCY
                               PLH             JANUARY 1988
```

APPENDIX

B.2

PFRSIM Subroutines for User Input

```
      SUBROUTINE PRPRTYS(NEQ,NC,Y,NT,X,MW,MWM,A,B,C,D,CPM,MUM,RHOM,
     $                   KFM)
      INTEGER NEQ,NR,NC
      REAL Y(*),X(*),MW(*),A(*),B(*),C(*),D(*),E
      REAL NT,MUM,CPM,RHOM,P,T,CPJ,MWM,RG1,TF,KF(6),KFNUM,KFDEN,KFM
      DATA RG1/82.059/
      NT=0.0
      DO 10 J=1,NC
         NT=NT+Y(J)
   10 CONTINUE
      T=Y(NEQ-1)
      P=Y(NEQ)
      MWM=0.0
      CPM=0.0
      KF(1)=5.374E-4+1.240E-6*(T-400.0)
      KF(2)=4.547E-5+2.646E-7*(T-400.0)
      KF(3)=5.126E-5+2.976E-7*(T-400.0)
      KF(4)=1.166E-4+3.635E-7*(T-400.0)
      KF(5)=8.557E-5+3.844E-7*(T-400.0)
      KF(6)=7.234E-5+3.721E-7*(T-400.0)
      KFNUM=0.0
      KFDEN=0.0
      E=1.0/3.0
      DO 20 J=1,NC
         X(J)=Y(J)/NT
         MWM=MWM+X(J)*MW(J)
         CPJ=A(J)+B(J)*T+C(J)*T**2+D(J)*T**3
         CPM=CPM+X(J)*CPJ
         KFNUM=KFNUM+X(J)*KF(J)*MW(J)**E
         KFDEN=KFDEN+X(J)*MW(J)**E
   20 CONTINUE
      KFM=KFNUM/KFDEN
      MUM=1.4E-4+1.8E-7*(T-400.0)
      RHOM=P*MWM/(RG1*T)
      RETURN
      END

      SUBROUTINE RATES(NEQ,NR,NC,Y,V,DP,RHOP,RHOB,ACT,EFF,R,RATE)
      INTEGER NEQ,NR,NC
      REAL Y(*),V(20,*),R(*),RATE(*),NT,T,P,ACT(*),EFF(*)
      REAL PH,PB,PC,RHOP,RG1,RG2,RP,DP,RHOB,KEQ,KP,KH,KB,KC,KT,AT
      DATA RG1,RG2/82.06,1.987/
      RP=DP/2.0
      T=Y(NEQ-1)
      P=Y(NEQ)
      NT=0.0
      DO 10 J=1,NC
         NT=NT+Y(J)
   10 CONTINUE
      PH=Y(1)/NT*P
      PB=Y(2)/NT*P
```

```
      PC=Y(3)/NT*P
      KP=EXP(-6693.5/T+16.86)
      KH=EXP(7800.7/T-16.05)
      KB=EXP(5636.6/T-11.63)
      KC=EXP(4479.1/T-9.76)
      KEQ=EXP(25980.0/T-46.87)
      KT=ACT(1)*KP*KB*KH**3
      AT=(1.0+KH*PH+KB*PB+KC*PC)**4
      R(1)=KT*(PB*PH**3-PC/KEQ)/(AT*3600.0)
      DO 20 J=1,NC
         RATE(J)=0.0
         DO 20 I=1,NR
            RATE(J)=RATE(J)+V(I,J)*R(I)*RHOB
   20 CONTINUE
      RETURN
      END

      SUBROUTINE DNDZ(NC,RATE,MWF,G,YAO,F)
      INTEGER NC
      REAL RATE(*),F(*),MWF,G,YAO,C
      C=MWF/(G*YAO)
      DO 10 J=1,NC
         F(J)=RATE(J)*C
   10 CONTINUE
      RETURN
      END

      SUBROUTINE DTDZ(NEQ,NC,Y,HF,A,B,C,D,RATE,NT,CPM,G,YAO,MWF,DP,
     $                DI,XW,KW,TJ,MUM,KFM,RE,HI,HO,UZ,MODE,MWM,F)
      INTEGER NEQ,NC
      REAL Y(*),HF(*),A(*),B(*),C(*),D(*),RATE(*),F(*)
      REAL NT,CPM,G,YAO,MWF,T,T1,T2,T3,T4,HRXN,SCPJDT,SENSHT,KW
      REAL TJ,DI,DO,DP,MUM,RE,HO,RF,UZ,HXFR,KFM,HI,MWM,MODE,PR
      F(NEQ-1)=0.0
      IF(MODE.LT.1.5)RETURN
      T=Y(NEQ-1)
      T1=T-298.0
      T2=(T**2-298.0**2)/2.0
      T3=(T**3-298.0**3)/3.0
      T4=(T**4-298.0**4)/4.0
      HRXN=0.0
      DO 10 J=1,NC
         SCPJDT=A(J)*T1+B(J)*T2+C(J)*T3+D(J)*T4
         HRXN=HRXN+(HF(J)+SCPJDT)*RATE(J)
   10 CONTINUE
      IF(MODE.GT.2.5)THEN
         DO=DI+2.0*XW
         RE=DP*G/MUM
         HO=1500.0/7373.26
         RF=0.001*7373.26
         PR=CPM*MUM/(MWM*KFM)
         HI=KFM/DI*1.40*RE**.95*PR**.33*EXP(-6*DP/DI)
         UZ=1.0/(1.0/HI+RF+DI/(DO*HO))
         HXFR=4.0*UZ*(TJ-T)/DI
      ELSE
         HXFR=0.0
      ENDIF
      SENSHT=NT*CPM*G*YAO/MWF
      F(NEQ-1)=(-HRXN+HXFR)/SENSHT
      RETURN
      END

      SUBROUTINE DPDZ(NEQ,EPS,MUM,DP,G,RHOM,FK,F)
      INTEGER NEQ
      REAL F(*),EPS,MUM,DP,G,RHOM,FK,CFACTOR
      CFACTOR=1.0/1013200.0
      FK=1.80+180.0*(1.0-EPS)*MUM/(DP*G)
      F(NEQ)=-FK*G**2*(1.0-EPS)/(DP*RHOM*EPS**3)
      F(NEQ)=F(NEQ)*CFACTOR
      RETURN
      END
```

APPENDIX B.3

Description of PFRSIM*

PFRSIM — PLUG-FLOW REACTOR SIMULATOR

This is a generalized interactive reactor simulation program that will apply to almost any well-defined reaction system in an axial or radial, homogeneous or heterogeneous, ideal plug-flow reactor with or without heat transfer. It was created for use by students in the Department of Chemical Engineering at the University of Texas at Austin as an educational tool in courses involving industrial reactor design.

DESCRIPTION OF VARIABLES

A	REAL(20)	HEAT CAPACITY EQN. CONSTANTS
ACT	REAL(20)	CATALYST ACTIVITIES
B	REAL(20)	HEAT CAPACITY EQN. CONSTANTS
C	REAL(20)	HEAT CAPACITY EQN. CONSTANTS
CONV	REAL	CONVERSION OF LIMITING REACTANT
CPJ	REAL	COMPONENT HEAT CAPACITY
CPM	REAL	MEAN HEAT CAPACITY, CAL/GMOL*K
D	REAL(20)	HEAT CAPACITY EQN. CONSTANTS
DI	REAL	INSIDE DIAMETER, CM
DO	REAL	OUTSIDE DIAMETER, CM
DP	REAL	CATALYST PARTICLE DIAMETER, CM
EPS	REAL	BED VOID FRACTION, CM^3/CM^3
F	REAL (22)	GRADIENTS FOR MOLES, TEMPERATURE AND PRESSURE WITH RESPECT TO Z
FK	REAL	COEFFICIENT IN PRESSURE DROP EQN.
G	REAL	MASS VELOCITY, G/CM^2*SEC

* Portions of the program are presented by permission of Doyle E. Broom and The Univeristy of Texas System.

HF	REAL(20)	STD. HEAT OF FORMATION, CAL/GMOL
HI	REAL	INSIDE HEAT TRANSFER COEFFICIENT, CAL/CM^2*SEC*K
HO	REAL	OUTSIDE HEAT TRANSFER COEFFICIENT, CAL/CM^2*SEC*K
KFM	REAL	THERMAL CONDUCTIVITY OF THE REACTION MIXTURE, CAL/CM*SEC*K
KW	REAL	THERMAL CONDUCTIVITY OF THE TUBE WALL, CAL/CM*SEC*K
L	REAL	RADIAL REACTOR LENGTH, CM
LR	INTEGER	COMPONENT# OF LIMITING REACTANT
MODE	REAL	CODE FOR REACTOR HEAT TRANSFER MODE
MUM	REAL	VISCOSITY OF REACTION MIXTURE
MW	REAL (20)	MOLECULAR WEIGHT OF COMPONENT
MWF	REAL	MOLECULAR WEIGHT OF FEED
N	INTEGER	NUMBER OF REACTION RATES PRINTED
NC	INTEGER	NUMBER OF COMPONENTS IN SYSTEM
NEQ	INTEGER	NUMBER OF DIFFERENTIAL EQUATIONS
NI	REAL (20)	INLET NUMBER OF MOLES/MOLE LR
NR	INTEGER	NUMBER OF REACTIONS DEFINED
NT	REAL	TOTAL NUMBER OF MOLES/MOLE LR
P	REAL	PRESSURE, ATM
PI	REAL	CONSTANT, PI=3.14159
PR	INTEGER	COMPONENT# OF DESIRED PRODUCT
Q	REAL	TOTAL HEAT DUTY, CAL/SEC
R	REAL(20)	REACTION RATE, GMOL/SEC*CM^3
RATE	REAL(20)	RATE OF PRODUCTION OF COMPONENT, GMOL/SEC*CM^3
RE	REAL	REYNOLDS NUMBER
RF	REAL	FOULING FACTOR, CM^{2}SEC °K/CAL
RHOB	REAL	CATALYST BED DENSITY, G/CM^3
RHOM	REAL	DENSITY OF REACTION MIXTURE G/CM^3
RHOP	REAL	CATALYST PARTICLE DENSITY, G/CM^3
T	REAL	TEMPERATURE, °K
TITLE	CHARACTER*70	TITLE OF REPORT
TJ	REAL	JACKET TEMPERATURE, K
TMAX	REAL	MAXIMUM TEMPERATURE IN BED, K
TMIN	REAL	MINIMUM TEMPERATURE IN BED, K
UZ	REAL	LOCAL HEAT TRANSFER COEFFICIENT, CAL/CM^2*SEC*K
UZAVG	REAL	AVERAGE HEAT TRANSFER COEFFICIENT, CAL/CM^2*SEC*K
V	REAL(20,20)	STOICHIOMETRIC COEFFICIENTS
X	REAL(20)	MOLE FRACTION OF COMPONENTS
XI	REAL(20)	INLET MOLE FRACTION OF COMPONENTS
XLMT	REAL	CONVERSION EXIT CRITERION
XW	REAL	TUBE WALL THICKNESS, CM
Y	REAL(22)	DEPENDENT VARIABLES FOR LSODE: MOLES, TEMPERATURE, AND PRESSURE
YAO	REAL	INLET MOLE FRACTION OF LR
YIELD	REAL	YIELD OF DESIRED PRODUCT
Z	REAL	BED LOCATION, AXIAL OR RADIAL, CM
ZPRINT	REAL	PRINTING INTERVAL, CM
ZSTEP	REAL	STEP SIZE, CM
ZTMAX	REAL	LOCATION OF MAX. TEMPERATURE, CM
ZTMIN	REAL	LOCATION OF MIN. TEMPERATURE, CM

Combinations of variables are defined in the subroutines given above.

PFRSIM consists of a main program that calls various subroutines, including those shown here. The user need only enter into the subroutines given here those equations unique to the system to be studied, as described below for the cyclohexane reaction. Please refer to the line numbers indicated on the subroutines.

Components	No.
Hydrogen	1
Benzene	2
Cyclohexane	3
Methane	4
Ethane	5
Propane	6

15–20 Enter thermal conductivity equations for each component.

33 Enter viscosity equation for mixture with hydrogen in excess. When mixture viscosity changes substantially with reaction progress, pure component equations should be given here followed by an averaging equation similar to that for thermal conductivity.

50–59 Enter equations for all kinetic and equilibrium constants. Define terms as real.

59–60 Enter rate equations and convenient combination terms. In this case there is only one equation. Multiple equations can be entered as required. The DO LOOP (61–64) will read each equation and the stoichiometric coefficient for each component and prepare rate equations for each component. In this case the result in terms of standard nomenclature is

$$\hat{r}_H = -3\hat{r}_1$$

$$\hat{r}_B = -\hat{r}_1$$

$$\hat{r}_C = \hat{r}_1$$

See Eq. EX-7.1 for the rate of reaction 1, the only reaction included. For multiple-reaction systems the DO LOOP would include all reaction rates involving each component along with the appropriate stoichiometric coefficients.

69–71 This is a general subroutine that requires no change. As can be seen, it writes the necessary differential equations, which in this case are

$$\frac{dn_H}{dZ} = -\frac{M_F}{\mathbf{G}Y_{A_0}} 3\hat{r}_1$$

$$\frac{dn_B}{dZ} = -\frac{M_F}{\mathbf{G}Y_{A_0}} \hat{r}_1$$

$$\frac{dn_C}{dZ} = \frac{M_F}{\mathbf{G}Y_{A_0}} \hat{r}_2$$

where H is hydrogen, B is benzene, and C is cyclohexane.

100 Enter value for outside heat-transfer coefficient, usually a constant value. But an equation may be inserted here if value changes with position.

101 Enter appropriate fouling factor.

110 No entry required. This is the heat-balance differential equation formed from the combination terms preceding it.

114–122 This last subroutine is for pressure drop and requires no user entry unless coefficients on the Ergun equation are to be changed based on experimental pressure drop studies for the particular catalyst to be used.

The user also prepares a data file giving the stoichiometric coefficients for each component in each reaction and the heat-capacity coefficients and heats of reaction for each component.

See Fig EX-7.3 for interactive displays that permit rapid study of operating and design variables and easy plotting of results together with useful tabulation of results (see Table EX-7.2).

APPENDIX C.1

Derivation of General Shell Balance for Fixed-Bed Reactor

BASIS: RADIAL VELOCITY ≈ 0; CYLINDRICAL REACTOR

Consider a cylindrical element with a radius r_s, thickness Δr_t, and height ΔZ such that the area normal to radial dispersion is $2\pi r_t \Delta Z$, the area normal to axial flow is $2\pi r \Delta r_t$, and the volume of the element is $2\pi r \Delta r_t \Delta Z$.

Component Balance for Component *j*

Axial flow (forced convection)

$$\underbrace{C_j u_s 2\pi r_t \Delta r_t \Delta t|_Z}_{\text{In}} - \underbrace{C_j u_s 2\pi r_t \Delta r_t \Delta t|_{Z+\Delta Z}}_{\text{Out}}$$

+ Axial dispersion (dispersion in z direction)

$$+\underbrace{\varepsilon N_{j_Z} 2\pi r_t \Delta r_t \Delta t|_Z}_{\text{In}} - \underbrace{\varepsilon N_{j_Z} 2\pi r_t \Delta r_t \Delta t|_{Z+\Delta Z}}_{\text{Out}}$$

+ Radial dispersion (dispersion in r direction)

$$\underbrace{+\varepsilon N_{jr} 2\pi r_t \Delta Z \Delta t|_{r_t}}_{\text{In}} - \underbrace{\varepsilon N_{jr} 2\pi r_t \Delta Z \Delta t|_{r_t+\Delta r_t}}_{\text{Out}}$$

+ Reaction

$$+\rho_b \sum_{n=1}^{n_r} \nu_{jn} \hat{r}_n 2\pi r_t \Delta r_t \Delta Z \Delta t$$

= Accumulation

$$= \varepsilon C_j 2\pi r_t \Delta r_t \Delta Z|_{t+\Delta t} - \varepsilon C_j 2\pi r_t \Delta r_t \Delta Z|_t \quad \text{(C-1.1)}$$

where C_j is the concentration of component j; u_s is the superficial velocity of the fluid flowing in axial direction; r_t is the radial distance; t is time; N_{ja}, N_{jr} are molar fluxes in the axial and radial directions, respectively, in moles/(fluid area)(time); ε is the bed void fraction.

Divide Eq. C-1.1 by $2\pi r\,\Delta r_t\,\Delta Z$ and rearrange to obtain

$$-\frac{C_j u_s|_{Z+\Delta Z} - C_j u_s|_Z}{\Delta Z} - \frac{\varepsilon N_{ja}|_{Z+\Delta Z} - \varepsilon N_{ja}|_Z}{\Delta Z}$$

$$-\frac{1}{r_t}\frac{\varepsilon r_t N_{jr}|_{r_t+\Delta r_t} - \varepsilon r N_{jr}|_{r_t}}{\Delta r_t} + \rho_b \sum_{n=1}^{n_r} \nu_{jn}\hat{r}_n = \frac{\varepsilon C_j|_{t+\Delta t} - \varepsilon C_j|_t}{\Delta t} \qquad \text{(C-1.2)}$$

Taking the limit as $\Delta r_t \to 0$, $\Delta Z \to 0$, $\Delta t \to 0$ and assuming ε is constant, gives

$$-\left[\frac{\partial(C_j u_s)}{\partial Z}\right]_{t,r_t} - \varepsilon\left[\frac{\partial(N_{ja})}{\partial Z}\right]_{t,r_t} - \frac{1}{r_t}\,\varepsilon\left[\frac{\partial(r_t N_{jr})}{\partial r_t}\right]_{t,Z} + \rho_b \sum_{n=1}^{n_r} \nu_{jn}\hat{r}_n = \varepsilon\left[\frac{\partial C_j}{\partial t}\right]_{Z,r_t} \qquad \text{(C-1.3)}$$

Substitute

$$N_{ja} = -\mathscr{D}_a\left[\frac{\partial C_j}{\partial Z}\right]_{r,t} \quad \text{and} \quad N_{jr} = -\mathscr{D}_r\left[\frac{\partial C_j}{\partial r_t}\right]_{Z,t} \qquad \text{(C-1.4)}$$

Also assume

$$\left[\frac{\partial \ln \mathscr{D}_a}{\partial Z}\right]_{t,r_t} \Bigg/ \left[\frac{\partial \ln\left|\dfrac{\partial C_j}{\partial Z}\right|}{\partial Z}\right]_{t,r_t} \approx 0 \quad \text{for all } 0 \leqslant r_t \leqslant R_t \text{ and } T \geqslant 0$$

$$\left[\frac{\partial \ln \mathscr{D}_r}{\partial r_t}\right]_{Z,t} \Bigg/ \left[\frac{\partial \ln\left|\dfrac{\partial C_j}{\partial r_t}\right|}{\partial r_t}\right]_{Z,t} \approx 0 \quad \text{for all } 0 \leqslant Z \leqslant L \text{ and } t \geqslant 0$$

where $\mathscr{D}_a$ and $\mathscr{D}_r$ are the effective axial and radial diffusivities defined by Eq. C-1.4.

$$-\left[\frac{\partial(C_j u_s)}{\partial Z}\right]_{t,r_t} + \varepsilon\mathscr{D}_a\left[\frac{\partial^2 C_j}{\partial Z^2}\right]_{t,r_t} + \varepsilon\mathscr{D}_r\,\frac{1}{r_t}\,\varepsilon\left[\frac{\partial\left(r_t \dfrac{\partial C_j}{\partial r_t}\right)}{\partial r_t}\right]_{Z,t} + \rho_b \sum_{n=1}^{n_r} \nu_{jn}\hat{r}_n$$

$$= \varepsilon\left[\frac{\partial C_j}{\partial t}\right]_{r,Z} \qquad \text{(C-1.5)}$$

The radial dispersion term can be simplified by applying the chain rule:

$$\left[\frac{\partial\left(r_t \dfrac{\partial C_j}{\partial r_t}\right)}{\partial r_t}\right]_{t,Z} = \left[r_t \frac{\partial^2 C_j}{\partial r_t^2} + \frac{\partial C_j}{\partial r_t}\right]_{t,Z} \qquad \text{(C-1.6)}$$

Substitute Eq. C-1.6 into Eq. C-1.5 and rearrange to get

$$\varepsilon\left[\frac{\partial C_j}{\partial t}\right]_{Z,t} + \left[\frac{\partial (C_j u_s)}{\partial Z}\right]_{t,r_t} - \varepsilon\mathscr{D}_a\left[\frac{\partial^2 C_j}{\partial Z^2}\right]_{t,r_t} - \varepsilon\mathscr{D}_r\left[\frac{\partial^2 C_j}{\partial r_t^2} + \frac{1}{r_t}\frac{\partial C_j}{\partial r_t}\right]_{t,Z}$$

$$= \rho_b \sum_{n=1}^{n_r} \nu_{jn}\hat{r}_n \qquad j = 1, \ldots, c \tag{C-1.7}$$

Equation C-1.6 can also be written in terms of moles of key component per mole of limiting reactant A:

$$\varepsilon\frac{\mathbf{G}_0 Y_{A_0}}{M_F}\left[\frac{\partial (n_{ka}\rho_f/\mathbf{G})}{dt}\right]_{Z,r_t} + \frac{\mathbf{G}_0 Y_{A_0}}{M_F}\left[\frac{\partial n_{ka}}{dZ}\right]_{t,r_t} - \varepsilon\mathscr{D}_a\frac{\mathbf{G}_0 Y_{A_0}}{M_F}\left[\frac{\partial^2 (n_{ka}\rho_f/\mathbf{G})}{dZ^2}\right]_{t,r_t}$$

$$-\varepsilon\mathscr{D}_r\frac{\mathbf{G}_0 Y_{A_0}}{M_F}\left[\frac{\partial^2 (n_{ka}\rho_f/\mathbf{G})}{\partial r_t^2} + \frac{1}{r_t}\frac{\partial (n_{ka}\rho_f/\mathbf{G})}{\partial r_t}\right]_{Z,t} = \rho_b \sum_{n=1}^{n_r}\nu_{jn}\hat{r}_n$$

$$k = 1, \ldots, k_n \tag{C-1.8}$$

ENERGY BALANCE

$$\frac{\textit{Axial flow of energy}}{\dfrac{\text{In} \quad - \quad \text{Out}}{(c_p\rho)_f u_s T 2\pi r_t\,\Delta r_t\,\Delta t|_Z - (c_p\rho)_f u_s T 2\pi r_t\,\Delta r_t\,\Delta t|_{Z+\Delta Z}}}$$

$$+\frac{\textit{Axial flow of energy}}{\dfrac{\text{In} \quad - \quad \text{Out}}{N_{qa} 2\pi r_t\,\Delta r_t\,\Delta t|_Z - N_{qa} 2\pi r_t\,\Delta r_t\,\Delta t|_{Z+\Delta Z}}}$$

$$+\frac{\textit{Radial conduction of energy}}{\dfrac{\text{In} \quad - \quad \text{Out}}{N_{qr} 2\pi r_t\,\Delta Z\,\Delta t|_{r_t} - N_{qr} 2\pi r_t\,\Delta Z\,\Delta t|_{r_t+\Delta r_t}}}$$

$$+\frac{\textit{Energy due to reaction(s)}}{+\rho_b \displaystyle\sum_{n=1}^{n_r}(-\Delta H_n)\hat{r}_n}$$

$$=\frac{\textit{Accumulation}}{(c_p\rho)_{fs} T 2\pi r_t\,\Delta r_t\,\Delta Z|_{t+\Delta t} - (c_p\rho)_{fs} T 2\pi r_t\,\Delta r_t\,\Delta Z|_t} \tag{C-1.9}$$

where c_p is heat capacity per unit mass; ρ is mass density; subscripts f, s, and fs refer, respectively, to fluid, solid, and the combined fluid and solid. N_{qa}, N_{qr} refer to heat flux due to axial and radial energy dispersion.

Divide Eq. C-1.9 by $2\pi r_t \, \Delta r_t \, \Delta Z$ and rearrange to obtain

$$-\frac{(c_p\rho)_f u_s T|_{Z+\Delta Z} - (c_p\rho)_f u_s T|_Z}{\Delta Z} - \frac{N_{qa}|_{Z+\Delta Z} - N_{qa}|_Z}{\Delta Z}$$

$$-\frac{1}{r_t}\frac{r_t N_{qr}|_{Z+\Delta Z} - r_t N_{qr}|_Z}{\Delta r_t}$$

$$+\rho_b \sum_{n=1}^{n_r} (-\Delta H_n)\hat{r}_n = \frac{(c_p\rho)_{fs}|_{t+\Delta t} - (c_p\rho)_{fs}|_t}{\Delta t} \qquad \text{(C-1.10)}$$

Taking the limit as $\Delta r_t \to 0$, $\Delta Z \to 0$, $\Delta t \to 0$ gives

$$-\left[\frac{\partial (c_p\rho)_{fs} u_s T}{\partial Z}\right]_{t,r_t} - \left[\frac{\partial N_{qa}}{\partial Z}\right]_{t,r_t} - \frac{1}{r_t}\left[\frac{\partial (r_t N_{qa})}{\partial r_t}\right]_{t,Z} + \rho_b \sum_{n=1}^{n_r} (-\Delta H_n)\hat{r}_n$$

$$= \left[\frac{\partial (c_p\rho)_f T}{\partial t}\right]_{Z,r_t} \qquad \text{(C-1.11)}$$

Define N_{qa} and N_{qr} in terms of effective conductivities (λ_a and λ_r), which in actuality must combine static, dynamic, and radiation terms.

$$N_{qa} = -\lambda_a \left[\frac{\partial T}{\partial Z}\right]_{t,r_t}$$

and (C-1.12)

$$N_{qr} = -\lambda_r \left[\frac{\partial T}{\partial r_t}\right]_{Z,t}$$

Substitute Eq. C-1.12 in Eq. C-1.11 and assume

$$\left[\frac{\partial \ln \lambda_r}{\partial r_t}\right]_{Z,t} \Big/ \left[\frac{\partial \ln T}{\partial r_t}\right]_{Z,t} \approx 0 \qquad \text{for all } 0 \leqslant Z \leqslant L \text{ and } t \geqslant 0$$

$$\left[\frac{\partial \ln \lambda_r}{\partial r_t}\right]_{Z,t} \Big/ \left[\frac{\partial \ln T}{\partial r_t}\right]_{Z,t} \approx 0 \qquad \text{for all } 0 \leqslant Z \leqslant L \text{ and } t \geqslant 0$$

The result is

$$
-\left[\frac{\partial(c_p\rho)_{\mathrm{f}}u_{\mathrm{s}}T}{\partial Z}\right]_{t,r_t}+\lambda_{\mathrm{a}}\left[\frac{\partial^2 T}{\partial Z^2}\right]_{t,r_t}+\frac{\lambda_{\mathrm{r}}}{r_t}\left[\frac{\partial\left(r_t\dfrac{\partial T}{\partial r_t}\right)}{\partial r_t}\right]_{Z,t}+\rho_{\mathrm{b}}\sum_{n=1}^{n_{\mathrm{r}}}(-\Delta H_n)\hat{r}_n
$$

$$
=\left[\frac{\partial(c_p\rho)_{\mathrm{fs}}T}{\partial t}\right]_{Z,r_t} \tag{C-1.13}
$$

But as for Eq. C-1.6, the third term in Eq. C-1.13 is

$$
\lambda_{\mathrm{r}}\left[\frac{\partial^2 T}{\partial r_t^2}+\frac{1}{r_t}\frac{\partial T}{\partial r_t}\right]_{Z,t}
$$

Substituting, rearranging, and assuming a constant value of $(c_p\rho)_{\mathrm{fs}}$ gives

$$
(\overline{c_p\rho})_{\mathrm{fs}}\left[\frac{\partial T}{\partial t}\right]_{Z,r_t}+\left[\frac{\partial(c_p\rho)_{\mathrm{f}}u_{\mathrm{s}}T}{\partial Z}\right]_{t,r_t}-\lambda_a\left[\frac{\partial^2 T}{\partial Z^2}\right]_{t,r_t}-\lambda_{\mathrm{r}}\left[\frac{\partial^2 T}{\partial r_t^2}+\frac{1}{r_t}\frac{\partial T}{\partial r_t}\right]_{Z,t}
$$

$$
=\rho_{\mathrm{b}}\sum_{n=1}^{n_{\mathrm{r}}}(-\Delta H_{\mathrm{n}})\hat{r}_{\mathrm{n}} \tag{C-1.14}
$$

Usually $(\overline{c_p\rho})_{\mathrm{fs}}$ is approximated by

$$
(\overline{c_p\rho})_{\mathrm{fs}}=\varepsilon(\overline{c_p\rho})_{\mathrm{f}}+(1-\varepsilon)(\overline{c_p\rho})_{\mathrm{s}} \tag{C-1.15}
$$

Nomenclature

For illustrative purposes three sets of units are given in brackets just after each definition of a symbol and in the order cgs, metric engineering, and English engineering. Modern simulator software usually provides a rapid means for converting to most frequently used systems of choice, and the issue of units is becoming a minor one.

Absolute temperature is used throughout. Obviously, differences in temperature may be expressed as centrigrade or Fahrenheit degrees as can fluxes given on a per degree basis.

Where necessary for clarity, length units with subscripts are used as follows:

cm_a, m_a, ft_a = axial distance in bed

cm_b, m_b, ft_b = thickness

cm_f, m_f, ft_f = distance in direction of flow

cm_n, m_n, ft_n = distances normal to direction of flow

cm_r, m_r, ft_r = distance in radial direction

A, B, . . . R, S	reactants and products
A, B	constants defined by equations where used
A	Arrhenius or pre-exponential factor [same units as rate constant]
ΔA	change in total work function of a system [cal, kcal, BTU]
A'	pre-exponential factor for reverse reaction [same units as reverse rate constant]
A''	reparameterized pre-exponential factor [same units as rate constant]
A_c	cross sectional area of bed [cm^2, m^2, ft^2]
A_i	inside surface area of reactor tube [cm^2, m^2, ft^2]
A_j, B_j, C_j, D_j	constants for empirical heat-capacity equations for ideal gases

$\Delta A_j, \Delta B_j, \Delta C_j, \Delta D_j$	summations of v_jA_j, v_jB_j, v_jC_j, and v_jD_j
A_O	observed pre-exponential factor [same units as rate constant]
$a, b, \ldots r, s,$	stoichiometric coefficients for indicated components $A, B, \ldots, R, S$, positive for products and negative for reactants
$a_A, a_B, \ldots, a_R, a_S$	activities of indicated components [dimensionless]
a_j	activity for component j [dimensionless]
a_m	exterior surface area of catalyst per mass of catalyst [cm^2/gcat., m^2/kgcat., ft^2/lbcat.]
a_s	exterior area per unit volume of a catalyst pellet [cm_n^2/cm^3, m_n^2/m^3, ft_n^2/ft^3]
b	thickness of conducting solid [cm_f, m_f, ft_f,]
b_i	total number of atomic weights of element i in the system
b_v	fractional bed volume of inert solids mixed with catalyst [(cm^3 inerts) (cm^3 bed)$^{-1}$, (m^3 inerts) (m^3 bed)$^{-1}$, (ft^3 inerts) (ft^3 bed)$^{-1}$]
b_w	wall thickness of reactor tube [cm, m, ft]
C	corrosion allowance [cm, m, in.]
$C_A, C_B, \ldots, C_R, C_S$	concentration of indicated component [moles/cm^3, kmoles/m^3, lbmoles/ft^3]
$C'_A, C'_B, \ldots, C'_R, C'_S$	concentration of indicated adsorbed species [moles/gcat., kmoles/kgcat., lbmoles/lbcat.]
$C_{A_S}, C_{B_S}, \ldots, C_{R_S}, C_{S_S}$	concentration at catalyst exterior surface for components $A, B, \ldots, R, S$ [moles/cm^3, kmoles/m^3, lbmoles/ft^3]
$C_{A_b}, C_{B_b}, \ldots, C_{R_b}, C_{S_b}$	concentration at bulk conditions for $A, B, \ldots R, S$ [moles/cm^3, kmoles/m^3, lbmoles/ft^3]
C_{A_e}	exit concentration [moles/cm^3, kmoles/m^3, lbmoles/ft^3]
C_{A_o}	inlet concentration [moles/cm^3, kmoles/m^3, lbmoles/ft^3]
C_j	concentration of component j [moles/cm^3, kmoles/m^3, lbmoles/ft^3]
C_{j_b}	concentration of component j in the bulk fluid [moles/cm^3, kmoles/m^3, lbmoles/ft^3]
C_{js}	concentration of component j on the catalyst surface [moles/cm^3, kmoles/m^3, lbmoles/ft^3]
C'_l	concentration of vacant sites [moles/gcat., kmoles/kgcat., lbmoles/lbcat.]
C_k	concentration of key component, k [(moles k)/cm^3, (kmoles k)/m^3, (lbmoles k)/ft^3]

C_1	conversion factor for Ergun equation (see Eq. 4.15)
$C^{\dagger}$	concentration of activated complex [moles/gcat., kmoles/kgcat., lbmoles/lbcat.]
c	number of components
c_p	heat capacity at constant pressure on a mass basis [cal/g°K, kcal/kg°K, BTU/lb°R]
c_{p_1}	heat capacity at constant pressure of fluid leaving stage #1 on a mass basis [cal/g°K, kcal/kg°K, BTU/lb°R]
c_{p_F}	heat capacity at constant pressure of feed on a mass basis cal g^{-1}°K^{-1} [cal/g°K, kcal/kg°K, BTU/lb°R]
c_{p_j}	heat capacity of component j at constant pressure [cal/(mole j)(°K), kcal/(kmole j)(°K), BTU/(lbmole j)(°R)]
c_{p_m}	heat capacity at constant pressure on a mole of mixture basis [cal(mole of mixture)$^{-1}$ (°K)$^{-1}$, kcal(kmole of mixture)$^{-1}$ (°K)$^{-1}$, BTU/(lbmole)$^{-1}$(°R)$^{-1}$]
c_v	heat capacity at constant volume on a mass basis [cal/g°K, kcal/kg°K, BTU/lb°R]
D	diameter of bed or reactor (cm, m, ft)
D_I	inside diameter of shell [cm, m, ft]
D_m	mean diameter of reactor tube [cm, m, ft]
D_O	outside diameter of reactor tubes [cm, m, ft]
D_p	equivalent spherical diameter of a catalyst pellet, which is the diameter of a sphere having the same exterior area per unit volume as the pellet [cm, m, ft]
D_v	diameter of a sphere having same volume as average particle [cm, m, ft]
$\mathscr{D}_A$	molecular diffusivity of A [cm_f^2/sec, m_f^2/hr, ft_f^2/hr]
$\mathscr{D}_a$	effective axial diffusivity in a packed bed [cm_a^2/sec, m_a^2/hr, ft_a^2/hr]
$\mathscr{D}_{I_A}, \mathscr{D}_{I_B}, \ldots, \mathscr{D}_{I_R}, \mathscr{D}_{I_S}$	effective diffusivity in the catalyst pellet for components, $A, B, \ldots, R, S$ [cm_f^2/sec, m_f^2/hr, ft_f^2/hr]
$\mathscr{D}_{I_j}$	effective diffusivity of j in the catalyst pellet [cm_f^2/sec, m_f^2/hr, ft_f^2/hr]
$\mathscr{D}_j$	molecular diffusivity of component j [cm_f^2/sec, m_f^2/hr, ft_f^2/hr]
$\mathscr{D}_r$	effective radial diffusivity in a packed bed [cm_r^2/sec, m_r^2/hr, ft_r^2/hr]

d_c	diameter of a cylinder [cm, m, ft]
d_i	inside diameter of Rasching-ring catalyst form [cm, m, ft]
d_o	outside diameter of Rasching-ring catalyst form [cm, m, ft]
d_p	equivalent spherical diameter of a pellet, which is the diameter of a sphere having the same exterior area as the pellet [cm, m, ft]
d_s	sphere diameter [cm, m, ft]
E	activation energy of forward reaction [cal/mole, kcal/kmole, BTU/lbmole]
E'	activation energy for reverse reaction [cal/mole, kcal/kmole, BTU/lbmole]
E_o	observed activation energy [cal/mole, kcal/kmole, BTU/lbmole]
E	weld efficiency
e	extent of reaction (see Eq. 3.32) **e** is equal to conversion in any reaction in which the limiting reactant appears where $a = 1$
F_A	flow rate of A, the limiting reactant in the feed [moles A/sec, kmoles A/hr, lbmoles A/hr]
$F(1), F(2), \ldots$	function 1, 2, etc.
F_n	moles of stream n [moles, kmoles, lbmoles]
F_p	moles of product stream [g, kg, lb]
F_{PT}	mass of pure product [g, kg, lb]
F_T	total mass of feed to reactor [g, kg, lb]
$\mathscr{F}_j$	flow rate of component j [moles j/sec, kmoles j/hr, lbmoles j/hr]
$\mathscr{F}_1$	total moles leaving Stage No. 1 [moles/sec, kmoles/hr, lbmoles/hr]
$\mathscr{F}_Q$	total moles of feed quench added at exit of Stage No. 1 [moles/sec, kmoles/hr, lbmoles/hr]
f_j	pure component fugacity of component j [atm, atm, psia]
f_j^0	standard-state fugacity of component j [1 atm]
$\hat{f}_A, \hat{f}_B, \ldots \hat{f}_R, \hat{f}_S$	partial fugacity of indicated components ($Y_A \phi_A P$ for ideal solution; $Y_A \hat{\phi}_A P$ for nonideal solutions) [atm, atm, psia]
$\hat{f}_j$	partial fugacity of reacting component j ($Y_j \phi_j P$ for ideal solution; $Y_j \hat{\phi}_j P$ for nonideal solution) [atm, atm, psia]
f_k	friction factor for flow in packed bed [dimensionless]
f_m	factor in Eq. 7.26
f_p	fraction of bed inlet pressure lost to friction

$e, f, g, \ldots, e', f', g'$	exponents for empirical rate equation
$\mathbf{G}$	mass velocity [g/cm_n^2 sec, kg/m_n^2 hr, lb/ft_n^2 hr]
G	total free energy of a mixture relative to the elements at T and P [cal/mole, kcal/kmole, BTU/lbmole]
$\Delta G^\dagger$	free energy of activation [cal/mole, kcal/kmole, BTU/lbmole]
ΔG°	standard free energy of a reaction at temperature T and 1 atm [cal/mole, kcal/kmole, BTU/lbmole]
$G^\circ_{f_A}, G^\circ_{f_B}, \ldots, G^\circ_{f_R}, G^\circ_{f_S}$	standard free energy of formation relative to the elements at temperature T and 1 atm for indicated component [cal/mole, kcal/kmole, BTU/lbmole]
$G^\circ_{f_j}$	standard free energy of formation of component j at 1 atm and temperature T [cal/mole, kcal/kmole, BTU/lbmole]
$\bar{G}_j$	partial molal free energy of component j [cal/mole, kcal/kmole, BTU/lbmole]
ΔG	change in total free energy of a system [cal, kcal, BTU]
g	acceleration due to gravity [cm^2/sec, m^2/hr, ft^2/hr]
H	total enthalpy of a mixture relative to the elements at temperature T and pressure P [cal/mole, kcal/kmole, BTU/lbmole]
$H_A, H_B, \ldots, H_R, H_S$	enthalpy at temperature T relative to the elements at 298°K (i.e, the heat of formation at temperature T) for indicated components [cal/mole, kcal/kmole, BTU/lbmole]
$H^\circ_{f_j}$	standard heat of formation of component j from the elements at temperature T and 1 atm [cal/mole, kcal/kmole, BTU/lbmole]
H_j	partial enthalpy of component j [cal/mole j, kcal/kmole j, BTU/lbmole j]
H_j	enthalpy of j relative to the elements at 298°K [cal/mole j, kcal/kmole j, BTU/lbmole j]
H^*	ideal gas enthalpy [cal/mole, kcal/kmole, BTU/lbmole]
H_m	enthalpy per unit mass [cal/g, kcal/kg, BTU/lb]
ΔH	change in total enthalpy of a system [cal, kcal, BTU]
ΔH°	standard enthalpy or heat of reaction at temperature T and 1 atm [cal/mole, kcal/kmole, BTU/lbmole]

ΔH_A	heat of reaction per mole of A reacted at temperature T corrected for pressure if necessary [cal/mole A, kcal/kmole A, BTU/lbmole A]
$\Delta H_1, \Delta H_2, \ldots, \Delta H_n$	heat of reaction at temperature of interest (corrected if necessary for deviation from ideality due to very high pressure operation) for reactions, $1, 2, \ldots, n$—per mole of component upon which rate equation is based [cal/mole, kcal/kmole, BTU/lbmole]
$(\Delta H^\circ)_T$	standard enthalpy of reaction at temperature T and 1 atm [cal/mole, kcal/kmole, BTU/lbmole]
$\Delta H^\dagger$	enthalpy of activation [cal/mole, kcal/kmole, BTU/lbmole]
HP	horsepower
$\mathbf{h}$	Planck constant [6.624×10^{-27} erg-sec/molecule]
h	heat transfer coefficient [cal/cm^{2}sec °K, kcal/m^{2}hr °K, BTU/ft^{2}hr °R]
h_c	height of a cylinder [cm, m, ft]
h_i	inside heat-transfer coefficient for tubular reactor based on average bed temperature at each Z [cal/cm^{2}sec °K, kcal/m^{2}hr °K, BTU/ft^{2}hr °R]
h_o	outside heat-transfer coefficient for heat transfer fluid in the jacket [cal/cm^{2}sec °K, kcal/m^{2}hr °K, BTU/ft^{2}hr °R]
h_s	heat-transfer coefficient between catalyst exterior surface and bulk fluid [cal/cm^{2}sec °K, kcal/m^{2}hr °K, BTU/ft^{2}hr °R]
h_T	heat-transfer coefficient between heat-transfer fluid and edge (radius) of bed [cal/cm^{2}sec °K, kcal/m^{2}hr °K, BTU/ft^{2}hr °R]
h_w	heat transfer between edge (radius) of bed and wall of reactor [cal/cm^{2}sec °K, kcal/m^{2}hr °K, BTU/ft^{2}hr °R]
i	normal or periodic rate of return
i, j	refer to element and component, respectively
I_H	integration constant (see Eq. 3.18)
I_κ	integration constant (see Eq. 3.19)
J_d	J factor for mass transfer in packed bed between catalyst exterior surface and bulk fluid, $(k_{s_j} \rho_f Y_{j_f} / \mathbf{G})$ Sc, [dimensionless]
J_h	J factor for heat transfer in packed bed between catalyst exterior surface and bulk fluid, $(h_s / c_p \mathbf{G})\,\mathrm{Pr}^{2/3}$, [dimensionless]

K	reaction equilibrium constant [dimensionless in terms of activities] for partial pressures or fugacities: [(atm)$^{\alpha}$, (atm)$^{\alpha}$, (psia)$^{\alpha}$, where $\alpha = r + s, \ldots - a + b, \ldots$]
$K_1, K_2, \ldots$	reaction equilibrium constant for reactions $1, 2, \ldots$ [(atm)$^{\alpha}$, (atm)$^{\alpha}$, (psia)$^{\alpha}$, where $\alpha = r + s, \ldots - a + b, \ldots$]
K_A, K_B, K_R, K_S, K_I	apparent adsorption equilibrium constants for indicated components [dimensionless in terms of activities]
K'	surface reaction equilibrium constant
K_f	equilibrium constant of formation from the elements
K_T	equilibrium constant at temperature T [(atm)$^{\alpha}$, (atm)$^{\alpha}$, (psia)$^{\alpha}$, where $\alpha = (r + s, \ldots) - (a + b, \ldots)$]
K_v	vapor-liquid equilibrium constant
$K_{\phi}, K_{\phi_1}, K_{\phi_2}$	fugacity coefficient quotients for equilibrium expression (see Eq. 3.25), with numeral indicating reaction number [dimensionless]
$K^{\dagger}$	equilibrium constant for formation of the activated complex
k	Boltzman constant [1.3805×10^{-16} erg/°K molecule]
$k_1, k_2, \ldots, k_n$	apparent reaction rate constants for indicated reactions on a mass of catalyst basis and power-law kinetics [moles/gcat. sec atm^n, kmoles/kgcat. hr atm^n, lbmoles/lbcat. hr psia^n. Where n is the reaction order. When used with H & W equations having adsorption constants as multipliers, omit pressure units].
k_A, k_B, k_R, k_S	adsorption rate constants for indicated components based on activities of components [moles/(moles sites) sec, kmoles/(kgmoles sites) hr, lbmoles/(lbmoles sites) hr]
k_a	concentration mass-transfer coefficient [cm$_f$/sec, m$_f$/hr, ft$_f$/hr]
$\hat{k}_c, \hat{k}'_c$	forward and reverse rate constants, respectively, based on concentrations and mass of catalyst [moles/gcat. sec(moles/cm^3)^n, kmoles/kgcat. hr (kmoles/m^3)^n, lbmoles/lbcat. hr(lbmoles/ft^3)^n, where n is the reaction order]
k_d	catalyst deactivation rate constant [atm^{-m} sec^{-1}, atm^{-m} hr^{-1}, (psia)$^{-m}$ hr^{-1}, where m is order of reacting component or impurity]

$k^{s}_{g_j}$	mass-transfer coefficient between catalsyt exterior and bulk fluid with partial pressure driving force [moles/cm_n atm sec, kmoles/m_n atm hr, lbmoles/ft_n psia hr]
k_n	number of key components
k_o	apparent psuedo first-order rate constant [moles/gcat.sec atm, kmoles/kgcat.hr atm, lbmoles/lbcat.hr psia]
k_p, k'_p	forward and reverse reaction rate constants, respectively, based on partial pressures and mass of catalyst for power law kinetics [moles/gcat.hr atm^n, kmoles/kgcat.hr atm^n, lbmoles/lbcat.hr psia^n], where n is the reaction order. When used with H & W equations having adsorption constants as multipliers, omit pressure term]
k_s, k'_s	surface reaction rate constants of forward and reverse reactions, respectively, based on activities of components [moles/(moles sites) sec, kmoles/(kmoles sites) hr, lbmoles/lbmoles sites) hr]
$k_{s_A}, k_{s_B}, \ldots, k_{s_R}, k_{s_S}$	mass-transfer coefficient for $A, B, \ldots, R$ and S between catalyst and bulk fluid with concentration driving force [cm_f/sec, m_f/hr, ft_f/hr]
k_{s_j}	mass-transfer coefficient between catalyst exterior and bulk fluid with concentration driving force [cm_f/sec, m_f/hr, ft_f/hr]
k_T	rate constant based on bulk fluid temperature [see k_p for units]
k_{T_s}	rate constant based on temperature at catalyst exterior surface [see k_p for units]
L	bed length [cm, m, ft]
$\mathscr{L}$	Lagrangian
ℓ	active site
L_R, L'_R	total number of active sites [moles/gcat, kmoles/kgcat, lbmoles/lbcat], prime designates used catalyst
M	molecular weight
M_{AB}	molecular weight factor for diffusivity (see Table 4.1)
M_F	molecular weight of feed
M_j	molecular weight of j
M_1	molecular weight of flow from Stage No. 1
M_s	mass of catalyst sample [g, kg, lb]
m	mass [g, kg, lb]

m	factor for particular system for correcting either inlet or outlet rate constant to produce a value of an equivalent isothermal temperature
m_{A_1}, m_{A_2}	mass of A in streams 1 and 2, respectively [g, kg, lb]
m_a	mass flowing per mole of A fed [g/mole A fed, kg/kmole A fed, lb/lbmole A fed]
m_f	total mass flowing per unit time [g/sec, kg/hr, lb/hr]
m_{T_1}, m_{T_2}	total mass in streams 1 and 2, respectively [g, kg, lb]
N_A	diffusion flux of A [moles A/cm_n^2 sec, kmoles A/m_n^2 hr, lbmoles A/ft_n^2 hr]
N_j	flux of j [moles j/cm_n^2 sec, kmoles j/m_n^2 hr, lbmoles j/ft_n^2 hr]
N_{ja}	axial dispersion flux of j [moles j/cm_n^2 sec, kmoles j/m_n^2 hr, lbmoles j/ft_n^2 hr]
N_{jr}	dispersion flux of j in radial direction [moles j/cm_n^2 sec, kmoles j/m_n^2 hr, lbmoles j/ft_n^2 hr]
N_o	Avogadro's number [6.02252×10^{23} molecules/mole]
N_q	heat flux [cal/cm_n^2 sec, kcal/m_n^2 hr, BTU/ft_n^2 hr]
N_{qa}	axial heat dispersion flux [cal/cm_n^2 sec, kcal/m_n^2 hr, BTU/ft_n^2 hr]
N_{qr}	heat flux in radial direction [cal/cm_n^2 sec, kcal/m_n^2 hr, BTU/ft_n^2 hr]
n	reaction order; also carbon number in Example 9
$n_A, n_B, \ldots, n_R, n_S$	moles of indicated component [moles, kmoles, lbmoles]
$n_{Aa}, n_{Ba}, \ldots, n_{Ra}, n_{Sa}$	moles of indicated component per mole of limiting reactant, A, fed [moles/(mole A fed), kmoles/(kmole A fed), lbmoles/(lbmole A fed)]
n_{Ba_0} or n_{Ba_0}	moles of B fed per mole of A fed
n_D	number of tubes at tube bundle diameter
n_i	components other than j [moles, kmoles, lbmoles]
n_j	moles of any component j [moles, kmoles, lbmoles]
n_{ja}	moles j per mole of A fed [moles/(mole A fed), kmoles/(kmole A fed), lbmoles/(lbmole A fed)]
$(n_{ja})_s$	moles j per mole of A fed at exterior surface of catalyst [moles/(mole A fed), kmoles/(kmole A fed), lbmoles/(lbmole A fed)]

n_{ka}	moles key component k per mole of A fed [moles/(mole A fed), kmoles/(kmole A fed), lbmoles/(lbmole A fed)]
$n_1, n_2, \ldots, n_c$	moles of component 1, 2, up to number c [moles, kmoles, lbmoles]
$n_{1a}, n_{2a}, \ldots, n_{ca}$	moles of components 1, 2, up to c per mole of A fed [moles/mole A fed, kmoles/kmoles A fed, lbmoles/lbmoles A fed]
n_o	observed order
n_p	polytropic exponent for compression
n_t	total number of tubes
n_r	total number of reactions
n_T	total moles in a mixture [moles, kmoles, lbmoles]
n_{T_a}	total moles per mole of A fed [moles/(mole A fed), kmoles/(kmole A fed), lbmoles/(lbmole A fed)]
P	total pressure [atm, atm, psia]
$P_A, P_B, \ldots, P_R, P_S$	partial pressure of indicated component [atm, atm, psia]
$P_A^*, P_B^*, \ldots, P_R^*, P_S^*$	partial pressure of indicated components at chemical equilibrium [atm, atm, psia]
P_c	pseudo-critical pressure [atm, atm, psia]
$(P_c)_A, (P_c)_B, \ldots$	critical pressure for components A, B, . . . , respectively [atm, atm, psia]
P_j	partial pressure of any component j [atm, atm, psia]
P_{j_b}	partial pressure of component j in the bulk fluid [atm, atm, psia]
P_{j_s}	partial pressure of component j at the catalyst surface [atm, atm, psia]
$(\mathrm{Pe})_a$	axial Peclet number, $D_p u_s/\mathscr{D}_a$
P_p	partial pressure of poison [atm, atm, psia]
P_r	reduced pressure
Pr	Prandtl number, $c_p \mu/\lambda_f$
Pw	present worth
p	pitch of tubes [cm, m, ft]
Q	heat added to the system [cal, kcal, BTU]
q	heat-transfer rate [cal/sec, kcal/hr, BTU/hr]
q_p	heat generation potential (see Chapter 6) [cal, kcal, BTU]
R	gas-law constant in PVT units [82.056 cm^3 atm/mole °K, 0.082056 m^3 atm/kmoles °K, 10.73, ft^3 psia/lbmole °R]

R	gas constant [1.987 cal/mole °K, 1.987 kcal/kmol °K, 1.987 BTU/lbmol °R]
R_b	radius of bed [cm, m, ft]
R_I	inside radius of shell [cm, m, ft]
R_{n}	cash flow in period n
R_{p}	catalyst radius [cm, m, ft]
Re	Reynold's number, $D_p \rho_f u_s/\mu$ or $D_p \mathbf{G}/\mu$ at steady state
R_{t}	radius of tube [cm, m, ft]
$\hat{r}_{\mathrm{A}}$	rate of reaction of feed component A, the limiting reactant [moles A reacted/gcat.sec, kmoles A reacted/kgcat.hr, lbmoles A reacted/lbcat.hr]
$(\hat{r}_{\mathrm{A}})_{\mathrm{b}}$	intrinsic rate based on bulk fluid conditions [moles/gcat.sec, kmoles/kgcat.hr, lbmoles/lbcat.hr]
$(\hat{r}_{\mathrm{A}})_e$	effective (observed or global) rate [moles/gcat.sec, kmoles/kgcat.hr, lbmoles/lbcat.hr]
$(\hat{r}_{\mathrm{A}})_{\mathrm{s}}$	intrinsic rate based on catalyst exterior surface conditions [moles/gcat.sec, kmoles/kgcat.hr, lbmoles/lbcat.hr]
$(\hat{r}_{\mathrm{A}})_0$	initial rate [moles/gcat.sec, kmoles/kgcat.hr, lbmoles/lbcat.hr]
r_{f}	fouling factor [cm_n sec °K/cal, m_n hr °K/kgcal, ft_n hr °R/BTU]
r_I	radial distance in the catalyst sphere or equivalent sphere [cm_f, m_f, ft_f]
$\hat{r}_j$	rate or reaction of component j per unit mass of catalyst [moles/gcat.sec, kmoles/kgcat.hr, lbmoles/lbcat.hr]
$(\hat{r}_j)_e$	effective (observed or global rate of reaction of component j [moles/gcat.sec, kmoles/kgcat.hr, lbmoles/lbcat.hr]
$\hat{r}_{j\mathrm{m}}$	mass rate of reaction of A [g/gcat.sec, kg/kgcat.hr, lb/lbcat.hr]
$(\hat{r}_n)_{\mathrm{s}}$	rate of reaction no. n at catalyst exterior-surface conditions [moles/gcat.sec, kmoles/kgcat.hr, lbmoles/lbcat.hr]
$\hat{r}_0$	initial or inlet rate [moles/gcat.sec, kmoles/kgcat.hr, lbmoles/lbcat.hr]
$\hat{r}_1, \hat{r}_2, \ldots, \hat{r}_n$	rate of reactions $1, 2, \ldots, n$ [moles/gcat.sec, kmoles/kgcat.hr, lbmoles/lbcat.hr]
$\bar{r}_{\mathrm{p}}$	average pore radius [cm, m, ft]
r_t	radial distance in a packed bed [cm, m, ft]

S	entropy or, where noted, total entropy of a mixture at T and P relative to the elements at 298° and 1 atm [cal/mole °K, kcal/kmole °K, BTU/lbmole °R]
$\mathbf{S}$	allowable stress [psi]
$S^\circ_{f_j}$	standard entropy of formation of component j from the elements at temperature T and 1 atm [cal/mole °K, kcal/kmole °K, BTU/lbmole °K]
S_g	total surface area per unit mass of catalyst [cm^2/g, m^2/kg, ft^2/lb]
S_j	entropy of component j [cal/mole °K, kcal/kmole °K, BTU/lbmole °R]
Sp	sphericity (see Table 4.3)
$\mathbf{S}_R$	fractional selectivity of R
ΔS	change in total entropy of a system [cal/°K, kcal/°K, BTU/°R]
ΔS°	standard entropy of reaction at temperature T and 1 atm [cal/mole °K, kcal/kmole °K, BTU/lbmole °R]
$\Delta S^\dagger$	entropy of activation [cal/mole °K, kcal/kmole °K, BTU/lbmole °R]
Sc	Schmidt number, $\mu/\mathscr{D}_j\rho_f$
s	number of equidistant active sites [moles/gcat., kmoles/kgcat., and lbmoles/lbcat.]
T	reaction temperature, understood to be bulk fluid temperature in models. [°K, °K, °R]
T_b	temperature of bulk fluid [°K, °K, °R]
T_c	pseudo-critical temperature [°K, °K, °R]
$(T_c)_A, (T_c)_B, \ldots$	critical temperatures of components A, B, . . . , respectively [°K, °K, °R]
ΔT_e	temperature error [°K, °K, °R]
T_{eq}	temperature at equilibrium for a given composition or conversion [°K, °K, °R]
T_h	temperature of heat-transfer fluid [°K, °K, °R]
TJ	heat-transfer medium temperature or jacket temperature. Used in programs and equivalent to T_h used in text [°K, °K, °R]
T_M	temperature of mixture [°K, °K, °R]
T_m	mean temperature in the reactor or portion thereof or base temperature for reparameterization [°K, °K, °R]
T_0	inlet temperature [°K, °K, °R]
T_{op}	optimum or operating temperatures [°K, °K, °R]

T_Q	temperature of quench [°K, °K, °R]
T_{R_t}	temperature at edge (radius) of the bed [°K, °K, °R]
T_r	reduced temperature
T_s	temperature at catalyst exterior surface [°K, °K, °R]
T_w	wall temperature [°K, °K, °R]
T_1	temperature at exit of Bed No. 1 [°K, °K, °R]
ΔT_a	adiabatic approach [°K, °K, °R]
t	time [sec, hr, hr]
$t^{\dagger}$	time required for passage of activated complex
t_v	vessel thickness [cm, m, ft]
U	general overall heat-transfer coefficient [cal/cm^{2}sec, kcal/m^{2}hr, BTU/ft^{2}hr]
U_m	internal energy per unit mass [cal/g, kcal/kg, BTU lb]
U_z	overall heat-transfer coefficient for tubular reactor [cal/cm$_n^2$ sec °K, kcal/m$_n^2$ hr °K, BTU/ft$_n^2$ hr °R]
ΔU	change in internal energy for a system [cal, kcal, BTU]
u	velocity [cm/sec, m/hr, ft/hr]
u_s	superficial flow velocity in packed bed (i.e. based on empty cross section) [cm/sec, m/hr, ft/hr]
V	volume, also volume of reactor bed [cm^3, m^3, ft^3]
V°	volume of gas adsorbed, corrected to 0°C at 760 mm pressure [cm^3, m^3, ft^3]
V_j	volume of component j in an ideal solution [cm^3, m^3, ft^3]
$\bar{V}_j$	partial volume of component j [cm^3, m^3, ft^3]
V_m	volume of an adsorbed monolayer of nitrogen [cm^3, m^3, ft^3]
V_{PR}	product value
v	volume per unit mass [cm^3/g, m^3/kg, ft^3/lb]
W	mass of catalyst [g, kg, lb]
W_{A_1}, W_{A_2}	fractional masses of A in streams 1 and 2, respectively
W_a	work done on the system [cal, kcal, BTU]
W_F	feed flow rate [g/sec, kg/hr, lb/hr]
W_f	lost work [cal, kcal, BTU]
W_p	total desired production [g/sec, kg/hr, lb/hr]
W_t	production per tube [g/tube sec, kg/tube hr, lb/tube hr]

X_A	fractional conversion of limiting reactant, A
$X_{A_1}, X_{A_m}, X_{A_Q}$	fractional conversion level in discharge from stage No. 1, mixture and quench, respectively
$X_{1a}, X_{2a}, \ldots, X_{na}$	extent of reaction for Reaction number $1, 2, \ldots, n$ on a per mole of A fed basis (See Eq. 3.32A)
Y_A	mole fraction of A in reacting mixture
Y_j	mole fraction of component j
Y_{A_f}	film factor, $(1/\tau_A)\,\mathrm{LM}\,(\tau - Y_A)$ where $\tau_A = a/(a+b-r-s)$
Y_{A_o}	mole fraction of A in the feed
y_{A_s}	mole fraction of A at catalyst exterior surface
$\mathbf{Y}_R$	fractional yield of R
$Y_{R_1}, Y_{R_m}, Y_{R_Q}$	mole fraction of product R in the discharge from stage No. 1, the mixture, and the quench, respectively
Z	axial distance in fixed-bed reactor [cm, m, ft]
z	compressibility factor
z_c	critical compressibility factor

Greek Letters

α	activity factor, (rate for used catalyst)/(rate for fresh catalyst), both at same conversion and operating conditions, also as an exponent and an empirical constant
α_{ij}	number of atoms of element i in component j
α_s	projected area of an adsorbed nitrogen molecule $[16.2 \times 10^{-16}\ \mathrm{cm}^2]$
α_t	thermal diffusivity, $\lambda/\rho c_p$
ε	bed void fraction
β	an empirical constant
ε_n	fractional error
ε_p	internal void fraction of a catalyst
η	effectiveness factor based on catalyst exterior surface conditions
η_o	effectiveness factor based on bulk conditions
λ	thermal conductivity $[\mathrm{cal\ cm}_b/\mathrm{cm}_n^2\ \mathrm{sec}\ °\mathrm{K},\ \mathrm{kcal\ m}_b/\mathrm{m}_n^2\ \mathrm{hr}\ °\mathrm{K},\ \mathrm{BTU\ ft}_b/\mathrm{ft}_n^2\ \mathrm{hr}\ °\mathrm{R}]$
λ_a	effective axial conductivity $[\mathrm{cal\ cm}_f/\mathrm{cm}_n^2\ \mathrm{sec}\ °\mathrm{K},\ \mathrm{kcal\ m}_f/\mathrm{m}_n^2\ \mathrm{hr}\ °\mathrm{K},\ \mathrm{BTU\ ft}_f/\mathrm{ft}_n^2\ \mathrm{hr}\ °\mathrm{R}]$
λ_c	critical conductivity $[\mathrm{cal\ cm}_f/\mathrm{cm}_n^2\ \mathrm{sec}\ °\mathrm{K},\ \mathrm{kcal\ m}_f/\mathrm{m}_n^2\ \mathrm{hr}\ °\mathrm{K},\ \mathrm{BTU\ ft}_f/\mathrm{ft}_n^2\ \mathrm{hr}\ °\mathrm{R}]$
λ_f	fluid thermal conductivity $[\mathrm{cal\ cm}_f/\mathrm{cm}_n^2\ \mathrm{sec}\ °\mathrm{K},\ \mathrm{kcal\ m}_f/\mathrm{m}_n^2\ \mathrm{hr}\ °\mathrm{K},\ \mathrm{BTU\ ft}_f/\mathrm{ft}_n^2\ \mathrm{hr}\ °\mathrm{R}]$

λ_I	effective thermal conductivity of catalyst pellet under operating conditions [cal cm_n^{-2} cm_f °K^{-1} sec^{-1}, kcal m_f/m_n^2 hr °K, BTU ft_f/ft_n^2 hr °R]
λ_i	Lagrange multiplier
λ_r	effective radial conductivity [cal cm_f/cm_n^2 sec °K, kcal m_f/m_n^2 hr °K, BTU ft_f/ft_n^2 hr °R]
λ_w	conductivity of reactor tube [cal cm_f/cm_n^2 sec °K, kcal m_f/m_n^2 hr °K, BTU ft_f/ft_n^2 hr °R]
κ	transmission coefficient
μ	fluid viscosity [g/cm sec, kg/m hr, lb/ft hr]
μ_j	chemical potential of component j [cal/mole, kcal/kmole, BTU/lbmole]
ν_j	stoichiometric coefficient for any component j, positive for products and negative for reactants [moles, kmoles, lbmoles]
ν_{jn}	stoichiometric coefficient for component j in Reaction n [moles, kmoles, lbmoles]
ν_{kn}	stoichiometric coefficient for key component k in Reaction n [moles, kmoles, lbmoles]
ρ	density [g/cm^3, kg/m^3, lb/ft^3]
ρ_b	bed bulk density [g/cm^3, kg/m^3, lb/ft^3]
ρ_c	critical density [g/cm^3, kg/m^3, lb/ft^3]
ρ_f	fluid density [g/cm^3, kg/m^3, lb/ft^3]
ρ_p	pellet density [g/cm^3, kg/m^3, lb/ft^3]
Φ	general power law modulus in terms of observables (see Eq. 5.21) [dimensionless]
ϕ	modified Thiele modulus for all shapes with a_s substituted for $R_p/3$ [dimensionless]
$\phi_A, \phi_B, \ldots, \phi_R, \phi_S$	pure-component fugacity coefficients for indicated components [dimensionless]
ϕ_g	modified modulus applicable to any simple power-law kinetics (see Eq. 5.19) [dimensionless]
ϕ_j	pure-component fugacity coefficient for component j [dimensionless]
ϕ_s	Thiele modulus for simple power-law kinetics (see Eq. 5.13) [dimensionless]
$\hat{\phi}_A, \hat{\phi}_B, \ldots, \hat{\phi}_R, \hat{\phi}_S$	partial fugacity coefficients for indicated components [dimensionless]
$\hat{\phi}_j$	partial fugacity coefficient for component j [dimensionless]

Index